MAKING DEMOCRACY COUNT

MAKING DEMOCRACY COUNT

How Mathematics Improves Voting, Electoral Maps, and Representation

Ismar Volić

Princeton University Press

Princeton & Oxford

Published by Princeton University Press
41 William Street, Princeton, New Jersey 08540
99 Banbury Road, Oxford OX2 6JX

press.princeton.edu

GPSR Authorized Representative: Easy Access System Europe - Mustamäe tee 50, 10621 Tallinn, Estonia, gpsr.requests@easproject.com

All Rights Reserved

Library of Congress Control Number for the cloth edition: 2023946898

First paperback printing, 2025
Paperback ISBN 9780691248813
Cloth ISBN 9780691248806
ISBN (e-book) 9780691248820

British Library Cataloging-in-Publication Data is available

Editorial: Diana Gillooly, Whitney Rauenhorst
Jacket/Cover: Heather Hansen
Production: Danielle Amatucci
Publicity: Matthew Taylor (US); Kathryn Stevens (UK)

This book has been composed in Arno Pro and Stratum2 with Industry

To Andrei and Iris:

may they one day live in a numerate democracy.

TABLE OF CONTENTS

MAKING DEMOCRACY COUNT

Introduction

ONE DAY MY daughter's fourth grade teacher announced that the following Friday was going to be a movie day. Everyone would come to school in pajamas, bring their favorite stuffed animal, and curl up to watch one of three options: *Bolt, Incredibles 2,* or *Coco*. To pick the movie, a vote would take place at the beginning of the day.

That morning, as my daughter was getting ready for school, I asked her to try to remember how the vote turned out. When she returned home, she duly reported that *Bolt* received 7 votes, *Incredibles 2* got 6 votes, and *Coco* got 4 votes. The teacher declared *Bolt* to be the winner and the class settled in for a movie afternoon.

Nothing against *Bolt*, but this was a terrible way to determine the winner. Most of the kids, ten of them, didn't give *Bolt* as their first choice. The will of the minority (7) was imposed on the remaining majority (10).

What could the teacher have done differently? She could, for example, have told the four kids who voted for *Coco* that their movie didn't make it, but they could cast another vote, this time between *Bolt* and *Incredibles 2*. The four new votes would have been added to the existing tallies for those two movies, with the upshot that the winner would now necessarily have majority support. If any two of the four kids who originally voted for *Coco* had voted for *Bolt*, that would have been the winner with at least 9 votes, but if—in a nail-biter twist—three had voted for *Incredibles 2*, that's the movie all seventeen kids would have watched, with *Bolt* dethroned after a 9–8 loss.

We will never know what would have happened. The plurality vote the teacher conducted asked only for the kids' top choice and nothing else. When so little information is asked for, only the coarsest tallying method is possible: count up the votes and the candidate with the most votes wins. The nuance of any preferences beyond the first choice is lost, resulting in a winner who does not necessarily represent the true will of the people.

And yes, this was just a bunch of kids choosing what movie to watch, so what's the big deal? But several months earlier, in the 2018 Democratic primary election in the 3rd District of Massachusetts, a few miles north of my daughter's school, you would have seen the same scenario playing out. Lori Trahan carried the nomination with 21.7% of the vote. Fast forward to the 2020 Republican primary in Florida District 3, far to the south, and you would see Kat Cammack winning with only 25.2% of the vote. Fast forward again, zagging back north to the 2022 Ohio Republican primary in the U.S. Senate race, and you would see J. D. Vance winning with 32.2% of the vote. You get the picture—all around us, people who have earned the support of only a minority of voters represent all of the voters.* This scenario is replicated all over the United States and the world in elections of all magnitudes, at all levels, deciding matters big and small.

What we're seeing is, at its root, a problem in mathematics.

Matters of politics have become mired in personalities and partisanship. Although we recognize that problems exist, we're getting worse at identifying them and increasingly paralyzed when it comes to constructing and assessing solutions. However, democracy is not just a human forum, it is also a system, a piece of civic infrastructure that runs on mathematics. Mathematics powers our basic democratic processes in ways that spread well beyond the seemingly simple matter of voting. Determining the size of representative bodies, distributing legislative seats, districting, and gerrymandering—all of these procedures rest on mathematical foundations.

* In heavily partisan districts, as most of them are, primaries are typically the real contests. All of these victors went on to be elected to office in their general elections.

Just as camera filters and lenses can reproduce an image faithfully or manipulate it intentionally or output a garbled mess, the mathematics of democracy can give the people a voice or silence some and amplify others or lead to results too fragile to trust. And indeed, a closer look at the manifestations of mathematics in our democracy reveals that the ways we use it are flawed, and archaic, and often serve discriminatory intent. They have murky, dubious, or politically motivated origins that few know about and even fewer remember.

The good news—the hopeful news—is that mathematics is also transparent, with no agenda or spin. It lets us see what's under the hood—we just have to look. If our politics are a screaming toddler and we are a parent incapacitated by the severity of the tantrum, then the math of those politics is the deep breath, a grounding mechanism that helps us understand that the child is just tired or hungry and we actually know how to fix that. Math is a clarifying way of looking at the world. It provides empowering confidence and is accessible to anyone. It is ready to reveal the deficiencies of our current democratic processes and recommend which new or updated ones can work better.

I have proof. For several years, I have witnessed the transformative effect of political numeracy education through teaching a college-level Math and Politics course. Students come to the class intrigued by the odd couple in the course title and hoping to earn a math credit needed for graduation. By the end, they are outraged that no one ever showed them how terrible our voting methods are, how blatantly devious gerrymandering is, how dysfunctional the U.S. Electoral College is. They are fired up about all the inequalities and discriminatory practices built mathematically into our system and are ready to get out there and do something about it. This book aims to bring my classroom to you, to empower you with knowledge (as well as outrage) that rests on a firm foundation of objective mathematics and that will give you the confidence to make a difference.

The time is right. There is growing awareness of the faults in our voting systems, and I don't mean fantasies of widespread voter fraud or conspiratorial voting machines. Initiatives to address inequities in representation and to implement something smarter are proliferating. (At the time of writing, at least ninety U.S. municipalities are trying to enact ranked choice voting.) After the 2016 election, the inadequacies of the Electoral College and its incompatibility with the popular vote have come front and center. As has gerrymandering, especially after the 2020 census and the many legal challenges to redistricting that followed. Politicians are starting to pay attention. More schools are building political quantitative literacy into their curriculum in recognition of its pedagogical appeal and relevance. Now is the time to get on the math and democracy bandwagon and join the movement to restore a functioning democracy.

———

It would of course be foolish to think that mathematics is the panacea for all of our political dysfunction. The role of politics, religion, community, emotion, greed, and power in democracy is undeniable and apparent to even the most detached of mathematicians. I tend to be even more sensitive to these things as an immigrant from Bosnia-Herzegovina. My life has to a significant extent been determined by that country's terrible war of the early 1990s, a horrific and bloody demise of democracy far removed from anything rational—and hence from anything mathematical.

But this book will intentionally ignore these things. Its scope and its intent are not to stretch into all things democracy. Everything you will read here is grounded in the quantitative. The motivation and the examples will come from a messy reality, but the analysis will proceed in a mathematically impartial way, without political commentary. The political context will be used only to inform the math. My guiding principle is that using the best version of mathematics in democracy is of benefit to everyone, regardless of all those extraneous factors. Using a voting method that best captures the will of the people; electing our officials in a way that respects the basic one person, one vote axiom of democracy; creating conditions so that underrepresented groups have a voice should be universal aspirations, and their implementation should be steered by tools that are equally all inclusive. Mathematics is one of those tools.

On the other hand, democracy is about people, and even the math of democracy is a story of human idealism, shortsightedness, and above all compromise. This means that we'll have to engage with the messiness on occasion. As definitive and unwavering as math is supposed to be, it doesn't do so well when it must proclaim itself the "best," the "most fair," or the "least biased." We'll see these words a lot because they're naturally woven into any discussion of politics and democracy, but they belong to a nonmathematical realm, one occupied by humans, in which opinions, preferences, and interpretations are allowed. For mathematics, these notions turn out to be too elusive. As a result, it will be easy for us to spot bad math (and there will be lots of it), but it will be trickier to find a replacement we can endorse. When considering math in the abstract, a diversity of definitions, theorems, and theories about a single subject can coexist simultaneously and independently (and they can all be equally true and valid), but because we will force them into competition for real-world application to democracy, we will sometimes have to be content with speculative outcomes. But we'll make the best of this. We'll figure out how to embrace the mathematical uncertainty.

This book is also not about the (mis)use of math and statistics in politics. I won't even address, let alone pick apart, the troubling ease with which politicians manipulate numbers, graphs, and charts or the cavalier way with which they bandy about cooked or carefully selected statistics. I have much respect for those who are waging the good fight of educating the public about the exploitation of statistics in politics, but this book is about the mathematics *behind* democratic processes, not in front of them. Of course, the two ends are but two tentacles of the political innumeracy kraken, and those of us who fight it stand shoulder to shoulder, math spears in hand, trying to flank the beast from different angles.

Finally, there is growing recognition that math curricula at all education levels need to be updated in a way that reflects the injustices, discriminations, and intolerances of the world. In this way, the optimistic educator reckons, we might even be able to use mathematics to tackle those issues. Many amazing people are fighting this good fight, writing and speaking about the archaic way we teach math and producing curricular materials that are relevant and timely.* As worthy and necessary as this effort may be, it is also outside the scope of this book. Our interest here is in the mathematical mechanics of democracy and not how mathematics can be used to explain or analyze specific social justice issues. But that's not to say that the content here has nothing to do with social justice. On the contrary—and as I'll argue repeatedly—implementing better math practices in democracy can lead to more equitable, less discriminatory outcomes.

So what does the math of democracy look like? We'll invest some time in unpacking concrete examples to get a feel for things—what goes right, what goes wrong—and then take on some formalism and abstraction to bring the big picture into focus. With only modest mathematical machinery, we'll be able to synthesize, extrapolate, generalize, and look for patterns in search of a cohesive framework that will support recommendations for better policies and mechanisms of democracy. We'll establish axioms, make definitions, and state theorems. We'll also encounter a surprising number of limitations and trade-offs, which will often manifest in paradoxical behavior, counterintuitive outcomes, and apparent inconsistencies—but we'll celebrate these. Probing strange outcomes can tell us a lot about the system.

On the other hand, the math of democracy is fairly straightforward: basic arithmetic is all you'll need. The focus will be on simple examples. If there is a more complicated or more abstract idea lurking around, I'll mention it in a footnote to avoid interrupting what I hope will be a comfortable, even cozy flow. You won't even notice I've slipped in some legit math!

As we move along, the mathematics will enable you to engage confidently in restoring our democracy by demystifying the systems that

---

* Examples include *Mathematics for Social Justice* by Gizem Karaali and Lily S. Khadjavi and *Rethinking Mathematics: Teaching Social Justice by the Numbers* by Eric Gutstein and Bob Peterson.

power it and examining how close they come to embodying our ideals. You'll be equipped to reject the prevailing refrain that things are just too complicated. You won't defer to history or tradition. You won't fear that something terrible is lurking in the details of an unfamiliar method that makes it secretly partisan. Math will offer a path to true progress, to tangible improvements and resolutions of impasses. You'll understand how the engine of democracy works, and you'll be ready to make your own judgments and take action.

PART I

Voting

VOTING IS the atom of democracy—individuals making choices about how they want to be governed and who will represent them. It sounds straightforward, but the political simplicity of the act of voting conceals the remarkable mathematical complexity of discerning the "correct" outcome. The devil is in the mechanics of tallying, in how the winners and losers are determined. Does the system embody our ideals about how democracy should function?

Taking a mathematical point of view helps us address this question. Along the way we'll confront the challenges of collective preferences, fairness, and the (im)possibility of a perfect democracy as an expression of individual desires. But we won't just shrug and say, "Nothing's perfect." Math will equip us to understand the trade-offs and how we can bring the system closer to those democratic ideals.

CHAPTER 1

The Best Way to Choose the Winner

ONE OF the first things I do when I teach my Math and Politics class is conduct an important poll. If only two kinds of pizza existed on the planet, pepperoni and vegetarian, which would my students get? The last time we did this, pepperoni got 13 votes and vegetarian got 12. (The vote is usually close, which I don't understand, since in my objective nonvegetarian view, pepperoni is obviously superior.) Everyone agreed that pepperoni was the winner and if we could only order one type of pizza, that is what we would get.

But then I add another option, barbecue chicken, and conduct the poll again. In that same class, 10 students still wanted pepperoni, 8 now chose barbecue chicken, and 7 stuck with vegetarian. We all again agreed that pepperoni had won and the experiment concluded with the observation that if we could only order one type of pizza, pepperoni would be it.

But then I pause. The silence becomes awkward and the students start realizing something is off. The gears begin to crank . . . and then from somewhere in the room comes the inevitable revelatory cry: "But wait! Most of us don't want pepperoni!" Witnessing lightbulb moments like this is why I do what I do.

MAJORITY, DICTATORS, AND MONARCHS

To appreciate what is happening with the pizza and the toppings, let's start with the first poll. Suppose a group of people must make a choice between two options, A and B. I'll typically refer to the options as *candidates*, although we won't always be talking about political candidates running for office. That will be the situation we want to examine most of the time, but we could also be discussing other scenarios in which people have to make a decision about a binary choice—a jury deciding whether or not to convict a defendant, a board of directors voting for or against a measure, a family trying to decide which of two movie options to watch, or students expressing their preference for one of two pizza toppings. So instead of "candidates," we could interchangeably use words such as "options," "choices," or "alternatives."

As you would expect things to proceed, all voters are first asked to say whether they prefer A or B. Nothing controversial there. But then, with everyone's preference in hand, how should the winner be decided? Most reasonable people would say that the candidate with more votes should win. For that to happen, a candidate necessarily has to receive more than half, or the *majority* of the votes. This procedure for selecting the winner has an unexciting name, *simple majority*. This is an example of a *voting method*, a generic term for any kind of process or algorithm for selecting a winner (or group of winners) from a collection of votes cast by any number of individuals.* With a simple majority method, a tie is possible only if A and B receive the same number of votes, which can happen if there is an even number of voters. Ties would have to be resolved with a separate mechanism, which we'll talk about later. In the first poll I conducted with my class, pepperoni was the simple majority winner.

But even if simple majority is an obvious way to select one of two options, is it the only way? Far from it. Here are three alternatives.

---

* Mathematicians sometimes study theoretical voting methods with *infinitely* many voters. As cool as that is, in this book we will always assume that the number of voters is finite.

- **Parity:** If one candidate gets an even number of votes, they are the winner. If both candidates get an odd number of votes or both get an even number of votes, it's a tie.
- **Dictatorship:** One of the voters decides the election, so whichever candidate this person votes for is the candidate that wins. We'll call this voter a *dictator*.*
- **Monarchy:** One of the two candidates wins no matter how the group votes. We'll call this candidate a *monarch*.

If these strike you as strange, there's a good reason. Under the parity method, if 98 of 100 people vote for A while 2 vote for B, it's a tie because both candidates have an even number of votes. If 99 people are voting and 2 vote for A while 97 vote for B, A wins because 2 is an even number and 97 is not.

If dictatorship is used and one voter is designated the dictator, then if only that person votes for A and 1 million other voters choose B, A still wins.

If monarchy is used and A has been designated the monarch, then the entire planet could vote for B but A would still win. Again the election is meaningless.

I hope you'll agree that these are all awful methods. If they were used in real life, the entire voting process would be farcical and wouldn't amount to much more than a performance. They disregard any sense of fairness or consensus. If these are our alternatives, we should definitely stick to simple majority.

But the question remains: Is simple majority the right method to use? Are these four methods all that's out there? What if we haven't yet discovered a method that's superior even to simple majority?

This is where it pays to bring math into the picture. Math is structured precisely for questions like this. It provides a framework for

---

* Although this is the standard terminology in social choice theory, it doesn't correspond to common usage. We usually think of a dictator as the ruler who continues to rule, often through rigged elections. Such a ruler is in this setting called a monarch to differentiate them from a dictator, who is a *voter* but not a candidate. Of course, a monarch could vote for themselves and thus also be the dictator, but in this book, I will regard candidates and voters as separate.

handling all possibilities without having to identify or examine each one. By anchoring itself in formal and often abstract concepts and by giving itself clear rules of engagement, math can extract substance from seemingly formless situations.

Mathematics is indeed able to say something about *all* two-candidate voting methods, even if we don't know how many methods there are or whether we have discovered them all. And even better, we get an answer that can be stated in plain language:

> **Theorem:** Simple majority is the best possible voting method for elections with two candidates.

This *theorem* confirms our intuition. Theorems are the pinnacles of mathematical knowledge. They are immutable truths—they aren't subject to opinion or feelings and they aren't susceptible to politics. They are true anywhere on this planet and beyond and they are forever. Intuition can be faulty; a theorem is not.

But as you apply your mathematical eye, you will not be entirely happy with the way this theorem is stated. What does "best" mean? How do we define "best"? Who decides what's "best"? It would be easy to end up chasing words in a circle. We could agree that "best" should mean "fairest" and "fairest" should mean "best for most people," and "best" should mean . . .

Deciding what "best" or "fairest" means for a voting method is interesting, complicated, and frustrating. It is the genesis of an entire field of study called *social choice theory* that lives at the intersection of math, statistics, economics, history, political science, psychology, and assorted other disciplines. Social choice theory is rightfully central to the practice of democracy and hence to this book, so we'll be digging into it. For a start, it will tell us what's really behind the theorem above. As we pursue it further, it will quickly lead to unexpected complications and some Nobel Prize–winning mathematics.

Is it surprising to see math playing a central role in studying something as messy and human as democracy? History tells us to be ready to find math where we don't expect it.

THE UNREASONABLE EFFECTIVENESS
OF MATHEMATICS

In 2020, a sixteen-year-old TikToker, @gracie.ham, posed some questions during one of her makeup tutorial videos: "Who came up with a concept like algebra?" and "What would you need it for?" The derisive fangs of internet groupthink, dipped in sexist venom, quickly plunged into Gracie, proclaiming her questions "dumb." But mathematicians immediately came to her defense, explaining that Gracie was in fact asking deep and substantive questions about the origins of mathematics, about conundrums mathematicians and philosophers have been grappling with for centuries. Gracie had scratched the surface of a puzzle that is far from prosaic: Is math discovered or invented? In other words, would math exist if people didn't? This is a fundamental enigma that's still up for debate. The best answer (or maybe it's a cop-out) is that math is both.

Most people think of math as arising from experience, usually observed in nature. To most, math is eager to explain what's around us, what governs the real world. That's not false, but math also doesn't need this world to thrive. It uses the real as a springboard to break free from the worldly, to cross into the abstract, to make itself universal. If along the way it becomes useful when it is restricted to our limited 3D world, that's great. But that's not the reason it exists. Its dual personality is one of the things that make the pursuit of mathematics so attractive.

Take, for example, my own field of research, *algebraic topology*. One of its branches, *knot theory*, draws inspiration from physical phenomena such as DNA knotting, chemical engineering, and the physics of fluid flow. The motivating idea is that a knot is basically a malleable one-dimensional loop of string in familiar three-dimensional space, and using that idea, you try to study its properties. But once you make this idea precise with the right definitions, it turns out that the theory can be expanded to allow both the string and the space it lives in to be of *any* dimension. This generalization gives rise to *embedding theory*, which studies the way n-dimensional geometric objects can be placed into m-dimensional geometric objects, where n and m are any positive integers. At this point, we're untethered from reality. It's all pure, abstract, free-form mathematics. Sometimes, though, we'll prove a theorem (which means we discover—or is it create?—a new rule or truth in this abstract game) that holds valid for some combinations of dimensions n and m and that might in particular hold in the case when n equals 1 and m equals 3. But this combination of dimensions is the setting of tangible knot theory, and the theorem suddenly has something to say about the tangling of DNA or the structure of a molecule used in drug design. The inquiry has taken us full circle back to the real via a detour into the abstract.

Mathematics is a long-term game, and this full circle can take decades or centuries and land in contexts that a theorem's creator/discoverer could not possibly have anticipated. For example, the *Perron-Frøbenius theorem*, a result proved in 1907 about the structure of matrices (two-dimensional arrays of numbers) of a certain type, is what powers Google's PageRank algorithm. There would be no Google empire without this theorem.

A more extreme instance of the delayed power of math is *Euler's theorem* from 1736. This seminal piece of mathematics, which talks about divisibility properties of numbers, turns out to be precisely what makes RSA, one of the first and the most widely used encryption algorithms, work. The ability to encrypt data flowing through the internet is what allows us to transfer private information securely. The internet would forever stay 1.0 without Euler's theorem. There would be no Facebook or Amazon (for better or worse) without encryption. There would be no personal accounts on any website or service, no private conversations, no online shopping. It's safe to say that Euler, illuminating his quill-written math with a candle, never imagined that private Snapchat conversations between two iPhones would be an eventual application of his theorem.

This uncanny ability of mathematics to show up when least expected isn't confined to digital infrastructures such as the internet. Mathematics is surprisingly good at explaining the physical world as well, even when that isn't its original function. This phenomenon was dubbed the *unreasonable effectiveness of mathematics* in a 1960 essay by the Nobel Prize–winning physicist Eugene Wigner. The standard example is Newton's law of universal gravitation, formulated to describe falling bodies on Earth but eventually unexpectedly explaining the motion of planets. Some physicists, such as Max Tegmark of MIT, go so far as to argue that the universe is just a giant mathematical construct and that we are merely chipping away at it with each new discovery.

Wigner confined his attention to physical applications of mathematics, so he didn't mention an arena that seems even more unreasonable in which math might hold some answers: how to achieve a desirable society given the preferences of its individual members. This is the subject of social choice theory, which grapples with the problem of consensus and, in particular, the problem of aligning voting methods with the stated values of a society. As I hope you're now expecting, mathematics proves to be a highly effective tool in tackling this challenge.

IT'S AS SIMPLE AS SIMPLE MAJORITY

A recap: we have a two-candidate voting method that appears to be decent (simple majority) and three terrible methods (parity, dictatorship, monarchy). To start, we'd like to characterize more precisely what we mean by "decent" and "terrible." Our ultimate goal is to establish the "best" voting method—and not just the best among the four we've described here but the best among all possibilities.

The standard—and (unreasonably?) effective—math approach is to first come up with properties we can all agree are desirable and then try to produce a voting method that satisfies them. This is a lot like crafting the perfect job ad that appeals to qualified candidates but is at the same time not too specific and not too broad so we won't have either zero applicants or 10,000 applicants. Our voting criteria should likewise be formulated in a way that attracts all potentially useful voting systems and eliminates the bad ones. And just as we hope to find the one job candidate who checks all the boxes, we're hoping that a single voting method will emerge as the sole survivor.

So what should these criteria be? All we know at this point is that we have some methods we don't like. We can work with that. Let's articulate some conditions in a way that's informed by examples we already have in hand. By digging into the essence of what is wrong with the bad methods, we can formulate properties that a good method must have.

Dictatorship is problematic because, well, there is a dictator. The election result is whatever that voter's ballot looks like. This ballot is special, but only in the dictator's hands. If the dictator were to trade ballots with another voter, then the new ballot in the dictator's hands would suddenly become special. That is the essence of the dictatorship problem.

Monarchy doesn't care about the ballots at all. It is not sensitive to the choices voters make on their ballots. If a voter were to switch their vote from one candidate to the other, nothing would change. Even worse, if *all* voters were to switch their votes, nothing would change. This observation captures the gist of monarchy.

The counterintuitive behavior of parity is that a candidate can gain votes yet go from winning to losing. Because odd and even numbers alternate and the parity flip-flops regardless of how high or low the numbers are, the actual ballot count does not matter at all. Doing better in the ballot tally does not mean doing better in the election. For example, if 9 people are voting and 6 people vote for candidate A and 3 people vote for candidate B, then candidate A is winning. But if one of the B voters switches their vote to A, then the score is 7–2 and now B wins, because they have an even number of votes, even though A gained a vote (and widened their lead over B).

Based on this breakdown, we can state some basic criteria against which to test voting methods:

- **Anonymity:** If any two voters trade ballots, the election result does not change.
- **Neutrality:** If all voters switch their votes from one candidate to the other, the election result switches.
- **Monotonicity:** It is impossible for the winning candidate to become the losing candidate by gaining additional votes.

The anonymity criterion means that a voting method does not treat any voter as special. As you would expect, dictatorship is not anonymous because if the dictator initially votes for A, then A is the winner, but if the dictator swaps ballots with someone who voted for B, B becomes the winner.

Neutrality is similar in flavor but is about candidates. It means that all candidates should be treated equally. By contrast, if A is the monarch, they always win. Even if all voters were to switch their votes (those who voted for A now vote for B and vice versa), A would stay the winner. Monarchy fails the neutrality criterion.

Monotonicity means that a candidate who is winning isn't hurt if someone decides to change their vote to support them or if a new ballot is cast for them. This criterion rules out the parity method or anything similar to it.

These three criteria should seem reasonable, but how do we know there aren't other ones we should apply? Why these three and only these three? The answer isn't obvious. What you're getting here is a distillation of years of academic research by mathematicians and economists—years of peer-reviewed papers, years of talks at seminars and conferences—that led to producing and honing just the right set of criteria. And now I can present the result to you as if it had been plain to see from the start. But it's not and it shouldn't be.

Math is typically passed down in a polished state that conceals the struggle of its creation, and I've done that here. There's not much benefit to seeing all the dead ends, seeming contradictions, and faltering steps on the way to the final result. I think it's useful to know about the messiness, though. It's useful to know that knowledge isn't self-evident, that we mathematicians spend most of our careers confused, that we have more questions than answers, that it's mostly about perspiration and little about inspiration. Only after many hours put in by many people can one hope for an orderly picture to emerge. I'm supplying such a picture and asking you to believe that it's the correct one.

Now, assuming you trust me that anonymity, neutrality, and monotonicity are the right criteria, we ask: Is there a voting method that satisfies all three criteria? Is there a method that treats all voters and candidates equally and never penalizes a candidate for gaining votes? The answer brings us back to the theorem from the previous section, but this time we can state it in more precise, mathematical language:

> **May's theorem:** Simple majority is the only voting method for elections with two candidates that satisfies anonymity, neutrality, and monotonicity and can produce a tie only if the same number of people vote for each of the candidates.

The theorem is named after mathematician Kenneth May (1915–1977), who wrote it down in his 1952 paper "A Set of Independent Necessary and Sufficient Conditions for Simple Majority Decision." To prove the theorem, May used a classic "sieve" method. The idea is to let all the possibilities fall through figurative meshes that represent criteria we want the falling things to satisfy. Only those that have the desired properties get to pass through. Imagine a job candidate pool that's progressively whittled down as candidates who don't have a college degree are sifted out, then those who have no experience are removed, and so on. It's the same here—voting methods are filtered through a series of meshes that correspond to each of the three criteria. In some sieve arguments, none of the possibilities survive the fall through all the meshes, and in others, infinitely many do. In the case of May's theorem, one and only one method manages to squeeze through—simple majority. We've found the perfect job candidate.

The cool part is that in order to do the sifting, the math does its abstract thing, no longer paying attention to the real-life motivating problem. Because we formalized how we want a voting method to perform, the process doesn't know (and doesn't care) how many possibilities are passing through the sieve at any stage. It's like the Twenty Questions toy my kids used to play with. The player thinks of a person, place, or object. The machine asks yes-no questions, running each answer through an algorithmic sieve, eliminating objects it has stored in its memory as it receives the answers, until there is only one possible thing the player could have had in mind. The yes-no questions in May's theorem are whether the method we're looking at satisfies anonymity, neutrality, and monotonicity. And out pops simple majority as the only possible option.

Although May's theorem says that simple majority is the best voting method according to the criteria we defined, we should also be mindful of the context. The setting is that of democratic elections where each voter has an equal say and each candidate has an equal chance. Given that assumption, it should not be controversial that anonymity, neutrality, and monotonicity are of paramount importance.

However, other considerations might sometimes outweigh these criteria. The considerations are not mathematical but depend instead on context, on tradition or civics or politics. For example, it might be important for a group of people to use a voting method that never produces a tie. Many a time my wife and I have regretted conducting a show-of-hands vote in my family that resulted in a 2–2 tie. The subsequent agony of trying to resolve it in a way that remained "fair" often made dictatorship look pretty good. According to May's theorem, we can't have a system that simultaneously satisfies the three criteria and never produces a tie. In terms of the sieve, demanding that ties never occur amounts to adding another filter, and this time no method survives. If we wanted to eliminate the possibility of ties, we would have to make a compromise and use a method that fails at least one of the criteria—in other words, a method that might not regard voters or candidates as equal or that might produce strange situations in which winning fewer votes might be a favorable strategy.

In a more consequential setting than my family's quarrels, there might be groups of voters among which one voice bears the ultimate weight, like that of an elder who will consider individual preferences but ultimately arrive at a decision on their own. Such a group would not care that their method fails neutrality because preserving other values, such as placing the opinion of their elders above all else, would be of greater importance.

May's theorem is a consequence of a broader result that says that if we require anonymity, neutrality, and monotonicity, then the *supermajority method*, the subject we'll look at next, is the only eligible method. As we'll see, supermajority can produce a tie even if one of the candidates receives more than half the votes. So if we add the requirement, as May's theorem does, that a tie should occur only when an equal number of people vote for each candidate, the only thing that's left is simple majority. Math has done its abstract dance to give us a definitive and actionable result. All is good with democracy so far.

QUOTA AND SUPERMAJORITY

Choosing a candidate (where, again, "candidate" can mean various things, such as a candidate in an election, a version of legislation, the adoption of a proposal, conviction of a defendant, and so on) is sometimes important enough that an agreement might be reached among the voters (or might be imposed by the rules of an institution) that a candidate can be the winner only if they get a certain large percentage of the votes. The hope is that if a greater majority of voters agree on an option, it will be harder to contest the results and the winning candidate will benefit from a broad base of support after the vote. However, reaching that level of consensus among voters can be challenging. Opportunities for scheming and coalition building that can block institutions and even governments abound.

The number that is required for a win, called the *quota*, depends on the situation or the type of vote in question. It can be expressed as a fraction, such as 2/3, that indicates that an option wins with the support of at least 2/3 of the voters, or as a percentage, such as 75%, that means that that percentage of voters have to approve of it. The quota should be between 1/2 and 1, which corresponds to 50% and 100%.

The voting method in which more than half the votes are required for a win is called *supermajority*. When the quota is exactly 1/2, or 50%, that reduces the method to the usual simple majority. Another important special case is when the quota is 1, or 100%. This is called *unanimity* (or *consensus*) because this requires *all* voters to select an option in order for it to win. For example, NATO is an institution that operates by unanimity at all levels (although no formal voting takes place), including decisions about the admission of new members.

An amendment to the Constitution of the United States is subject to a supermajority system because it requires at least 2/3 of the votes in both the House and the Senate, plus it must be ratified by at least 3/4 of the fifty states. This means that the number of representatives in the House who would have to vote for an amendment is 2/3 of 435:

$$\frac{2}{3} \times 435 = 290.$$

For the Senate, this number is $\frac{2}{3} \times 100 = 66.66\ldots$ (the ellipses mean that 6 repeats forever).

Now senators are people, so they come in integer quantities, namely whole numbers, and we can't chop them up into decimal values. We therefore must round up to 67 to give us the smallest integer that's larger than the quota.

For the required number of states, we get

$$\frac{3}{4} \times 50 = 37.5,$$

which we then must round up to 38. As innocent as this rounding process looks, it will create problems later when we talk about the apportionment of seats in the House of Representatives.

The U.S. legislative system also uses a supermajority method, although it's a little more layered. If 2/3 of the House and 2/3 of the Senate approves a bill, then it passes and becomes law. If the legislation has the backing of a simple majority but not the 2/3 supermajority in either chamber, then the president has the choice of vetoing it or signing it into law.

However, even when legislation has simple majority support from both chambers and the support of the president, it can fall victim to a particularly contentious supermajority rule called the *filibuster*: a senate vote can be delayed indefinitely by a senator who holds the floor. (The word "filibuster" comes from the Dutch word for "pirate," which is appropriate for such a hostage situation.) The filibuster is regularly used to block legislation. Strom Thurmond's stonewalling of the Civil Rights Act of 1957 still holds the record, clocking in at over twenty-four hours. A filibuster can be overridden if the legislation has 3/5 of the votes, meaning that sixty senators must support it (this is called *cloture*). This number was 2/3, or sixty-seven senators, until 1970, when it was reduced to 3/5 because of the increase in the number of filibusters.

Juries in U.S. criminal cases also use supermajority to establish a verdict of guilty or not guilty. Some states even require unanimity, as do all federal trials. If a jury cannot reach the percentage of guilty votes a state requires, it is declared hung and the judge can order another trial or declare a mistrial.

The most ceremony-laden supermajority election is that of the pope. Of the 203 cardinals from sixty-nine countries, those under the age of eighty are eligible to vote. They are summoned to Rome and locked in the Sistine Chapel until a 2/3 majority has voted for a new pope. The pope does not have to be a cardinal but usually is (the last noncardinal pope, Urban VI, was elected in 1378). The process is called the *conclave*, meaning "with a key," because the cardinals are literally in custody behind locked doors for however long it takes them to reach a decision. No phones, TV, newspapers, or other means of communication or contact with the outside world are allowed. Various *Da Vinci Code*–style rituals, oaths, and threats of excommunication hang over the process. The cardinals vote four times a day and burn the ballots after each vote. The color of the smoke communicates to the world whether a decision has been reached—black if it has not, white if it has. The quota has changed over time. In 1996, John Paul II decreed that a simple majority of 1/2 would suffice after twelve days of unsuccessful voting, but in 2007, Benedict XVI switched it back to 2/3 supermajority.*

A distinctive feature of the supermajority method is that a candidate can receive a majority of the votes but still lose the election. For example, if the quota is 3/4 and 200 people are voting, then a candidate needs 150 votes to win. So a candidate could collect as many as 149 of 200 votes and still not win. In this case, a tie might be declared—even though one candidate has 149 votes and the other has only 51. The other

---

* The Berlin Philharmonic uses a conclave election to choose its chief conductor and its artistic director. The 124 tenured musicians of this self-governing institution meet at an undisclosed location and cut themselves off from the outside world until they have decided who to offer the jobs to. The first round is conducted by a secret ballot. Any of the voters can write any names down and a committee somehow tallies these votes to narrow the field for the second-round vote. The tallying methods are for some reason not disclosed to the public. Majority support is required. If the musicians do not reach a decision, they must conclave again at a later date.

candidate might even be pronounced the winner if, for example, they are the incumbent and the rules say that only a supermajority can remove them.

This possibility helps us formulate another useful test that we can inspect methods against.

◆ **Majority criterion:** Whenever a candidate receives a majority of votes, the method declares that candidate to be the winner.

Supermajority thus does not satisfy the majority criterion because a candidate can receive more than half of the votes but still not win the election. May's theorem can now be restated as saying that the only system that satisfies the anonymity, neutrality, monotonicity, *and majority* criteria is simple majority.

We've sliced two-candidate elections many ways by now. What we've learned is that simple majority is the best system, and we've formalized in precise, mathematical language what we mean by best. With this experience under our belt, we're ready to tackle a more complicated situation: elections with more than two candidates.

This is where democracy will start to get in trouble.

But first, let's talk numbers.

THE TYRANNY OF CONVENIENT NUMBERS

May's theorem is unusual in that it gives us a single clear answer to the question of the best method for a two-candidate election. More often mathematics will help us by eliminating unacceptable options and clarifying which choices remain without going the last mile and prescribing exactly what we should do. We've already seen an example of this in the broader theorem that singled out supermajority as the method satisfying the three basic criteria. But "singled" is a misleading word in this context. Supermajority is not one method; it's a family of methods with simple majority at one extreme and unanimity at the other. The quota can be anything between 1/2 and 1, with each fraction giving the rule for an election. In addition to choice of methods, the choice of fractions for a supermajority is a rich source of dysfunction in democratic processes.

2/3, 3/4, 3/5 . . . Do these quotas seem arbitrary? Pulled out of a hat? What mathematical process decided that 2/3 is the right supermajority for papal elections? What quantitative analysis decided that 60 senators is the right number to end the filibuster? Why not 57 or 62? Why does the UN Security Council have 10 rotating members, and not 9 or 11? To a mathematician, 3/5 or 10 reeks of convenience—or even worse, politics—devoid of mathematics. When a simple number acts as the determinant for an inherently difficult problem, a numerate person should be skeptical about whether enough thought has gone into choosing that number or wonder whether those who chose it were politically motivated. A particularly troublesome offender is 435, the number of seats in the House of Representatives; we'll dissect this at length later.

Here is another number whose origin and appropriateness are questioned by few: twelve jurors. This is the most common jury size in the United States. (Some states have smaller juries, but the Supreme Court has ruled that six is the minimum.) This number was adopted from the English system, which, in turn, uses it because King Morgan of Glamorgan, Wales, said in 725 AD that twelve was the magic number. Why? Because he thought the judge and the jury were supposed to mirror Jesus and his twelve apostles. So twelve is an arbitrary, biblical number that was embedded in our criminal judicial system by inertia and without scrutiny.*

A cavalier attitude toward the politics of numbers is not an American or British invention. Politicians in the country I am from, Bosnia (I'll use this shortened version of the full name, Bosnia-Herzegovina), were recently engaged in the contentious process of revising the election laws. One option that, unfortunately, was considered in earnest, was to mimic the U.S. Electoral College in the ten cantons that make up one of Bosnia's entities. That proposal would have given some number of electoral votes to each canton based on its population, but it would then have given each canton twenty more votes. Where did the number twenty come from? Why would every canton get twenty electors for

---

* An excellent mathematical account of how jury size can affect trial outcomes is found in Jeff Suzuki's book *Constitutional Calculus*.

free, regardless of population size? In the United States, the extra two electoral votes for each state (which come from the number of senators each state has) make for some ugly math, and twenty makes things just about ten times uglier. But a closer look at who would benefit most from these twenty extra votes quickly revealed that this number was motivated entirely by politics and did not involve in-depth mathematical analysis. This proposal was eventually rejected, but only because a different kind of political scheming presented itself as more appealing.

The world doesn't explain or resolve itself in easy numbers. Ask π or the Planck constant. The convenient numbers floating around the world's democracies are largely unjustified. They have a concrete bearing on societies, laws, and lives, but they're more often the product of tradition or manipulation or happenstance than of deep thinking. When they're procedurally or politically inconvenient, like the 2/3 cloture, they are replaced by something equally unexamined that makes life easier, like 3/5. No matter how much the population, demographics, politics, or geopolitical conditions might change around them, these numbers are seldom revisited to ensure that they reflect the circumstances or work better for the people. We'll examine many more examples of convenient, lazy numbers and show how a mathematical lens can lead to more sensible choices.

CHAPTER 2

The Worst Way to Choose the Winner

MAY'S THEOREM says that if a group of voters has to choose between two candidates, then simple majority is the method to use. Each voter selects one of the two options, and the option with more votes wins. If the group feels the winner should have the support of more than half the voters to win, then it can agree on a quota and proceed with a supermajority vote. Easy enough.

But what should we do if there are more than two candidates?

MOST OF NOT MANY

Most people would say that the natural generalization to be drawn from simple majority is that everyone should again select their favorite candidate and the one with the most votes should win. And they wouldn't be wrong. It's certainly the simplest thing that comes to mind. This method is called *plurality* (or *relative majority* or *first past the post* or *winner take all*). Its simplicity is one of the main reasons it's the most widely used voting method in the world, including in the United States.

Plurality has been used in U.S. elections since the founding of the country. Of the roughly 520,000 elected officials in the United States, only a handful have been elected by a method other than plurality. This is how we choose our school boards, city councils, mayors, governors, senators, and scores of other representatives who perform

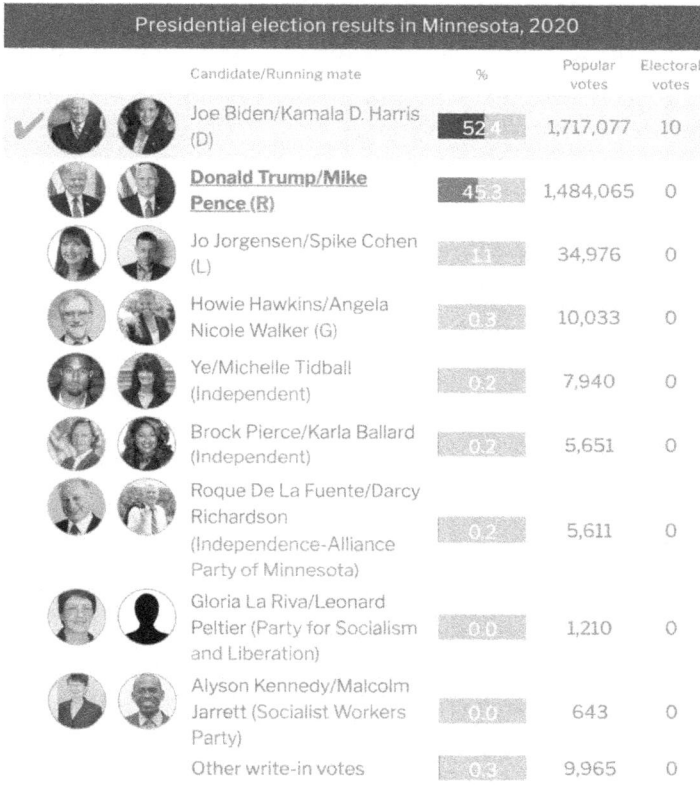

Candidate/Running mate	%	Popular votes	Electoral votes
Joe Biden/Kamala D. Harris (D)	52.4	1,717,077	10
Donald Trump/Mike Pence (R)	45.3	1,484,065	0
Jo Jorgensen/Spike Cohen (L)	1.1	34,976	0
Howie Hawkins/Angela Nicole Walker (G)	0.3	10,033	0
Ye/Michelle Tidball (Independent)	0.2	7,940	0
Brock Pierce/Karla Ballard (Independent)	0.2	5,651	0
Roque De La Fuente/Darcy Richardson (Independence-Alliance Party of Minnesota)	0.2	5,611	0
Gloria La Riva/Leonard Peltier (Party for Socialism and Liberation)	0.0	1,210	0
Alyson Kennedy/Malcolm Jarrett (Socialist Workers Party)	0.0	643	0
Other write-in votes	0.3	9,965	0

Presidential election results in Minnesota, 2020

FIGURE 2.1. Minnesota presidential election results, 2020.

hundreds of functions at all levels of government. It is also how all states except Maine (and Alaska, starting with the 2024 elections) decide the winner of the presidential election. For example, figure 2.1 shows the results of the 2020 presidential elections in Minnesota. Joe Biden was declared the winner because he got the most votes. End of story.

Plurality makes sense—shouldn't the candidate favored by the most people win? Yes and no. It depends on what we mean by most. If someone wins more than half the votes, then yes, they should be the winner. Majority rule in action.

But most doesn't have to mean majority. As soon as there are three or more candidates, plurality can select a candidate who has received fewer than half the votes. In other words, plurality can select a winner who is not the top choice for most people. Remember my daughter's fourth-grade class movie and my pizza-toppings poll? That's what happened.

To drive this crucial point home, consider the results of this election:

Candidate	Percentage of votes
A	30%
B	25%
C	20%
D	10%
E	10%
F	5%

The plurality method would declare candidate A to be the winner. However, a vast majority of the voters, 70% of them, did not choose A. So now A, who received only 30% of the votes, represents 100% of the people. Maybe those 70% don't mind being represented by A and everything is dandy. But maybe they despise A's positions and would rather leave the country and renounce their citizenship than have A represent them. Maybe in a head-to-head contest with A, B would have been the representative a significant portion of voters preferred. Or C. Maybe that significant portion is more than half the total number of voters, which means that the majority of voters would prefer B over A. Or C over A.

With plurality, none of this is knowable. Because the method requires the bare minimum of input from voters, math can do little more than add up the total number of votes each candidate receives. But that's not good enough. Plurality does not capture the people's will in any sense of that loaded phrase. It does not create broad agreement. It does not find a consensus that most people can get behind.

It gets worse. In an election with nine candidates, as in the Minnesota presidential example, it is possible for a candidate to win with barely over 11% of the vote, or one vote more than 1/9 of the total if the remaining votes are split equally among the other candidates. When there are one hundred candidates, it's possible to win with barely over 1% of the votes.

Sure, this is an exaggerated example, but the fact that it *could* happen is still shocking and points to a systemic flaw in the plurality method.

And that flaw is on display all the time. In 2019, the Center for Election Science looked at 154 mayoral elections in America's biggest cities. They found that on average, 6.8 candidates ran for mayor in each race and that 55.6% of people voted for candidates who did not win. The 2021 mayoral election in my own city of Boston corroborates this: the winner was a candidate who received 33.4% of the votes in the first round of voting.

And that's just the mayoral races. Examples proliferate at all levels of public service. Of the eleven elections for governor of New Hampshire before 2018, nine winners received fewer than 50% of the votes. In two of the most recent Democratic primaries for U.S. House races in districts adjacent to where I live, the winners carried only 21.7% and 22.4% of the votes. Donald Trump became the 2016 Republican nominee for president with 44.9% of the votes.

Time and again plurality gives us *minority rule*. People elected by fewer than half the voters represent us all.

It's a simple observation—*most* doesn't mean *majority*—and the math will soon tell us that its consequences are far reaching. Plurality is vulnerable to anomalies such as vote splitting and the spoiler effect, it favors two-party systems, it discourages diversity of candidates and ideas, and it breeds authoritarianism. Its crudeness and lack of insight into voters' preferences beyond their top choice have turned this instrument of democracy into an implement of its demise. Most countries have understood this. Even though plurality can be found all over the world, the only democracies that rely on it extensively are the United States, Canada, the United Kingdom, and India.

In fact, the only advantage of plurality is that it's easy. Everyone can understand it. Find one name on the ballot and fill in the bubble next to it. Done. But if we want to enjoy a democracy where election outcomes are the most satisfactory for most people, then plurality is inadequate.

The rest of this chapter will be exasperating. Irritating. The refrain will be that plurality is deeply flawed. But it's important to break it down until it gives up all its tricks and secrets. Then we'll rebuild it into something better.

A SPLITTING HEADACHE

It will not be surprising if posterity declares the 2016 Clinton versus Trump presidential election the most significant in United States history. It marked the ascendance of a new sort of populism whose fallout has prompted many a first draft of democracy's obituary and whose full consequences are still unclear. The ideological and cultural rifts this election exposed were deep. They came to a head with the 2020 Biden-Trump election and the insurrectionist attack on the Capitol on January 6, 2021.

But before the 2016 presidential election were the Republican primaries, when the chatter of democracy entering hospice care began in earnest.

As a quick reminder, the way we elect the president is that the Democratic and Republican Parties each choose their nominee and then people vote for those candidates or for candidates from other, smaller parties who have no chance of winning (but are still important, as we'll see). People vote in primary and caucus elections, state by state, around the country (over a period of weeks, which creates a slew of other problems), essentially to tell each party's delegates how they should vote in the two national conventions where the nominees are actually selected.*

In 2016, Donald Trump emerged as the Republican Party's surprise nominee, securing the majority of its delegates. Other major contenders were Ted Cruz and Marco Rubio, U.S. senators from Texas and Florida, and John Kasich, the former governor of Ohio. The results, obtained by plurality voting in most states, were as follows:

---

* Read all about the bewildering mathematics of delegate allocation in the book *Delegate Apportionment in the US Presidential Primaries* by Michael A. Jones, David McCune, and Jennifer Wilson.

Candidate	Percent of votes won*
Trump	44.9%
Cruz	25.1%
Rubio	13.8%
Kasich	11.3%

As these primaries and caucuses were taking place, polls suggested that Cruz and Rubio would each have defeated Trump in one-on-one contests. That is, if Rubio and Kasich voters had been asked to pick between Trump and Cruz, a significant majority of them would have opted for Cruz. The same is likely true for Cruz and Kasich voters, who would have picked Rubio had they been given a choice between him and Trump. Put differently, if Rubio and Kasich had dropped out, Cruz would have won the Republican nomination. If Cruz and Kasich had dropped out, Rubio would have won the nomination. Trump's win did not reflect how most voters felt about him. According to most Republican voters, he should not have been their party's nominee.

What happened here was an example of *vote splitting*, which occurs when similar candidates (which Cruz, Rubio, and Kasich were, at least relative to Trump) divide a majority of the votes, allowing the less popular candidate to rise to the top. Not only does such a candidate lack the support of the majority, but they can also misrepresent the consensus values of the group by espousing fringe ideologies, as Trump did.[†]

Another famous election that brought the perils of plurality to the national stage was the 1998 Minnesota gubernatorial race. (This was the first time I started paying attention to the math of politics.) The main candidates were Norm Coleman (Republican), Skip Humphrey (Democrat), and Jesse "The Body" Ventura (Reform), a former World Wrestling Federation wrestler and an actor whose political résumé consisted of a four-year post as mayor of Brooklyn Park, a small Minnesota

---

[*] The numbers don't add up to 100% because other candidates won the remaining votes, but excluding them doesn't affect this discussion.

[†] I will use the term "fringe" loosely to mean a candidate or political platform that is outside the mainstream. Of course, the term is temporally relative; what is fringe at one moment can quickly morph into the prevailing position, as was the case with Trumpism.

town. Ventura was a populist who flaunted his ignorance of the ways of politics and urged voters to vote against "politics as usual."

Ventura won with 37% of the votes. Coleman was second with 34%, followed by Humphrey with 28%. Much was made of the results, which were billed as a revolt by young voters, a rejection of the same old stale politics. But polls revealed a simpler, purely mathematical explanation for the upset: vote splitting. A huge majority of the Humphrey voters would rather have seen Coleman than Ventura as the governor. In other words, nearly 62% of the voters preferred Coleman over Ventura. This might not seem like classic vote splitting by similar candidates because Coleman and Humphrey came from different parties, but it does when viewed as a contest between "standard" and "nonstandard" politics, between the mainstream and the fringe, in which the mainstream candidates ended up dividing the votes.

Ventura was a one-term governor who is mostly remembered for "Jesse checks," consumer rebates on sales taxes that he pushed through the Minnesota legislature. For many of us of a certain generation, his most memorable achievement is his role in the movie *Predator*, a classic 1980s testosterone fest of action badassness. Coincidentally, the lead role in that movie was played by Arnold Schwarzenegger, another future governor who was elected in 2003 in a recall election that replaced Governor Gray Davis of California. Like Ventura, Schwarzenegger won with fewer than half the votes.

SPOILS OF PLURALITY

Even if the 2016 and 2020 presidential elections are the ones that portend the end of U.S. democracy in historical memory, math books will forever speak of a different one.* In 2000, Republican candidate George W. Bush and Democratic candidate Al Gore were locked in a bitter battle for the White House that came down to Florida. The winner in that state would secure enough Electoral College votes to make him the new president. The Electoral College is a topic for later, but I'll briefly describe it here.

---

* Forever and repeatedly: I will use it as an example nine more times.

In U.S. presidential elections, there are a total of 538 electoral votes. Each state is allocated a certain number of them. A candidate who wins by plurality vote in a state (except in Maine and Nebraska) wins the electoral votes of that state, and a cumulative majority of electoral votes—at least 270—must be secured to win the presidency.

In 2000, this was the Florida vote count:

Candidate	Number of votes
George W. Bush	2,912,790
Al Gore	2,912,253

Bush won the state by 537, or 0.009%, of the votes. The narrow margin prompted Gore to request a hand recount that was contested all the way up to the United States Supreme Court. As the nation waited with bated breath and tried to distract itself by learning new terminology such as "dimpled chad,"* the highest court finally ruled in favor of Bush. The Supreme Court blocked the recount, reversing the decision of the Florida supreme court. Bush claimed all of Florida's twenty-five electoral votes, giving him a total of 271 and hence the presidency.

But that's not where the interesting math lives. There was another candidate on the Florida ballot, Ralph Nader from the Green Party, who had gained traction nationally by positioning himself to the left of Gore, promising to promote environmental justice, universal health care, and affordable housing while fighting corporate greed. Including Nader, the Florida tally really looked like this:

Candidate	Number of votes
George W. Bush	2,912,790
Al Gore	2,912,253
Ralph Nader	97,488

* Some ballots are marked by punching a hole through them, train conductor–style. The small circle of paper that is punched out is a *chad*. When the chad is still attached to the paper, it can be *hanging* or *dimpled*, depending on how fiercely it's clinging to the paper. Much of the legal battle in the 2000 Bush-Gore election had to do with whether ballots that had such chads should be counted.

Polls showed that about 45% of Nader voters would have voted for Gore if Nader hadn't been an option, which is unsurprising because Nader was politically closer to Gore than to Bush. About 25% would not have voted at all and the rest would have voted for Bush or other third-party candidates. This means that if Nader hadn't been running for president and we redistributed his votes according to the polling—45% to Gore and (generously) 30% to Bush—more Floridians would have chosen Gore for president than Bush. In fact, Gore would have won by about 14,000 votes. But because the plurality method can't read the minds of voters, George W. Bush became the forty-third president of the United States.

What happened in this election is called the *spoiler effect*, which occurs when a candidate who has no chance of winning—in this case, Nader—takes away enough votes from a major candidate—in this case, Gore—to flip the election in someone else's favor—in this case, Bush.* Spoiler candidates are typically on the same side of the political spectrum as the spoilees, but farther away from the center.

U.S. presidential election spoilers are not a rare occurrence. Up to seven elections, or about 12% of them, might have been decided by a spoiler:[†]

- Trump-Clinton, 2016
- Bush-Gore, 2000
- Clinton-Bush, 1992
- Wilson-Roosevelt, 1912
- Harrison-Cleveland, 1888
- Taylor-Cass, 1848
- Polk-Clay, 1844

"Might" is an important word because the polls that ask people who they would have voted for had they not been able to vote for their favorite candidate are not necessarily complete, representative, or reliable. It is still debated whether Ross Perot, a Texas billionaire who collected an

* This definition of a spoiler is not precise enough to be mathematically useful. We'll tighten it up when we talk about ranked voting.

† William Poundstone's book *Gaming the Vote* has more about these elections and lots more examples of vote splitting and spoiler.

impressive 18.9% of votes nationwide (but no electoral votes) spoiled the election for George H. W. Bush and helped elect Bill Clinton in 1992.

An even more disputable instance is the 2016 election in which the Green Party's candidate, Jill Stein, might have acted as a spoiler for Hillary Clinton. In three of the key states that Donald Trump won— Michigan, Pennsylvania, and Wisconsin—Stein took more votes than Trump's margin of victory. If we assume that Stein voters would have voted for Clinton over Trump had Stein not been in the race, Clinton would have won those states and the presidency. However, it's not so clear that this assumption is justified. Stein voters might have turned to her because of their dissatisfaction with the Democratic Party, so some might have chosen Trump or the Libertarian candidate Gary Johnson instead. Even more likely, they might simply have sat out the election.

If you consider the possibility of spoilers a flaw, then we have a presidential voting system that might have malfunctioned 12% of the time. Does that sound like a lot? Does plurality have enough merit for us to tolerate the spoiler defect? Would you buy a gallon of milk if there was a 12% chance it was spoiled?

How big a problem is the spoiler effect more broadly? Remember the study of 154 mayoral elections that concluded that most people vote against the winning candidate? That finding likely extrapolates to other types of elections, simply because American politics is dominated by two parties whose candidates (or subsets of candidates) are bound to be politically proximate, creating fertile ground for spoilers at the primary, within-party stage. Third-party spoilers in general elections might not be as pronounced; some estimates say that only about 1.5% of general races are decided by such candidates. But when you toss in nonpartisan races, such as most local school board elections, spoilers might not be so uncommon.

A PARTY OF TWO

In 2019, the mayor of Fall River, Massachusetts, Jasiel Correia, was indicted for defrauding investors. He was accused of spending their money on a lavish lifestyle that included fancy cars and adult entertainment.

After he refused to resign, a recall election was held. The first part of the ballot asked whether Correia should be recalled. The people's choice was clear: 61.5% of voters said yes and 38.6% said no. The second part of the ballot asked who the new mayor should be. Correia won with 35.5% of the votes (yes, he was one of the candidates).* No other candidate received more than 35.4% of the votes. He remained the mayor of Fall River, a town where the large majority of voters thought he should have been ousted. In 2021, Correia was sentenced by a federal jury to six years in prison.

A similar dynamic received national attention in the media-frenzied 2021 recall election of California governor Gavin Newsom. The recall, which was initiated by conservative forces who had long been annoyed by Newsom's progressive politics, found broader support in the public's dissatisfaction with his handling of the COVID-19 pandemic. California voters first had to say whether they wanted to recall Newsom and then to select one of forty-six(!) replacement candidates who made it onto the ballot. Unlike in Correia's case, Newsom was not one of them. With a long, colorful cast of characters befitting a reality TV spectacle that included a shaman, a YouTube influencer, and a former Olympic athlete, the fear was that a small plurality would guarantee a win, possibly less than 30%. But if Newsom had been recalled by a tight margin, say 52%–48%, that would have meant that more people wanted him to stay as governor than voted for his replacement. Newsom avoided becoming a mathematical anomaly; Californians kept him in office, with 61.9% affirming their confidence in him.

These examples illustrate that problems with plurality go deeper than just vote splitting and the spoiler effect. As we'll see next, they can manifest themselves in subtle and peculiar ways that can thwart the views of the electorate, whether they are expressed explicitly or not.

---

* This number is less than the number of people who wanted Correia to remain mayor. Either some people answered only the first question or they thought Correia should not be removed but would rather not have him as mayor or some combination of the two.

Duverger's Law

In his 1951 book *Political Parties,* French political scientist and attorney Maurice Duverger posited that plurality elections lead to two-party systems. One reason, he argued, is that it is hard for smaller parties to reach the threshold number of votes required to win. Because they stand virtually no chance, candidates from smaller parties are disinclined to enter races or end up dropping out. These decisions are potentially reinforced by the concern that they'll act as spoilers. This was the case, for example, when Michael Bloomberg opted not to run for president in 2016 as an independent to avoid taking votes away from the Democratic nominee. In Maine's 2014 gubernatorial election, the independent candidate Eliot Cutler even told his supporters not to vote for him and to instead choose the Democratic candidate, Mike Michaud. That state had been particularly traumatized by the spoiler effect, which had decided the outcome of the gubernatorial election four years earlier in favor of the radical right-wing Republican Paul LePage.

The second reason, Duverger claimed, is a phenomenon called *favorite betrayal.* A voter who does not want to "waste" their vote on a minor candidate will instead cast their ballot for one of the major candidates even if this is not their favorite. The reasoning is that the minor candidate has no chance of winning, so why use the vote in a way that will have no bearing on the outcome? The vote is better used to support one of the main candidates who will wage the ultimate battle. Anyone who decided to vote for Al Gore instead of Ralph Nader in 2000 or for Hillary Clinton instead of Jill Stein in 2016 understands this concept. I am guilty of this practice—I cast a ballot in a recent presidential primary election in my state based on who I thought might win against the candidate from the other party and not on how I really felt.

The outcome of favorite betrayal is that minor candidates who choose to stay in a race receive fewer votes than the actual amount of support they have. Polling sees them as insignificant, and the media follow suit with lack of coverage. All this, in turn, makes such candidates unlikely to run again because they feel the participation barriers are insurmountable.

In a system that encourages minor candidates to exit races and voters to betray those who stick around, political diversity is diminished. The civic system is ultimately reduced to a duopoly. Nonfringe, reasonable minor candidates are forced to shift their positions toward one of the two major parties as they try to peel voters away. And the major candidates can simply counter by cherry-picking the messaging from minor candidates that resonates with voters and incorporating it into their own platforms, thereby preventing the exodus of their supporters. In short, there is no incentive to remain independent or to appeal to a smaller slice of the electorate because the race is all about beating every other candidate.

It's important to note that both candidate *and* voter are compelled to alter their conduct. The candidate chooses not to participate in the political process and the voter casts their ballot for someone other than their top choice. A voting system that dissuades fresh voices from entering the political arena and offers voting dishonesty as a necessary option should undoubtedly be discarded as undemocratic.

Despite its name, Duverger's law is not a law or a theorem. Duverger himself described it as something that "approaches most nearly perhaps to a true sociological law." It is empirical, far from deterministic, and contested academically. While the United States is an obvious example of Duverger's law in action, plurality elections lead to multiparty systems in many other countries. For example, Canada, which uses plurality for its parliamentary elections, currently (August 2023) has five parties in the House of Commons. The United Kingdom has three major parties; India has seven national and fifty-four state parties. In fact, one might argue that the reason Duverger's law holds in the United States is not plurality at all but a plutocratic system that is legislatively enshrined by the Supreme Court's *Citizens United* decision of 2010 that makes it financially unfeasible—and hence practically impossible—for minor candidates to reach enough voters.*

---

* *Citizens United* reversed campaign finance restrictions, giving an upper hand to wealthy candidates.

Duverger himself was aware of the limitations of his "law" and allowed for exceptions. For example, it is possible that certain political
parties do not have a strong, broad national presence but have concen-
trations of local support. In that case, they might claim a few seats in the
national legislature by carrying a few districts where they have sufficient
following. This persistence of support in a number of enclaves might
explain the continued presence of the Liberal Democrats in the UK. In
the Canadian House of Commons, local two-party systems—but different pairs of parties in different districts—aggregate to a multiparty
system at the national level. The point is that Duverger's law is still in
evidence locally despite its absence in federal arenas.

The ongoing debate notwithstanding, Duverger's law is a clear force
in analysis of elections and of the relationship between election outcomes and voting methods. A large 2012 study of 6,745 single-member
districts from fifty-three democracies found that the outcome, on average, is indeed a duopoly, with third (and rarely fourth, fifth, etc.) parties
occasionally managing to break the mold and eke out representation.
In the United States, instances of breaking the duopoly are rare in recent
history. The few third-party or independent members of Congress typically align strongly with one of the two major parties, caucusing with
them and often even becoming their members during their legislative
tenure.*

A hypothesis of Duverger's law that is harder to dispute is that *proportional representation* systems are considerably better at stimulating
diversity. We'll talk about these in chapter 13, but the focal idea is that a
legislature should reflect the voters' wishes in a proportional manner:
the percentage of seats won should be the same or close to the percentage of votes received. So one might, for example, have voting districts
in each state from which multiple candidates are elected to the House.
That way, if, say, a voting district elects five members and the Alien Party

---

* The grip of Duverger's law on the U.S. political system has tightened steadily over the
decades. Until the 1960s, members of various parties—including Greenback, Labor, Populist,
Silver, Progressive, Socialist, Prohibitionist, and Minnesota Farmer-Labor Parties—made regular appearances in the House of Representatives. This trend has all but disappeared in recent
decades.

wins 20% of the vote, then one Alien goes to the House because that is 20% of the allocated seats for that district. This possibility encourages campaigning on narrower platforms that might appeal to only a small percentage of people. This is a viable strategy because the system does not require a win over all other candidates; it only requires crossing some reasonable threshold.

Countries that use proportional representation rather than plurality, as many in Europe do, are on the whole more politically diverse (although there are exceptions here as well, such as Ireland and Malta). The reason is that the bar for representation is lower than it would be for plurality. Yes, the system might give a voice to people from the fringes, like the adult film actress Cicciolina who won on a libertarian platform in Italy or more recently the members of the One Nation Party in Australia, whose members' values range from Islamophobia to climate change denial. But that is the price of democracy. These fringe figures are elected because more voters are truly enfranchised. When fringe candidates are elected using plurality, it's because vote splitting has disenfranchised the majority of the electorate. According to Duverger's law, the plurality voting system stifles political diversity, and in the final analysis, that is the more dangerous outcome, the surest path to tyranny, dictatorship, and autocracy.

Center Squeeze

Let's say there are three candidates, Left, Center, and Right, running in a plurality election. On Election Day, 36% of people vote for Left, 34% for Right, and the remaining 30% for Center. Left wins and everyone goes home. But what if all the Right voters prefer Center to Left? Then we'd have the familiar situation in which all the Center and all the Right voters, 34% + 30% = 64%, would rather have Center be the winner.

Of course, the roles of Left and Right can be flipped in this scenario. The point is that the Center candidate, whom we imagine to be more moderate, is pushed out of the election by one of the more partisan, possibly extreme candidates. This smells like vote splitting, but it is not because we do not have candidates who are politically similar attracting portions of votes from the same voter pool. Center and Right voters are different, and neither of them has enough votes to win. The subtlety is in the second-tier analysis, namely in the Right's second choice, because this is where we see that the selection of Left does not really reflect the will of the majority.

This situation is called the *center-squeeze effect*. It occurs when moderate candidates are squeezed out of the election by polarizing candidates, even if they might be preferred over each of those candidates in individual matchups and have the potential to build consensus among the majority of the voters. The center-squeeze effect was evident in the Republican primaries leading up to the 2022 midterm elections, when many extreme candidates secured nominations over more moderate ones.

Negative Campaigning

Here's a scene you might suspect comes from a children's book about citizenship. Two candidates are running against each other for an elected seat, but they are campaigning together. When one speaks, the other listens. When it is the other one's turn to speak, she praises her opponent but politely emphasizes her own additional qualities and policy ideas, making the case that you should vote for her. At the end of the discussion, they shake hands, smile, and leave the stage together amicably. As icing on this unicorn cake, they endorse each other in the race.

Ridiculous, right? Isn't campaigning all about digging up dirt, defaming, and trash talking? It certainly feels like it. Character assassination has been a mainstay of U.S. politics since the infamous 1964 TV ad for Lyndon Johnson's campaign that cut dramatically from a girl picking a daisy to a screen-filling mushroom cloud, suggesting that Barry Goldwater would lead the nation into nuclear war.*

Even though the volume of negative ads seems to be decreasing (it peaked in 2012), mudslinging still works. It energizes the base and

---

* That ad feels like a PBS Kids video these days. Campaign ads have become much more vitriolic. For example, look up the 2011 *Give us your cash* ad on YouTube.

makes elections seem more relevant by dramatizing them. It preys on fears and insecurities. Trashing the opponent projects an image of strength and leadership. Never mind that it tells us nothing about the policies of the mudslinger—who cares about that anyway?—the idea is to scare voters and convince them that the other candidate is to blame for bad things.

Although negative campaigning has developed into a vast and complex machine, there is a simple little engine that gives it wings: plurality voting. Plurality encourages negative campaigning because there is no penalty for slander. It's a zero-sum game. A no vote for one person amounts to support for the other, so the strategy focuses on making the voter *refuse* to vote for someone else. And what better way to do this than paint the opponent as incompetent, dangerous, or corrupt?

And what better way to turn people off from voting at all? So many Americans are disgusted with politics that they have removed themselves from the voting process entirely. The highest turnout for a presidential election in recent decades is 62% in 2020, a percentage that puts the United States at about thirtieth place among the OECD (Organisation for Economic Co-operation and Development) countries. This poor showing is especially embarrassing for a country that purports to be the world's leading democracy. Negative campaigning and the incivility that comes with it are some of the strong drivers of voter apathy. It's worth reiterating that plurality voting fuels that apathy.

By the way, the utopian scene of two candidates being civil to each other, focusing on policies and accomplishments instead of stabbing at each other, is not fiction at all. Mark Eves and Betsy Sweet, Democrats who ran for Maine governor in 2018, campaigned like this. So did Mark Leno and Jane Kim in the 2018 San Francisco mayoral elections. So did Andrew Yang and Kathryn Garcia in the 2021 New York mayoral elections. The list goes on. What did all these elections have in common? The voting system was not plurality. It was something smarter and mathematically more sophisticated, a method that among its many merits does not reward negative campaigning. Intrigued? Hang around until the next chapter.

SECOND TIME'S THE CHARM

Some general elections are conducted using two rounds of voting. The first is the usual plurality vote. But then, if no candidate wins over 50% of the votes, a second election is scheduled in which only the two top-scoring candidates compete. This round is called the *runoff* election and the system is sometimes called *plurality with runoff*. The idea is to avoid declaring a winner who has only minority support by forcing the second election in which one of the two candidates will necessarily collect the majority of the votes to win.

Both high-profile U.S. Senate races in Georgia went to a runoff in 2020. In one race, Jon Ossoff (Democrat) and David Perdue (Republican) were the top two scorers in the plurality round but were both shy of the 50% mark, with 47.9% and 49.7% of the votes, respectively. In the other, Raphael Warnock (Democrat) and Kelly Loeffler (Republican) won 32.9% and 25.9% of the votes in the first round. Georgia voters came out again two months later to cast their votes for one of the two candidates in each of the two runoff races. Ossoff and Warnock won, handing the Democratic Party the narrowest of advantages in the Senate. Two years later, in the 2022 midterm elections, Warnock was engaged in another high-profile runoff, this time with former professional football player Herschel Walker (and he won again).*

Louisiana is another state that requires a general election runoff under the same conditions as Georgia. Ten states use runoffs for primary elections in case no candidate wins a majority (except in North Carolina, where the quota is 30%). Hundreds of cities and towns use runoffs for mayoral elections.

---

* Georgia's runoff system was established in 1964 by state representative and segregationist Denmark Groover after he lost an election. Groover wanted to diminish the power of Black voters by asking people to set aside time, often during work hours, to get out and vote in yet another election. The turnout in runoff elections is typically lower, and that is especially true for minority voters. As he publicly acknowledged, Groover was banking on this to ensure that Georgia's voters would elect white candidates. Groover is no longer around to enjoy the irony that both Warnock and Walker are Black.

Several states now also use *top-two primaries* (or *jungle primaries*). Recognizing that partisan primaries are one of the drivers of extreme political polarization, these states have instituted a system in which all candidates, no matter what their party affiliation is, are listed on the same primary ballot. The top two finishers in the primary (usually a plurality election) then advance to a runoff. It is not uncommon for the runoff to be between candidates from the same party. Washington and California use this system for congressional and state primaries and Nebraska uses it for state races only. In 2020, Alaska went further and implemented top-four primaries for congressional and state elections. By moving four candidates forward, Alaska leaves the (mathematical) shelter of May's theorem. We'll see their solution in the next chapter: the *instant runoff* method.*

Runoff elections are common around the world. Over seventy countries use it for presidential elections, including Brazil, Chile, Finland, France, Nigeria, Portugal, and Turkey. The eyes of the world were fixed on France's 2022 runoff between the liberal incumbent Emmanuel Macron and the right-wing candidate Marine Le Pen, a rematch of the 2017 "battle for the soul of France." Macron won by a comfortable margin.

———

A runoff election with two candidates is better than a simple plurality because it tells us that the majority of voters prefer one candidate to the other. And we have adopted the principle that when the majority agrees on something, that's what we should go with. If an election does not have a majority winner, a runoff will produce one. Sounds great, right? We've solved the problems with plurality elections!

Nope. Runoff does not eliminate any of the problems we've identified. Consider, for example, the results of the 2021 Boston *nonpartisan* (meaning all candidates run at the same time) preliminary mayoral election in figure 2.2.

---

* The same system was approved in Nevada in 2022, but by state law it must be approved again in 2024 in order to take effect in 2026.

Michelle Wu

33.4%	36,060 votes

Annissa Essaibi George

22.5%	24,268 votes

Andrea Joy Campbell

19.7%	21,299 votes

Kim M Janey

19.5%	21,047 votes

John F Barros

3.2%	3,459 votes

All others

1.7%	1,839 votes
	107,972 votes

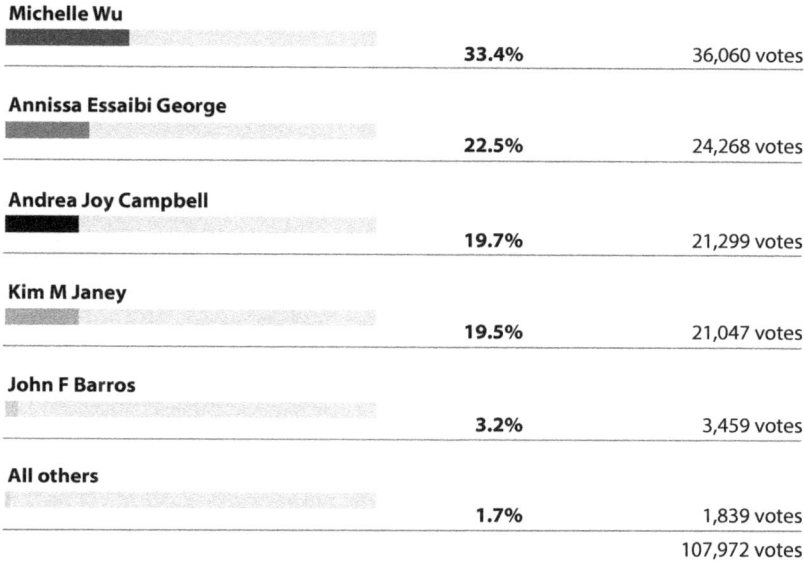

FIGURE 2.2. Boston mayoral preliminary election results, 2021.

As Boston election rules mandate, the two top scorers, Michelle Wu and Annissa Essaibi George, competed again in a runoff election seven weeks later. Wu won and became the first elected woman and a person of color to lead the city. But what if, for example, Campbell and Janey had split votes? That would have meant that had one of them dropped out, say Janey, Campbell would have carried 19.7% + 19.5% = 39.2% of the votes, handily qualifying for the runoff instead of George and perhaps winning the entire race. Or maybe Barros was a spoiler for Janey. With his 3.2% of the votes added to hers, Janey would have captured 22.7%, edging out George for a spot in the runoff.

The first round of the 2023 Chicago mayoral elections looked similar. All sorts of possible vote-splitting and spoiler scenarios could have sent different candidates to the runoff. As is always the case with plurality, we will never know.

One of the more famous examples is the 1991 Louisiana gubernatorial race. It had three leading contenders: the incumbent Buddy Roemer, a Republican who had switched from Democrat while in office; Edwin Edwards, a Democrat generally considered to be corrupt (probably a founded suspicion because he was sentenced to ten years in prison for racketeering in 2001); and . . . drumroll . . . David Duke, a Republican who also happened to be a white supremacist and a former grand wizard of the Knights of the Ku Klux Klan.

In the first round, Edwards won 33.8% of the vote, the highest percentage. Then came the shocker: Duke was in second place with 31.7% of the votes. Roemer followed in third with 26.5%. So Edwards and Duke advanced to the runoff, leaving Roemer voters tearing their hair out. Between the crooked rock and the racist hard place, they ultimately chose the lesser evil. Edwards won the runoff with 61% of the vote. The agony of the Roemer voters was immortalized in bumper stickers that said "Vote for the Lizard, not the Wizard" and "Vote for the crook—it's important."

The mathematical curiosity in this election is that polls showed Roemer beating both Edwards and Duke in one-on-one contests. That is, when voters were asked who they would pick between Roemer and Edwards or Roemer and Duke, Roemer came out ahead. In addition, most Duke voters had Roemer as their second choice. Neither the initial plurality round nor the subsequent runoff could detect any of this and the election ultimately produced the "wrong" winner. The only solution that plurality offers would be for Duke voters who also liked Roemer to engage in favorite betrayal, ensuring that at least their second choice was elected.

———

Runoff issues are not confined to the United States. Let's look more closely at the 2022 French presidential election (figure 2.3).

Almost any of the candidates below Jean-Luc Mélenchon could have been his spoiler. If he had received just 1.46% of the vote from somebody else, he could have bumped Marine Le Pen into third place and gone into the runoff election with Emmanuel Macron.

Or look at the 2019 Croatian presidential election (figure 2.4). All sorts of possible vote-splitting and spoiler scenarios could have sent different candidates to the runoff.

Percentage of votes per candidate

Candidate	Percentage
Macron	27.84%
Le Pen	23.15%
Mélenchon	21.95%
Zemmour	7.07%
Pécresse	4.78%
Jadot	4.63%

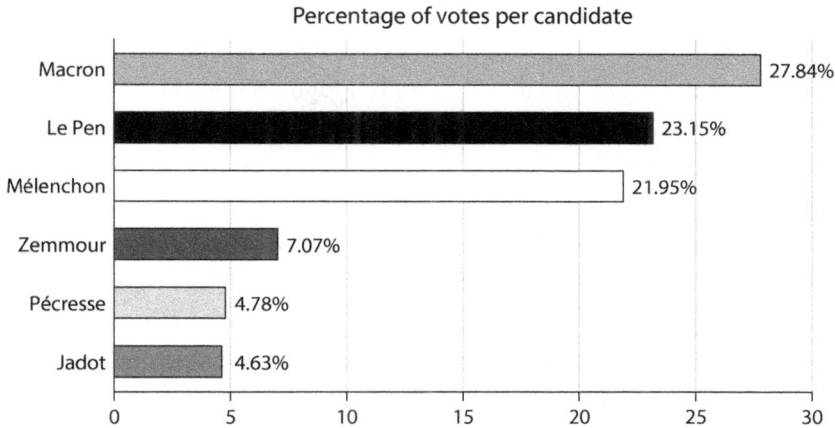

Other candidates:
Jean Lassalle 3.13%, Fabien Roussel 2.28%, Nicolas Dupont-Aignan 2.06%, Anne Hidalgo 1.75%, Philippe Poutou 0.77%, Nathalie Arthaud 0.56%

FIGURE 2.3. First-round French presidential election results, 2022.

Candidate ⇕	Party ⇕	First round		Second round	
		Votes ⇕	% ⇕	Votes ⇕	% ⇕
Zoran Milanović	Social democratic party	562,783	29.91	1,034,170	52.66
Kolinda Grabar-Kitarović	Independent (HDZ)[a]	507,628	26.98	929,707	47.34
Miroslav Škoro	Independent	465,704	24.75		
Mislav Kolakušić	Independent	111,916	5.95		
Dario Juričan	Independent	87,883	4.67		
Dalija Orešković	Independent	55,163	2.93		
Ivan Pernar	Party of ivan pernar	44,057	2.34		
Katarina Peović	Workers' front	21,387	1.14		
Dejan Kovač	Croatian social liberal party	18,107	0.96		
Anto Đapić	DESNO	4,001	0.21		
Nedjeljko Babić	HSSČKŠ	3,014	0.16		

FIGURE 2.4. Croatian presidential election results, 2019.

What we see is that the usual plurality unknowables persist even into the runoff election for the simple reason that the first round is conducted using plurality, with all of its baggage. While the runoff does ultimately allow voters to give more input, namely to disregard their initial decision and declare their preference between the two candidates still standing, the choice they are offered is flawed at the root. May's theorem ratifies the use of simple majority for the two-candidate round, but the two candidates vying for it are not necessarily the right options. The math of the second round checks out, but the math of the first round is still bad. And because the second round depends on the first, the entire process is defective.

Runoff elections also have significant practical drawbacks. Elections are organizational nightmares. The cost can be enormous, and I am not talking about the millions typically spent on campaigning in major races. I am talking about the money spent on the machines, the army of personnel, the polling places, and so on. In 2017, an MIT Election Data and Science Lab study concluded that nationwide election administration costs about $8.10 per voter and adds up to about $2 billion per year. Researchers say that this is likely a lowball estimate. And then there's the cost to voters. By the time the runoff rolls around, people are weary of politics. If they also dislike both runoff candidates or have to take time off work to vote, the chances are lower that they'll show up a second time. The turnout for runoff elections is on average 35% lower than for the first round.

One of the better ways to vote, instant runoff, has runoff elections built into the system without the need to drag voters back to the booths. It performs runoff elections automatically in a gradual and more nuanced fashion instead of hacking away all but the top two candidates. But first, we must digress and talk about poker and ping-pong.

TIES AND TOP HATS

In January 2018, an official of the Virginia State Board of Elections, adhering to the rules set forth in 1705, dipped his hand into a ceramic bowl made by a local artist and drew one of the two pieces of paper in it. The paper said "David Yancey," indicating that Yancey had won a seat in the Virginia House of Delegates. This was the final resolution of an election in which he and his opponent Shelley Simonds each received 11,608 votes.

In June 2009, Thomas McGuire and Adam Trenk tied in the race for a city council seat in Cave Creek, Arizona. They each got 660 votes. The town judge pulled a deck of cards from a cowboy hat, shuffled it, and asked McGuire and Trenk to draw. McGuire drew a six of hearts. Trenk drew a king of hearts and got the seat on the council.

Such random, sometimes senseless, and often comical procedures for breaking ties litter the American election landscape. A 2014 Florida election was resolved by a series of steps involving a top hat, a deck of cards, and ping-pong balls. A 2006 Alaska election was decided by a coin toss. New Mexico law mandates that a tie should be resolved by a game of chance, such as poker, while Nevada caucus ties call for drawing cards from a deck that has been shuffled at least seven times, with the highest card winning (not a big surprise considering Nevada's main claim to fame). More than half the states have some such lot-drawing or gaming procedure for breaking ties. Some require a runoff election to be held between the tying candidates, while some leave it to the governor or the Board of Elections to decide the winner. In all cases, the course of action is either random, left to personal or political whims, or leads to a new, expensive, and exhausting (for both voters and candidates) election.

The most dramatic example of a tie in U.S. history is the 1800 presidential election in which Thomas Jefferson and Aaron Burr each received 73 Electoral College votes. According to the Constitution, a vote in the House of Representatives shall in such a case break the tie. Each state's delegation carries one vote, so, for example, if this were to happen again today, California, with its 39.7 million inhabitants, would have the same voting weight as Wyoming, with its 0.6 million people. In 1800, there were sixteen states in the Union and thus sixteen votes to break the Jefferson-Burr tie. Thirty-six times the vote in the House was a tie. Jefferson was finally elected president on the thirty-seventh vote.

In the Senate's one-hundred-person structure, the vice-president acts as the tie breaker in case of a 50-50 split. In recent years, with the country and consequently the Senate split just about down the middle, the vice-president's vote has played an increasingly crucial role, perhaps an inappropriately outsized one.

If all this seems arbitrary, it is. Is it troubling that chance decides who gets to act as the representative of the people? Definitely. On the other hand, what else could the Virginia election officials have done in the Yancey-Simonds election? What else can the Senate do other than defer to the vice-president (and their political agenda)?

These election impasses and the subsequent need for a creative resolution are corollaries of plurality voting. Plurality elections contain no information that could be used to break ties in some informed way with input from the voters. What's left is dependence on centuries-old traditions, rules, and laws that are at best inappropriate for our time and are frequently just plain ridiculous.

There is nothing more to say here because there is nothing to be done. Mathematics will not tell us whether drawing cards is better than flipping a coin. Its advice is to turn to better voting methods.

CHAPTER 3

From Best to Worst

IF I'VE DONE MY JOB, you're now sprinting in the opposite direction from plurality. When we care about democracy, we can't be indifferent to the perils of this simplistic way of electing our officials, our representatives, our presidents, or, for that matter, our pizza toppings.

The fundamental problem with plurality is that it does not collect enough information; it asks the voter only for their top choice. There isn't much math can do with that. But what if more data is solicited? Then we can try to feed this additional information into more sophisticated math to find the winner who is the most satisfactory to the most people—a candidate who represents the broadest consensus.

CANDIDATE PROFILING

One way to gather more information is to ask the voters to *rank* the candidates in order of preference. To smooth the discussion, let's establish some terminology.

A voting method that asks for a ranking is called a *ranked choice* (or a *preferential*) *method*. Such a method presents each voter with a *preference ballot* (or *preference order*), as in figure 3.1.

Figure 3.2 shows an actual preference ballot from the 2018 Republican primary elections in Maine.

The voter does not have to fill out the entire ballot; they can just mark their top few choices in order and leave the rest blank. However, the more candidates each voter ranks, the more complete the picture of the

	Preference ranking			
	1st choice	2nd choice	3rd choice	4th choice
Candidate A	●	○	○	○
Candidate B	○	○	○	●
Candidate C	○	○	●	○
Candidate D	○	●	○	○

FIGURE 3.1. An example of a preference ballot with all choices marked.

collective preference will be. Examples here will assume that all voters ranked all candidates and will at times address what happens if that is not the case.

The collection of all the ballots is called a *profile* of the election. I'll summarize profiles as tables. For example, suppose an election had four candidates—A, B, C, and D—and twenty-seven voters. A profile might look like this:

9	6	4	3	3	2
A	B	A	B	D	C
C	D	D	C	A	D
D	C	B	D	C	B
B	A	C	A	B	A

The first column tells us that nine people ranked A first, C second, D third, and B fourth. From the second column, we know that six people ranked B first, D second, C third, and A fourth. We can abbreviate by employing the greater-than symbol (>), summarizing the columns as nine people choosing the ranking A > C > D > B, six people choosing B > D > C > A, and so on.

Not every possible ranking appears in the table. For example, you don't see A > C > B > D. This means that no voter expressed such a preference in this election. If every preference was expressed by at least one voter, our table would have twenty-four columns.*

* For n candidates, the number of possible ways to order them is n!, or n *factorial*, the product of all positive integers up to and including n. To see why, think of arranging n candidates behind lecterns on a stage for a debate. Any of them could go behind the first lectern, which

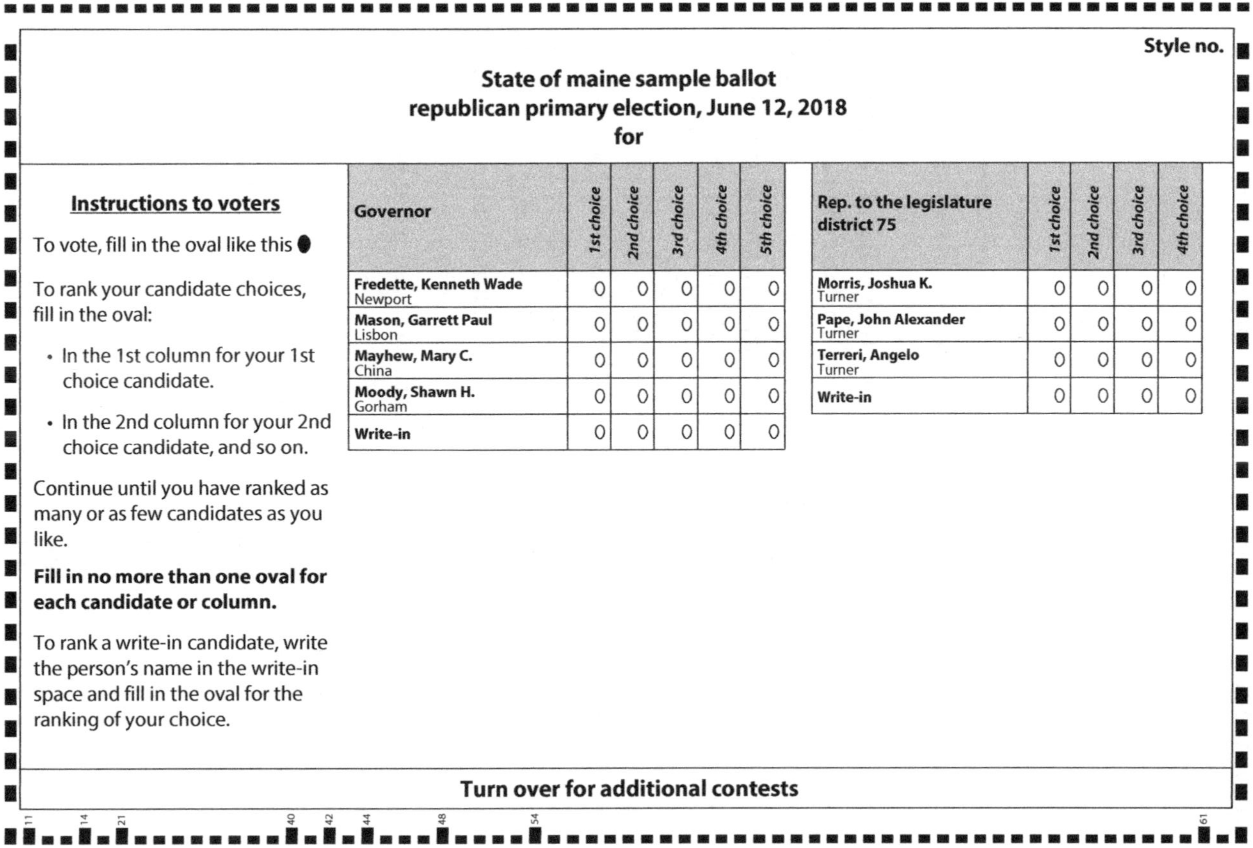

Style no.

State of maine sample ballot
republican primary election, June 12, 2018
for

Governor	1st choice	2nd choice	3rd choice	4th choice	5th choice
Fredette, Kenneth Wade Newport	O	O	O	O	O
Mason, Garrett Paul Lisbon	O	O	O	O	O
Mayhew, Mary C. China	O	O	O	O	O
Moody, Shawn H. Gorham	O	O	O	O	O
Write-in	O	O	O	O	O

Rep. to the legislature district 75	1st choice	2nd choice	3rd choice	4th choice
Morris, Joshua K. Turner	O	O	O	O
Pape, John Alexander Turner	O	O	O	O
Terreri, Angelo Turner	O	O	O	O
Write-in	O	O	O	O

Instructions to voters

To vote, fill in the oval like this ⬤

To rank your candidate choices, fill in the oval:

- In the 1st column for your 1st choice candidate.

- In the 2nd column for your 2nd choice candidate, and so on.

Continue until you have ranked as many or as few candidates as you like.

Fill in no more than one oval for each candidate or column.

To rank a write-in candidate, write the person's name in the write-in space and fill in the oval for the ranking of your choice.

Turn over for additional contests

FIGURE 3.2. Maine Republican primary ranked ballot, 2018.

From the profile, we can extract the results of the types of elections we saw in the previous chapter. Had this been a plurality election, all the information would be concentrated in the first row of the table:

9	6	4	3	3	2
A	B	A	B	D	C

This says that A won 9 + 4 = 13 votes, B won 6 + 3 = 9, C got 2, and D got 3. There is no majority winner (with twenty-seven voters, fourteen is the requisite majority). Candidate A is the plurality winner with the most first-place votes.

The profile can also be used to determine the winner of a runoff election. The real question in the runoff is this: If a voter did not have one of the top two candidates as their first choice already, which of those two would they vote for? The profile has the answer. We can look at such a voter's ballot and check which of those two candidates appears higher in their ranking.

In our example, A and B are the top scorers, so they advance to the runoff. The voters in the last two columns didn't have A or B as their top choice. Their favorites, C and D, are eliminated, so we need to ask these voters who they would choose between A and B. Each of the three voters whose ballots appear in the next-to-last column ranked A higher than B, so we assume that they would vote for A in the runoff. In the last column, the two voters ranked B higher than A, so we infer that they would vote for B.

This reasoning process can be visualized all at once by erasing C and D from the table and shifting cells up into the empty slots. In the original profile, first identify the cells to be eliminated (these cells are shaded):

gives n options. For the second lectern, there are $n-1$ options because one candidate has already been placed. For the third lectern, there are $n-2$ options. By the time we get to the last lectern, there is only one candidate left and we have no option but to put them there. Since for each placement, any of the subsequent placements could occur, the total is then the product of all the options, which is $n \times (n-1) \times (n-2) \ldots 2 \times 1 = n!$. This number gets very large as n gets bigger. In Gavin Newsom's 2021 recall election, which had forty-six candidates, the number of possible orderings would have been $46! = 5.5 \times 10^{57}$.

9	6	4	3	3	2
A	B	A	B	D	C
C	D	D	C	A	D
D	C	B	D	C	B
B	A	C	A	B	A

Then eliminate the data in those cells:

9	6	4	3	3	2
A	B	A	B		
				A	
		B			B
B	A		A	B	A

Finally, shift A and B up everywhere (and get rid of the bottom two rows of the table, which are now empty):

9	6	4	3	3	2
A	B	A	B	A	B
B	A	B	A	B	A

Now A has 9 + 4 + 3 = 16 first-place votes while B has 6 + 3 + 2 = 11, so A wins the runoff.

The plurality and the runoff winner happen to be the same, but this did not have to be the case. If the three voters in the next-to-last column had ranked B above A, then the final runoff tally would have been 14–13 in favor of B.

We're already seeing the value of preference ballots. If plurality is the method of choice, the data is there. If runoff is the method of choice, the data is there.

But neither of these methods uses *all* the information in the profile. Plurality hacks away all but the first row, while runoff eliminates all but the top two candidates in one fell swoop, without regard for how the voters judged the relative merits of the unfortunate losers. By looking at methods that actually use all the data collected from the voters— the *instant runoff,* the *Borda count,* and the *Condorcet method*—we can develop a general but nuanced picture of the benefits (and limitations) of ranked choice voting.

SUCCESSIVE ELIMINATION

Anyone who lives and votes in San Francisco, Oakland, Minneapolis, St. Paul, Santa Fe, New York City, or about fifty other jurisdictions across eighteen states has heard of *instant runoff*. So too those living anywhere in Maine and Alaska or those who vote from overseas in Alabama, Arkansas, Georgia, Louisiana, Mississippi, and South Carolina, including military personnel stationed abroad. These pioneers are at the forefront of a sea change driven by the recognition that plurality must be replaced by something better.

Reporters often call the system being used by these voters ranked choice voting, and that's true, but there is more to it. What's really being used is the instant runoff way of *tallying* ranked profiles. There are various ways to tally, and this is just one of them. But because it's the most popular, it has become synonymous with ranked choice voting.

Instant runoff is also called *sequential runoff*, or *Hare's method*, because it was proposed in 1859 by Sir Thomas Hare (1806–1891), an English lawyer, in *A Treatise on Election of Representatives, Parliamentary and Municipal*.* It is essentially a more refined version of the runoff method, the major difference being that losing candidates are removed one at a time instead of all but the top two being eliminated at once.

Here's the algorithm: with a ranked profile in hand (by which I mean loaded onto a computer),

- If there is a candidate who received the majority of first-place votes, that's the winner.
- If that is not the case, find the candidate with the least number of first-place votes.

---

* He actually proposed the method now called *single transferable vote*, which he designed to elect multiple candidates (see chapter 13). Instant runoff is a special case. It was developed by the MIT professor William Ware (1832–1915) in the 1870s. Ware implemented instant runoff at MIT, where it is used to this day.

- Remove this candidate from each ballot and give the votes for them to the next candidate on that ballot.
- Repeat until there is a candidate with the majority of first-place votes.

A couple of complications are worth acknowledging, although neither is terribly relevant for the development of the theory and they don't impair the smooth running of the method. We'll mostly ignore them:

- If a ballot is incomplete, meaning the voter ranked some but not all of the candidates, use it until there are no more candidates left on it. The partial ranking will still contribute to the process up to the point where the candidates appear on the ballot. A ballot that has no effect after some point is called *exhausted*.
- If two candidates tie for the least number of first-place votes at some stage of the elimination, who should be removed first? This tie can be broken in various ways, one of which removes the candidate with fewer second-place votes. If there is a second-place tie, look at the third-place votes, and so on. If there is a tie for the winner, look at who had the most first-place votes in the previous round or the one before it if that's a tie as well. Keep going like this if necessary. In the unlikely event that this doesn't work, then some other mathematical tie-breaking method based on the rankings can be used.

The second point warrants a brief reflection. Remember all those strange and ridiculous ways ties are broken in the plurality system? No need for them here. The information that is collected from the voters can be used to break ties in an informed way, not an arbitrary one.

An example is worth a thousand votes. Here's a ranked election profile with five candidates and forty voters:

12	9	8	6	3	2
A	C	B	E	D	A
B	D	C	B	A	C
D	B	A	A	C	E
E	A	E	D	B	D
C	E	D	C	E	B

From the first row, candidate A has the most top votes, $12 + 2 = 14$, but that's fewer than 21, the necessary majority, so we proceed with the instant runoff algorithm. B has 8 top votes, C has 9, D has 3, and E has 6. Because there is no majority winner and D has the fewest first-place votes, D is eliminated from the election and gets erased from the entire table:

12	9	8	6	3	2
A	C	B	E		A
B		C	B	A	C
	B	A	A	C	E
E	A	E		B	
C	E		C	E	B

This is now a four-candidate profile. In the first column, E is now the candidate who is ranked after B for those 12 voters, so we move E up into the empty slot. But then an empty slot opens where E was, so we move C into that slot. We repeat this in all the columns. In the next-to-last column, A is moved into the top slot formerly occupied by D. This means that A now has 3 first-place votes that used to belong to D: the votes of the candidate who is removed are transferred to the next candidate on the ballot. The table now looks like this:

12	9	8	6	3	2
A	C	B	E	A	A
B	B	C	B	C	C
E	A	A	A	B	E
C	E	E	C	E	B

What just happened was essentially a runoff between all candidates except D. Because D is no longer in the race, the runoff would ask the voters who is their next choice. But that's already encoded in the table; shifting candidates into the empty slots above them has the effect of everyone declaring who their choice is after D.

Now we look at first-place votes again. A has $12 + 3 + 2 = 17$, B still has 8, C still has 9, and E still has 6. No one has crossed the 21-vote threshold yet. Because E has the fewest first-place votes, they are the next to be eliminated.

12	9	8	6	3	2
A	C	B		A	A
B	B	C	B	C	C
	A	A	A	B	
C			C		B

Everybody is bumped up, and B inherits E's 6 first-place votes. And we have the result of the next automatized runoff among the candidates left standing after E is eliminated:

12	9	8	6	3	2
A	C	B	B	A	A
B	B	C	A	C	C
C	A	A	C	B	B

Candidate A still has 17 first-place votes, B now has $8 + 6 = 14$, and C has 9. No majority winner yet. Rinse and repeat, with C getting removed.

12	9	8	6	3	2
A		B	B	A	A
B	B		A		
	A	A		B	B

Shifting candidates up into empty slots gives the final table, the results of the runoff without C.

12	9	8	6	3	2
A	B	B	B	A	A
B	A	A	A	B	B

This is now a two-candidate profile, which must produce a majority winner (unless there is a tie). And sure enough, A still has 17 votes, while B now has $9 + 8 + 6 = 23$, a majority that wins the election.

The instant runoff winner, B, is not the same as the plurality winner, A, indicating that instant runoff is indeed doing something different from plurality. It's also apparent that it does something different from regular runoff. In the original profile, A and C were the top scorers, so they would have competed in the second round of a regular runoff. B, the instant runoff winner, would not have been in the picture at all. This is okay. It simply reinforces the fact that more information can produce a different result. Plurality, runoff, and instant runoff are like looking at a galaxy with the naked eye, through the Hubble telescope, or the Webb telescope. With each new instrument, we receive and process more information, producing an image that is more intricate and colorful—and more faithful—than the one before.

Even though B did not initially have a majority, they received an *eventual majority*. If this process had been conducted in person, voters would have been informed at each stage which candidate didn't make it because of their low score and asked who they would vote for next now that that candidate had been eliminated. If nobody won a majority again, the process would repeat. This is, for example, how host cities for the Olympics are chosen. Each member of the International Olympic Committee casts a single vote. If no city gets a majority, the city with the fewest votes is tossed out. Then members are asked to pick their favorite among the remaining cities and so on until a city receives a majority of the votes. Instant runoff does all this automatically, without having to repeatedly poll the voters. That is the "instant" part. And that is how a majority consensus is eventually reached.*

---

* Some proponents of instant runoff say that it ensures that a majority of voters will elect the winner. But this wording can be misinterpreted as saying that the winner is the first choice for a majority of the voters, which is not necessarily true. This is why I prefer the term "eventual majority." In addition, even an eventual majority might not be achieved because of ballot exhaustion. If enough people do not rank enough candidates and their ballots become exhausted, the winner might have only plurality and not majority support. But this is acceptable; all those exhausted ballots participated up to a point and helped influence the final outcome. The winner is still the product of a better-informed process than plurality, even if they do not receive an eventual majority.

For three candidates, runoff and instant runoff are the same because eliminating the candidate who places last is the same as advancing the top two candidates to the runoff.

Reaping the Benefits

Instant runoff knows enough to be robust against vote splitting and the spoiler effect, or at least some of their most obvious incarnations. (We'll see in a minute why it doesn't always avoid them.) For example, suppose the infamous 2000 Bush-Gore-Nader election was conducted using instant runoff in Florida. The profile of this version of the election might have looked something like this:

2,912,790	2,912,253	97,488
Bush	Gore	Nader
Gore	Nader	Gore
Nader	Bush	Bush

This is of course a tremendous oversimplification of the situation. For instance, it is not true that all Bush voters would have listed Gore next and Nader last. Many voters would probably have just circled one person and left incomplete ballots. And other candidates were on the ballot, such as Pat Buchanan, who won a respectable 17,484 votes. But whatever the profile might have been, one thing is certain: Nader would have been eliminated before Bush and Gore, and instant runoff would have transferred a large portion of his votes to Gore. At that point, Gore would have become the majority winner, carrying Florida and winning the presidency.

Had Republican primary voters been able to rank the candidates in 2016, Trump would not have won the nomination because instant runoff preempts vote splitting. If, say, Kasich had the fewest first-place votes, his votes would largely have been allocated to Cruz and Rubio, perhaps enough to put one of those candidates above Trump in terms of delegate count. If not, then either Cruz or Rubio would have been eliminated next. Let's say it was Rubio. The majority of Rubio votes would have gone to Cruz, who would then have overtaken Trump.

Instant runoff also reveals when vote splitting and the spoiler effect *didn't* occur. Take, for example, the August 2022 special election for Alaska's single seat in the House of Representatives. Two years earlier, Alaska had voted to enact instant runoff in all its elections, becoming the second state to do so after Maine. The candidates were Mary Peltola, a Democrat, and Sarah Palin and Nick Begich, both Republicans. The election profile looked like this:

27,053	15,467	11,290	34,049	3,652	21,272	47,407	4,645	23,747
Begich	Begich	Begich	Palin	Palin	Palin	Peltola	Peltola	Peltola
Palin	Peltola		Begich	Peltola		Begich	Palin	
Peltola	Palin		Peltola	Begich		Palin	Begich	

The empty spaces mean that those voters either did not submit fully ranked ballots or their ballots were eliminated after the first round for other technical reasons. The first-place tally looked like this:

Peltola: 75,799 (40.19%)
Palin: 58,937 (31.27%)
Begich: 53,810 (28.53%)

If this had been the usual plurality election, the story would have ended here. Peltola would have won the race. But the question of potential vote splitting between the two Republican candidates would naturally have emerged and would have remained unresolved. The simple first-place numbers, without the profile, give support for the claim that if only one of the Republicans had run, they would have won with about 60% of the vote. Textbook vote splitting, it seems, cast a shadow over Peltola's mandate to represent her constituency.

But this was an instant runoff election, so the story doesn't end here. Because Begich received the fewest first-place votes, he was eliminated first. The second round tally was

Peltola: 91,266 (51.48%)
Palin: 86,026 (48.52%)

There was no vote splitting! Enough Begich Republicans chose Peltola, a Democrat, over Palin, the other Republican, to seal the deal and keep Peltola on top. There was no monolith of Republican voters who put party above all. Instant runoff revealed nuance in the way Alaskans regarded the three candidates: enough of them didn't like what Palin had to offer that they parted ways with party-line voting.

After the election, Palin issued a statement/call to action/plea for donations. Among the usual incendiary sound bites like "freedom," "election integrity," and "the Left's corrupt agenda," she tossed this out: "Ranked choice voting is corrupt: 60% of Alaskans voted Republican, yet a Democrat won." Either she does not understand ranked choice or she is being deliberately misleading, but this is precisely an argument *for* ranked choice. Instant runoff shows that even though 60% of the voters voted for *a* Republican, those voters were unwilling to rally behind *one* of those Republicans. It shows clearly that 51.48% of voters would rather see Peltola than Palin in the House. Palin's is also an unwitting argument *against plurality* because her grievance would have seemed plausible if plurality had been used. All we would have seen is that 60% of the voters chose a Republican, and Peltola's win might in that case have felt unrepresentative of the voters' true sentiment because of vote splitting.

Another benefit of instant runoff is that it prevents "wasted" votes. A voter does not have to worry about supporting a candidate who has no chance of winning because if that candidate is eliminated, their vote will be transferred to the next choice. The voter continues to participate in the process and still has a say in the outcome.

On the flip side, a third-party candidate or a minor candidate from one of the big parties need not worry that they will act as a spoiler in a ranked choice election. They can guiltlessly campaign and try to appeal to voters with genuinely new ideas and a distinctive platform. Such an alteration in political psychology could have serious consequences. Minor candidates could assert themselves in the political process and voters could simultaneously express their support for greater political diversity. The two-party system could be shaken and diversified to everyone's benefit, encouraging greater participation of women and members of underrepresented groups. Duverger's law would become a thing of the past.

Of course, any time the doors to participation are flung open, assorted cranks and zealots also rush in with the fervor of Best Buy shoppers on Black Friday. Political extremism begets voting extremism because it appeals to raw emotion, fear, and insecurity. But fringe candidates are typically not the top choice for a majority of voters, who tend to put such candidates near the bottom of the ranked ballots. In round after round, extreme or extremely polarizing candidates rarely gain more first-place votes and eventually are overtaken by contenders with broader-based support.

Instant runoff can also act as an antidote to the negative campaigning that has reduced our political discourse to venomous trash talk. Doing well in an instant runoff means landing near the top of the ballot for as many voters as possible, so candidates have a strong interest in appealing to voters from all over the political spectrum. Alienating voters from the other side with negative campaigning is shooting yourself in the foot. Remember those pairs of candidates from chapter 2 who ran against each other but campaigned together and treated each other with respect? They did this because their jurisdictions use instant runoff. Each focused on the common ground, not on (manufactured) differences, in the hope that voters who supported other candidates would be open to placing them high on their ballots.

An example of how such a strategy might pay off is the 2010 Oakland, California, mayoral race. Jean Quan did not campaign by demanding that voters reject their favorite candidates. Instead, she asked voters to place her *second* on their ballots while trying to earn the support of as many of them as she could. The voters responded positively and Quan became the mayor of Oakland (after nine rounds of elimination) because she secured almost 25,000 second- and third-place votes.

What Do the Naysayers Say?

What are the objections to instant runoff? The most common is that it's too complicated. Sure, it's more complicated than plurality, but it's not *that* complicated. It simply asks voters to rank some candidates, which is not a terribly onerous task. A machine can then take over and perform the algorithm that governs the rounds of tallying and elimination.

Another objection is that instant runoff requires voters to have an opinion on all the candidates and be able to sort them in order of preference. And yes, that is tough when there are thirteen people running for a slot on the town school board and you've never heard of eleven of them. But it's also not true. Instant runoff still works with incomplete ballots, which means that a voter is allowed to fill out as much of it as they want. If they know and care about two candidates, they can rank just those two and that's that. For that matter, a voter can simulate a plurality election by selecting a single person. That voter's ballot will count only in the first round of instant runoff, but that's fine; they still contributed and expressed their opinion. Many instant runoff jurisdictions are aware of the potential for mental paralysis when too many options are presented and allow only a small number of candidates to be ranked, maybe four or five, to relieve the voter of forming an opinion on a horde of contenders.

A complaint sometimes advanced by opponents of ranked choice methods is that these methods allow voters to "vote more than once," referring to the voters' capability of expressing an opinion about more than one candidate by ranking them. This, of course, would fly in the face of the venerable one person, one vote dictum, but the flaw in this argument is that it equates "one vote" with "voting for one person." These happen to be identical for the plurality method, but for other methods, "one vote" must be interpreted according to the method used. In ranked choice voting, the meaning is "one ranking of as many candidates as desired." A ranked choice ballot allows a voter to express their preferences across all the candidates, but the vote *counts* only once. Interpreting this as fraudulent would be like saying dribbling a ball in basketball is cheating because it's not allowed in soccer.

Critics point out that instant runoff is far from instant, and that is true. The "instant" part refers to the possibility of conducting successive runoffs without holding separate elections. But if tens or hundreds of thousands of ranked ballots are in play, the tallying machines can take a while to produce the results of each round. In the 2020 New York City mayoral elections, the largest instant runoff election held in the United States to date, it took two weeks and eight rounds of counting before Eric Adams finally reached 50.4% of the votes and was declared the winner.*

Reporting the results to the public at each round might also foment excitement and then disappointment because the front-runner can change from round to round. Thankfully, this does not happen too frequently. In the vast majority of instant runoff races, the initial plurality winner ends up being the winner. For example, in the sixty-three races that took place across New York City in 2020, only three produced winners who did not initially get the most first-place votes.

These three races could have faced the criticism that the three people who got the most first-place votes, the "most liked" candidates in the sense of plurality, weren't necessarily the ultimate instant runoff winners. But we know enough about voting by now to understand that the plurality winner might not represent the will of the electorate. The real favorite might have been obscured by vote splitting or the spoiler effect. Still, conceding an election to someone who won fewer first-place votes is tough, both for the candidate and for their supporters. Just ask the 2009 Burlington, Vermont, mayoral candidate Kurt Wright, a Republican, who lost the election to Bob Kiss, a Progressive, even though Wright was the plurality winner. The backlash was so intense that Burlington repealed instant runoff in 2010. In 2021, it was restored for city council elections, but not mayoral races.†

---

* The process can be sped up through *batch elimination*, in which all candidates whose total number of first-place votes is fewer than any other candidate's first-place vote count. So if, say, six candidates win 47%, 26%, 15%, 5%, 4%, and 3%, respectively, of the first-place votes, the last three can be removed all at once because their total, 12%, is less than the lowest percentage of the other three candidates, 15%. This elimination does not affect the outcome because there is mathematically no way any of these low-scoring candidates can possibly win the instant runoff.

† Another example of a nail-biter upset was the selection of the city to host the 2000 Summer Olympics. Beijing was consistently ahead in several rounds of instant runoff but lost at the end to Sydney.

The 2009 Burlington count exhibited some other strange behavior that is rare (failure of *monotonicity*, which we'll talk about shortly), but the retaliation against it was not on mathematical grounds; it was a natural revolt against the unfamiliar, discomfort with the new. We tend to cling to the habitual—"Why are we replacing something we've been using for hundreds of years?"—or worse, fabricate a suspicion that someone is trying to pull a fast one—"Party X must be trying to do this to put Party Y at a disadvantage." Looking at first-place votes is easy and deeply embedded in our voting muscle memory; there's hard work ahead to deprogram and break the habit. The political-cultural shift that would have to accompany the transition to instant runoff is possibly its greatest obstacle.

Are There Mathematical Grounds for Objections?

Instant runoff does have mathematical vulnerabilities. Center squeeze, for example, can still present problems. Taking the earlier example in which 36% of voters supported Left, 34% supported Right, and 30% supported Center, suppose the election was really an instant runoff contest with this profile:*

36% rank Left > Center > Right
34% rank Right > Center > Left
16% rank Center > Left > Right
14% rank Center > Right > Left

Center is eliminated first and Left wins with 52% of the vote after one round. However, Center is preferred both to Left (64%–36%) and to Right (66%–34%) in individual matchups. Center has been squeezed by the more partisan candidates.

Another pitfall of instant runoff is that a candidate might be the second choice for many voters yet be eliminated right away. In the profile

---

* This profile reflects the political assumption that Left > Right > Center and Right > Left > Center won't appear because each extreme prefers Center to the other extreme. So the labels here aren't just generic mathematical A's and B's; it really is the centrist candidate that is getting squeezed.

below, D is the first to be eliminated, even though they are the first choice for nine voters, almost as many as all the other candidates, and the second choice for everybody else!

10	10	10	9
A	B	C	D
D	D	D	A
B	C	A	B
C	A	B	C

This seems unfair. So let's figure out precisely what "unfair" means. Let's look at just A, who is the instant runoff winner, and D. The table below has those two candidates highlighted.

10	10	10	9
A	B	C	D
D	D	D	A
B	C	A	B
C	A	B	C

Only ten voters (first column) prefer A to D, while $10 + 10 + 9 = 29$ (all the other columns), a strong majority, prefer D to A. In other words, if the voters were asked to choose between A and D, the answer would be a resounding D. So a candidate can do well overall, hanging out near the top of the profile, but instant runoff might punish them for not having enough first-place votes. This phenomenon is reminiscent of center squeeze; D could be a solid, moderate candidate who is not as polarizing as some others and so ended up near the top for most voters but not as the first choice for enough of them.

Let's talk about an even weirder issue. Here's a profile:

6	6	6	4	3	2
A	C	B	B	A	C
B	A	C	A	C	B
C	B	A	C	B	A

We start with the usual total of the first-place votes: A has 9, B has 10, and C has 8, which means that C is out of the picture. The new profile is

6	6	6	4	3	2
A	A	B	B	A	B
B	B	A	A	B	A

A has 15 first-place votes and B has 12, so A wins.

But now suppose that the initial profile summarizes the results of a poll conducted before the actual election and that between this poll and the election something happens that brings more voters to candidate A's camp—maybe a leading newspaper endorses them. As a result, three of four people who have the ranking B > A > C change their minds and decide to vote A > B > C, joining the six people who already have this ranking. In addition, both voters who have C > B > A also decide to place A higher, so their ballots now look like C > A > B, the same ranking as the six people in the second column.

After these shifts, the actual election profile looks like this:

9	8	6	1	3
A	C	B	B	A
B	A	C	A	C
C	B	A	C	B

Now B is eliminated first. This gives a new profile:

9	8	6	1	3
A	C	C	A	A
C	A	A	C	C

C wins 14–13! Let's summarize what just happened: A was doing well. Then several people decided to rank them higher, which turned out to be the kiss of death for A and A lost. Furthermore, the way those people decided to change their rankings *had nothing to do with* C, yet C ended up winning. The relative rankings of C versus A or C versus B did not change from one profile to the other. The same number of people had the same pairwise preferences for C in relation to A and B.

This example demonstrates that instant runoff fails *monotonicity* (or exhibits the *paradox of positive association*), a concept we've encountered before. A method is *monotone* if a winning candidate cannot become a losing candidate by gaining more votes. In the example, A was winning, then they received more votes (in the sense that they got placed higher on some ballots) and lost. Doing better is not necessarily better.

We start to see hints of an important feature of instant runoff that I'll devote a later chapter to: susceptibility to strategic voting. Supporters of candidate C can scheme to place A above B even if that is not their true preference because that helps their candidate get elected.

The last lesson we can extract is this: contrary to popular perception, instant runoff does not eliminate spoilers—when you define spoilers carefully, that is. We've described a spoiler informally as a minor candidate who siphons away enough votes from a major candidate to flip the election. But we now want a more precise statement. And here is the commonly accepted one: a *spoiler* is a losing candidate who changes an election outcome if they are removed from the election. The occurrence of such a thing is called the *spoiler effect*.

All the instances of the spoiler effect we've seen so far meet this definition, including the most egregious one, the 2000 Bush-Gore-Nader election. But the formal definition encompasses a wider range of situations. Looking at the profile above, we see that removing A gives this result:

9	8	6	1	3
B	C	B	B	C
C	B	C	C	B

Now B has 16 top votes and C has 11. B wins the election. When A was in the race, C won; but when A was removed, B won. According to our definition, A is a spoiler for B.

The difference between how we talked about a spoiler in previous chapters and how we're defining it now is that a spoiler no longer has to be a minor candidate who is perceived to have no chance of winning.

You would not have thought of, say, Sarah Palin, as a weak candidate in the Alaska election we examined, yet she still managed to be a spoiler for Nick Begich. When she is removed, Begich has more votes than Peltola. The effect doesn't go the other way, however; Begich was not a spoiler for Palin.*

The Alaska election also exhibits the failure of monotonicity. If 6,000 of the voters who put only Palin down on their ballots had instead ranked Peltola first and Palin second, you'd think this would have been better for Peltola, right? Not exactly. Now Palin is eliminated first and Peltola loses the head-to-head contest to Begich. It would paradoxically have been bad for Peltola to have won over these 6,000 Palin voters. Big-time failure of monotonicity.

It gets even more interesting (or more distressing, depending on your mood). Suppose 6,000 of the voters who ranked Palin > Begich > Peltola decided not to vote at all. It seems that this shouldn't affect Peltola's victory because she is ranked last on these ballots anyway—why would it hurt her if those people didn't vote? But it does. With this change, Palin is again eliminated first and Peltola again loses to Begich. This is an example of the *no-show paradox*, which was first illuminated by Steven Brams and Peter Fishburn in 1983. Those 6,000 voters could have swayed the result in a direction they favored, namely electing the second-highest person on their ballots rather than the last, *by sitting out the election.*

You've probably had enough at this point, but I can't resist this final scenario in which instant runoff can act peculiar. Suppose there are three candidates and two groups of people who vote separately—maybe one group consists of overseas voters—and who have their profiles tallied separately using instant runoff. The first group's profile looks like this:

7	5	5	4
A	C	B	B
C	A	A	C
B	B	C	A

---

* This is spelled out in detail in Adam Graham-Squire and David McCune's paper "A Mathematical Analysis of the 2022 Alaska Special Election for U.S. House."

C is eliminated, so we get this profile:

7	5	5	4
A	A	B	B
B	B	A	A

A wins 12–9.

This is the profile for the second group:

5	4	4	4
A	C	B	C
C	A	A	B
B	B	C	A

Here B is eliminated, and we get this profile:

5	4	4	4
A	C	A	C
C	A	C	A

A wins again, 9–8. So A is the overall winner because they won in both groups.

But imagine we make an administrative change: instead of profiling and tallying the two groups separately, we aggregate all the ballots into a single profile:

12	9	9	4	4
A	C	B	B	C
C	A	A	C	B
B	B	C	A	A

Now A is in last place with 12 top votes, while B and C both have 13. So A is tossed out first, B and C go on to the second round, and C wins the election. Not a single voter changed their preference, yet the results of the election changed simply because two separate profiles became one.

This is an example of *Simpson's paradox*, a sneaky trap that also pops up in many places unrelated to voting. The common theme is the merging of groups. For example, let's say that one year Division A of a company hires twelve people, four of whom are women. The next year they hire forty-eight people and eighteen are women. They happily report that their percentage of female hires increased from 33% to 38%. In the same time period, Division B went from hiring twenty-two women of forty-eight new employees the first year and six women of twelve the second year. They're excited too because their percentage of female hires increased from 46% to 50%. The company is feeling good about itself.

But let's look at the total percentages for these two divisions. In the first year, 4 + 22 = 26 women were hired of 12 + 48 = 60, which is 43%. The following year, 18 + 6 = 24 of 48 + 12 = 60, or 40%. The overall percentage went down.

You might suspect that these numbers are carefully cooked, that the behavior is too freaky to happen in the real world, but it's not. It happened, for example, in the 1999 implementation of the incentive program in California's public schools. In that program, separate groups of students showed improvement while the statistics for all the groups combined indicated the opposite. A recent study demonstrated the role of Simpson's paradox in the University of California's admissions as it relates to applicants' SAT scores and socioeconomic background. Simpson's paradox is also exhibited in law enforcement and can lead to racial bias.

Remember how you were sprinting from plurality by the end of the last chapter? Don't take flight again. Instant runoff isn't perfect, but the value of collecting more information from voters in the form of rankings is clear. Let's move forward and check out some other methods.

ASSIGNING POINTS

If I could choose a job during a revolution, it would be the one Jean-Charles de Borda had in revolutionary France. Borda was born in 1733 in Dax, France, and showed great interest in math and science from an early age. Most of his older brothers pursued military careers, so Borda cleverly combined his passion for math with family tradition and studied military engineering. He became a naval officer and did important work on ballistics and fluid dynamics when he wasn't commandeering ships, fighting in the American Revolutionary War, or getting captured and released by the British. Borda invented or improved several surveying instruments and used them to do extensive hydrographic work during his seafaring years.* When the French Revolution's Constituent Assembly mandated in 1789 that a new system of weights and measurements be established—the metric system—those instruments proved to be precisely what was needed. Borda was part of a group of scientists charged with determining the exact length of a meter, which they did by figuring out the length of the Earth's meridian from the equator to the South Pole and dividing it by 10 million. Yes, that's my ideal revolutionary job; it sure beats manning the barricades.

Borda's achievements earned him many accolades. He has as namesakes a street in Paris, five French naval ships, and (does it get any cooler?) a crater on the moon. His name is one of seventy-two inscribed on the Eiffel Tower. More relevant for our purposes, however, is that his scholarly research earned him election to the French Academy of Sciences. That is where, in 1770, he delivered a lecture on fair voting practices. In his talk and in an essay he published in 1781, he criticized the plurality method the academy used to choose its members and proposed a new method now known as the *Borda count*.† The academy adopted this newfangled system in 1784, but the incessant bickering over its merits went on until someone by the name of Napoleon Bonaparte shut it down and brought plurality back in 1801, two years after Borda's death.‡

---

* If you're about my age or older, you might remember the tables of trig and log values in the backs of math textbooks. Those were an encounter with Borda, who created many of them to aid his surveying work.

† The method was not new; the Roman Senate used it in the second century. The Catalan logician and philosopher Ramon Llull described it in 1299, as did Cardinal Nicolaus Cusanus in 1433 (who was aware of Llull's writing). Borda does not appear to have been aware of either.

‡ Napoleon was a stalwart supporter of mathematics. Said he: "The advancement and perfection of mathematics are intimately connected with the prosperity of the State."

As instant runoff works through round after round, it rewards candidates for being ranked high enough on enough ballots. The Borda count has the same motivation but cuts to the chase by turning the election into a single giant calculation. To perform the Borda count on a ranked profile for an election with n candidates:

- Assign $n - 1$ points to each first-place ranking, $n - 2$ for each second-place ranking, and so on. Thus each next-to-last ranking receives 1 point and a last-place ranking receives 0 points.
- Add the points for each candidate.
- The candidate with the most points wins.

Let's calculate the Borda count for a profile with five candidates and nineteen voters. Because there are $n = 5$ candidates, the points to be assigned range from 0 to 4.

	6	4	4	3	2
4 points	B	A	C	C	E
3 points	D	B	A	A	A
2 points	C	D	D	B	B
1 point	E	E	B	E	D
0 points	A	C	E	D	C

From the first column, B was ranked first on 6 ballots. For each of those ballots, they receive 4 points, so $6 \times 4 = 24$ points in sum. Second-ranked D gets $6 \times 3 = 18$ points from those same six voters. C gets $6 \times 2 = 12$, E gets $6 \times 1 = 6$, and A gets $6 \times 0 = 0$ points. This calculation is made for every column and each candidate's total is computed. Here are the totals:

Candidate A: $6 \times 0 + 4 \times 4 + 4 \times 3 + 3 \times 3 + 2 \times 3 = 43$
Candidate B: $6 \times 4 + 4 \times 3 + 4 \times 1 + 3 \times 2 + 2 \times 2 = 50$
Candidate C: $6 \times 2 + 4 \times 0 + 4 \times 4 + 3 \times 4 + 2 \times 0 = 40$
Candidate D: $6 \times 3 + 4 \times 2 + 4 \times 2 + 3 \times 0 + 2 \times 1 = 36$
Candidate E: $6 \times 1 + 4 \times 2 + 4 \times 0 + 3 \times 1 + 2 \times 4 = 25$

Because B has the most points, they are the winner.

The Borda count seems like a reasonable system. Receiving more points for doing better is not a strange concept. It prevents vote splitting and spoilers and it eliminates Duverger's law as well as center squeeze (unlike instant runoff). It encourages moderate, compromise candidates and discourages negative campaigning because it is advantageous for a candidate to land near the top of many ballots and rack up points.* Like the instant runoff, the Borda count does not reward extremism. Even if some people place a fringe candidate at the top, more people will rank them near the bottom, punishing them with low point scores.

While the Borda count is used only for certain types of legislative elections in a handful of countries, including Slovenia and Iceland, it has a strong presence in organizations and institutions with at least some democratic governance. Many colleges and universities use it for student government and other elections. My own college elects faculty to certain committees by a two-round process. The first asks the faculty to rank up to five people out of all eligible colleagues and a Borda count advances the top five scorers to the second round. In the second round, faculty rank the five finalists and the winner is decided using instant runoff.

The Borda count is frequently used in sports, for instance in the selection of college football's Heisman Trophy winner and to determine NCAA rankings.

Some contests, like Major League Baseball's Most Valuable Player, Formula One and NASCAR racing, the Tour de France, and Association of Tennis Professionals rankings (where the points earned correspond to finishing positions and are aggregated over a season or stages of the race), use a more general class of voting methods called *positional*, which includes Borda as a particular example. Such methods assign points just like Borda does, but the point values do not necessarily decrease by one. For example, in Formula One, the point values for first through ninth place are 18, 15, 12, 10, 8, 6, 4, 2, and 1, respectively. The tallying process remains the same, namely that the person with the most points wins. The sequence of values often follows a mathematical pro-

---

* A 1999 article titled "Would the Borda Count Have Avoided the Civil War?" by A. Tabarrok and L. Spector argues that had Borda count been used in the 1860 elections, Lincoln would have lost to Stephen Douglas and the Civil War might have been avoided. The reason is precisely that Lincoln was a more polarizing figure than Douglas.

gression, either arithmetic (the difference between successive values is constant, as in Borda) or geometric (the ratio of successive values is constant). Medal counts and country rankings in the Olympics use a positional system. Usually three points are given for a gold medal, two for silver, and one for bronze. The points are added to produce each country's standing.

My favorite example of positional voting is Eurovision (see figure 3.3), an annual song contest between European countries plus countries such as Israel, New Zealand, and Australia—the more the merrier. This feast of absurdity and camp is the most watched live nonsports event in the world, best encountered with a Cosmo (or something more potent) in hand. Some familiar artists such as Olivia Newton-John, Bonnie Tyler, Enya, Julio Iglesias, and Celine Dion have competed in it.* The show is not entirely without merit; it has given humankind such gifts as ABBA, Netta, and Måneskin.

Each country holds an internal contest or puts together a committee of music experts to choose their representative. The Eurovision competition begins with two semi-final shows. Of the twenty-six songs that make it to the final event, ten come from each of the semi-finals and six are automatically given entry: the previous year's winning country, which also hosts the competition, and the "five great ones" (lest the rest of Europe forget who's really in charge)—France, Germany, Spain, Italy, and the United Kingdom. After the songs are performed, judges from each of the competing countries video call into the show to announce their ratings. They can assign points to ten songs/countries (and cannot vote for their own country): 1, 2, 3, 4, 5, 6, 7, 8, 10, and 12. The last two values do not follow the sequential order, so this is a positional voting method rather than a straight-up Borda count.

The voting by judges/countries is largely a farcical display of geopolitical cliquishness—most countries give the most points to their neighbors and allies. Norway favors Sweden and Serbia favors Montenegro and so on. The second round of the process allows viewers to

---

* You can probably tell from these names alone what kind of cheesy affair this is and yes, countries can hire performers from other places to compete for them.

Italy	524	Sweden	109
France	499	Serbia	102
Switzerland	432	Cyprus	94
Iceland	378	Israel	93
Ukraine	364	Norway	75
Finland	301	Belgium	74
Malta	255	Azerbaijan	65
Lithuania	220	Albania	57
Russia	204	San Marino	50
Greece	170	Netherlands	11
Bulgaria	170	Spain	6
Portugal	153	Germany	3
Moldova	115	United Kingdom	0

FIGURE 3.3. Screenshot of Eurovision voting, 2021.

televote by phone, text, or the Eurovision app from each of the competing countries. Each country then produces a ranking and assigns 1,
2, 3, 4, 5, 6, 7, 8, 10, and 12 points, according to the country's televoting
results. So each country's viewers act as providers of a single ranked
ballot scored according to a positional voting system. The number of
ranked Eurovision ballots is thus twice the number of contestants; one
for the official judges and one for the viewers from each participating
country. Of course, the viewer votes can throw off everything that happened in the first round of scoring and the final winner might end up
being a song that is not in the lead after the judges' tally—to what I am
sure is the producers' delight, as it keeps the suspense going until the
last moment.*

---

* Speaking of TV drama, most reality shows where contestants are successively eliminated
use combinations of voting methods—in a meticulously planned and staged way—that maximize the tumultuous theatrics and attract viewership. Plurality is a common method because it
promotes divisiveness and packs the most shock value. Toss in the option of viewer voting and
you have a Molotov cocktail of unpredictability and thrills. These elements can be found in
shows like *Survivor* (plurality, pairwise competition, public voting), *American Idol* (public voting by judges, viewer voting), *RuPaul's Drag Race* (public voting by judges, dictatorship, direct
contestant competition), and many others.

Not All Fun and Games

Although positional voting is widely used, it is not as innocuous as it seems. Donald Saari, a professor of mathematics and economics at University of California, Irvine, and one of the most prominent scholars of voting theory, has shown that different winners can arise from a single profile simply by changing the positional voting scheme.* Saari endorses the Borda count to avoid this and related kinds of volatility.

However, the Borda count is not immune to strange behavior. For example, let's look at this profile:

	3	2
2 points	A	B
1 point	B	C
0 points	C	A

The scores are

A: $3 \times 2 + 2 \times 0 = 6$
B: $3 \times 1 + 2 \times 2 = 7$
C: $3 \times 0 + 2 \times 1 = 2$

B wins—but A received a majority of first-place votes, 3 of 5. Remember the majority criterion that says that if a candidate wins a majority of the votes, a decent voting system should elect that candidate as the winner? Well, Borda count fails it.

Here we have a ready fix. Amend the Borda count method to introduce an initial step. If a candidate gets a majority of the votes, that candidate is the winner. Otherwise, proceed with the Borda count. However, this doesn't eliminate a broader related problem. A candidate can receive a solid plurality of the votes and do poorly with the Borda count. In 1994, the Associated Press preseason college football poll, which is tallied by the Borda count, placed Nebraska in fourth place even though it had the most votes, 18. Nebraska was wedged between Florida State

---

* The result appears in the ominously titled 1984 article "The Ultimate of Chaos Resulting from Weighted Voting Systems."

(10 first-place votes) and Michigan (2 first-place votes). We know plurality isn't a great system, but when a method ranks the plurality winner so poorly, it raises a red flag about what the results actually mean.

Another strange aspect of the Borda count appears in this profile:

	7	6	4
3 points	A	B	C
2 points	D	D	D
1 point	B	C	A
0 points	C	A	B

The Borda count gives

A: $7 \times 3 + 6 \times 0 + 4 \times 1 = 25$
B: $7 \times 1 + 6 \times 3 + 4 \times 0 = 25$
C: $7 \times 0 + 6 \times 1 + 4 \times 3 = 18$
D: $7 \times 2 + 6 \times 2 + 4 \times 2 = 34$

So D wins, which makes sense because this candidate hovers near the top of the table and Borda rewards that. (This is not necessarily the case with instant runoff. We've seen an example in which instant runoff punished an otherwise stellar candidate for not having enough first-place votes. If we had run Borda on that profile, we would have seen that D deservedly won.) But D is not the first choice for any of the voters. Nobody would be elated about the results of this election. On the other hand, this outcome might be fine, because in order to do so well with so many voters, D is probably decent, nice, moderate, and inoffensive, a crowd-pleaser. Maybe we should regard this as an advantage of Borda count—it elects true compromise candidates.

On the candidate side, an aspiring office seeker can exploit this aspect of Borda by pursuing a campaign strategy that attempts to please as many voters as possible by staying away from polarizing issues and rejecting negative campaigning. In other words, they might try to win by being bland. Whether being boring is good or bad for democracy is a question that depends on the time, the place, and the electorate.

But wait, there's more! Check out this profile:

	4	3	3
3 points	C	A	B
2 points	A	B	D
1 point	B	D	C
0 points	D	C	A

Borda count declares B the winner:

A: $4 \times 2 + 3 \times 3 + 3 \times 0 = 17$
B: $4 \times 1 + 3 \times 2 + 3 \times 3 = 19$
C: $4 \times 3 + 3 \times 0 + 3 \times 1 = 15$
D: $4 \times 0 + 3 \times 1 + 3 \times 2 = 9$

Now pretend that this profile instead reports the results of a preelection poll, and candidate D, seeing that they didn't stand a chance, dropped out of the race. When people vote, the election profile looks like this:

	4	3	3
2 points	C	A	B
1 point	A	B	C
0 points	B	C	A

Now the Borda count is

A: $4 \times 1 + 3 \times 2 + 3 \times 0 = 10$
B: $4 \times 0 + 3 \times 1 + 3 \times 2 = 9$
C: $4 \times 2 + 3 \times 0 + 3 \times 1 = 11$

Candidate C wins because D, a candidate who has nothing to do with B or C, dropped out. Stranger yet, the relative position of B and C has not changed for any of the voters. Yet the winner changed. This example shows that like instant runoff, Borda count does not eliminate spoilers. The election outcome changes depending on whether D is in the race. Because D is not the winner, they are a spoiler for C.

Closely related to the spoiler effect is the following phenomenon. Let's look at these two profiles:

	3	2	2
3 points	D	A	B
2 points	A	B	C
1 point	B	C	D
0 points	C	D	A

	3	2	2
3 points	D	A	B
2 points	A	C	C
1 point	B	B	D
0 points	C	D	A

The only difference between them is that voters in the middle column have switched their preference between B and C. The Borda count declares B the winner in the first profile and A the winner in the second profile. The fact that some voters switched their rankings of candidates C and B gave A the victory, even though nobody changed their relative preference between A and B from one profile to the other. From the point of view of A and B, C is irrelevant. That this candidate changed position in some of the ballots should not have affected how A and B relate to each other.

This example shows that the Borda count fails *independence of irrelevant alternatives*, or *IIA* for short. A method that satisfies IIA should behave like this: if candidate A is the winner, then modifying the ballots in a way that does not change A's position relative to another candidate B should not make B the winner. (We'll see a more precise statement shortly.)

One of the most embarrassing and public failures of IIA occurred at the 1995 World Figure Skating Championships. At the time, the scoring system used ranked ballots and a certain form of plurality to extract the player standings from the ballots. Near the end of the competition, the skater ranking was

1. Chen Lu (China)
2. Nicole Bobek (United States)
3. Surya Bonaly (France)

Then a U.S. favorite, Michelle Kwan, appeared on the ice. After she completed her routine, the ranking was

1. Chen Lu (China)
2. Surya Bonaly (France)
2. Nicole Bobek (United States)
4. Michelle Kwan (United States)

Let that soak in. . . . Michelle Kwan came in fourth, but that caused the second and third places to switch. Kwan should have been an irrelevant alternative as far as second and third places went, but she wasn't. This is the failure of IIA in all its glory.

This incident prompted the International Skating Union to change its scoring system in 1998 and to use ranked ballots to conduct pairwise comparisons and thus obtain the ordering of the skaters. This made the problem worse, as could have been foreseen; various examples can readily be constructed that exhibit the Bobek-Bonaly-style flip. At the 2002 Winter Olympics in Salt Lake City, a similar IIA failure occurred, again involving Michelle Kwan. She was ahead of Sarah Hughes until Irina Slutskaya skated, but afterward, Hughes was ahead of Kwan. In 2004 the International Skating Union changed the voting system again, hoping to put this turmoil (and an accusation that the pairs competition was fixed) behind it. Their choice was a version of range voting, the subject of chapter 5.*

Finally, the Borda count shares a flaw with instant runoff. Let's look back at an earlier profile, now highlighting candidates A and B.

	6	4	4	3	2
4 points	B	A	C	C	E
3 points	D	B	A	A	A
2 points	C	D	D	B	B
1 point	E	E	B	E	D
0 points	A	C	E	D	C

---

* IIA is not something we humans are entirely responsible for. It is found in nature; see Jordan Ellenberg's *How Not to Be Wrong* for an extraordinary tale of IIA and slime mold.

B is the Borda winner. What sank candidate A was their poor performance with the 6 voters in the first column. It was impossible to recover from that many zeros. But in the rest of the columns, A is at or near the top, so this is a candidate who is popular with many voters. In fact, if we look at the relative positions of A and B, we see that only six voters of nineteen, those in the first column, prefer B to A. In other words, if we held a runoff election between A and B, A would win 13–6.

This seems unfair, right? It is—so much so that it prompted enough consternation in French academic high society to give birth to yet another voting method.

COMPARING PAIRS

Like Borda, Marie-Jean-Antoine-Nicolas de Caritat, Marquis de Condorcet (1743–1794), was a brilliant mathematician who began publishing original mathematics in his early twenties. He was admitted to the French Academy of Sciences at the age of twenty-six, supported by superstars such as Étienne Bézout and Joseph-Louis Lagrange. Condorcet also has a moon crater and a Paris street named after him (but his name is not on the Eiffel Tower). He was a fierce advocate for social justice and argued for equality regarding race, gender, and religion. He hung out with the likes of Voltaire, Benjamin Franklin, and Thomas Jefferson and was one of the leaders of the French Revolution (and ultimately died by its hand because he was an aristocrat). Condorcet sought to apply mathematics to social issues and in 1785, in that spirit, he published *Essay on the Application of Analysis to the Probability of Majority Decisions*, considered to be the first rigorous work on mathematics and politics. It is also one of the most pretentious and convoluted pieces of academic writing ever composed.

In this essay, Condorcet proposed what we now call the *Condorcet method*.* Much of it was motivated by the disdain he felt for his contemporary and fellow academician Borda. Condorcet considered Borda a failed mathematician because he occupied himself with the worldly, applicable mathematics of engineering and physics. But most of all,

---

* Remember Ramon Llull, who really discovered the Borda count in 1299? He also discovered the Condorcet method.

Condorcet was annoyed that Borda's system can produce a winner who would be beaten in a head-to-head contest with another candidate.

Condorcet's method is based on this observation, jacked on steroids: compare *all* pairs of candidates in head-to-head competitions. If there is a candidate who beats everyone else, that's the winner. In other words, the Condorcet method conducts all possible pairwise runoffs between candidates and declares the candidate who always comes out on top to be the winner. Like the Borda count, the Condorcet method needs only the information already present in a ranked profile.

Here is an example:

15	11	9	6	5
A	B	C	B	D
B	A	B	A	C
C	D	A	C	B
D	C	D	D	A

To see who wins between, say A and B, we focus on only those two candidates and count how many voters placed A higher than B and how many placed B higher than A. It helps to highlight this pair:

15	11	9	6	5
A	B	C	B	D
B	A	B	A	C
C	D	A	C	B
D	C	D	D	A

A higher than B: 15 voters in the first column.
B higher than A: 11 + 9 + 6 + 5 = 31 voters in all other columns.
So B wins that pairwise contest.
Let's do another head-to-head contest, say between B and C.

15	11	9	6	5
A	B	C	B	D
B	A	B	A	C
C	D	A	C	B
D	C	D	D	A

B higher than C: 15 + 11 + 6 = 32 voters.

C higher than B: 9 + 5 = 14 voters.

So B wins that pairwise contest as well.

Running through all possible pairs, of which there are six,* gives the full result:

A versus B: A wins 15, B wins 31, so B is the winner.

A versus C: A wins 32, C wins 14, so A is the winner.

A versus D: A wins 41, D wins 5, so A is the winner.

B versus C: B wins 32, C wins 14, so B is the winner.

B versus D: B wins 46, D wins 0, so B is the winner.

C versus D: C wins 30, D wins 16, so C is the winner.

Because B beat all the other candidates in one-on-one contests, they are the *Condorcet winner* of the election.

There is a pleasant way to visualize a Condorcet election by drawing a graph in which nodes represent candidates and arrows indicate pairwise contests. Figure 3.4 shows the graph for the example we just analyzed. Each arrow points from the winner to the loser of that pairwise contest. If all the arrows point away from a candidate, as they do for B, that candidate is the Condorcet winner.

If a candidate has all arrows pointing at them, as D does, they have lost all their pairwise contests; such a candidate is called the *Condorcet loser*.†

We know from polling that had the Condorcet method been used in the 2000 presidential election in Florida, Gore would have been the Condorcet winner. In the 2016 Republican primaries, using the Con-

---

* If there are n candidates, the first can be matched with $n - 1$ remaining candidates. The second candidate can then be matched with $n - 2$ candidates (one fewer because they were already pitted against the first candidate). There are $n - 3$ battles left for the third candidate. Continuing in this vein, the total number of head-to-head contests ends up being $n - 1 + n - 2 + n - 3 + \ldots + 3 + 2 + 1$, the sum of the first $n - 1$ integers. There is a nifty formula for calculating this number: $n(n - 1) \div 2$.

† This is the beginning of the connection between voting theory and a branch of mathematics called *graph theory*, an enormously useful field that models many real-life phenomena. A satisfying theorem by McGarvey from 1953 says that any graph you can concoct represents the tallying of some ranked profile by the Condorcet method.

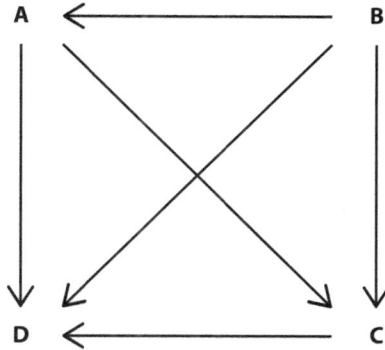

FIGURE 3.4. Representing a Condorcet method election with a graph.

dorcet method among the main contenders would likely have exhibited Trump as the Condorcet loser.

The Condorcet method is hard to argue against. In some sense, the Condorcet winner is the true majority winner because they defeat every other single candidate and they do so necessarily by a majority of the votes.* If you are a fan of May's theorem, this is the way to elevate that result to the setting of more than two candidates—reduce the election to a bunch of two-candidate contests where we know what to do. The winner will be a decent consensus candidate who is well liked by the electorate. Like other ranked choice methods, Condorcet prevents vote splitting and the spoiler effect.

Despite these strengths, the Condorcet method is not widely used. (Neither are its derivatives.) One conspicuous reason is that it can be computationally expensive. For each pair of candidates, each ballot must be checked to see which is ranked higher. For large numbers of candidates and lots of voters, this process is labor intensive. A few parties, such as the Libertarian Party in Washington state and the Pirate Party (yes, that pirate, as in Jack Sparrow) in many European countries, do use Condorcet, as do student governments at several universities

---

* We need a qualification here, which is that the Condorcet winner might not win all the pairwise contests by a majority of the total number of people who cast a ballot. Ballot exhaustion might mean that comparison of two candidates is possible for only a subset of the ballots, and the majority of that subset might not constitute a majority of all the voters.

around the world. The Wikimedia Foundation used it until 2013 to select its Board of Trustees. When the size of the election allows it, the Condorcet method provides a good reality check that another method has produced the "right" winner.

The real reason why Condorcet is relatively underutilized, however, is mathematical. Did you notice the caveat "*if* there is a candidate who beats everyone else" when I described the method? There's a good reason for it, as this profile will demonstrate.

10	8	7
A	B	C
B	C	A
C	A	B

Because A is preferred to B by seventeen voters and B to A by eight voters, A wins that pairwise contest. B wins against C, 18–7. So far, so good, A is better liked than B and B is better liked than C. But now comes the kicker. When we compare A and C, we find that C wins against A, 15–10. The voters, as a group, prefer A to B and B to C, but they also prefer C to A!

This situation is called a *Condorcet cycle*, represented by the graph in figure 3.5. Our discovery of this cycle demonstrates that the Condorcet method might not produce a winner, and even more disconcertingly, it exposes a kind of paradoxical behavior: a group can have circular preferences.

When we deal with an individual, we expect consistent preferences. We expect that if they prefer soup to salad and salad to pasta, they also prefer soup to pasta. A compact way of saying this is that their preferences are *transitive*. Someone whose preferences aren't transitive strikes us as irrational. Not so with groups. Most people in a group might prefer soup to salad, most might prefer salad to pasta, and most might prefer pasta to soup—and all of them would be models of rationality. The key reason is that *different* majorities of people prefer one dish to the next. A group's preferences are not necessarily transitive.*

---

* Condorcet himself talked about this paradox in his seminal essay. Because of this, though, it's not entirely clear what method he was actually advocating. Some have argued that he was

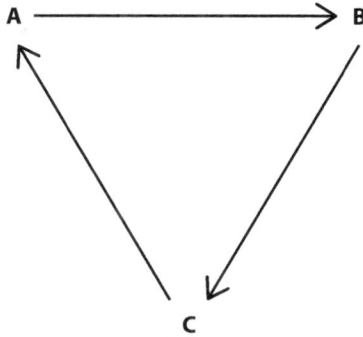

FIGURE 3.5. A Condorcet cycle.

In the face of this possibility, we can take comfort in research, both theoretical and empirical, that shows that Condorcet cycles are rare, meaning that the probability of there being a cycle in a generic election is small. But the fact that they can occur prompts the motivating insight of voting theory: extracting group preferences from individual preferences is a tricky business. The conversion of individual opinions into a collective position must be handled carefully, especially because we run democracies with the assumption that individual expressions cumulatively produce what is best for society. Social choice theory exists precisely because these are highly nontrivial and mathematically (and socially) volatile issues.

Condorcet cycles are not confined in size to three candidates; any number of candidates can form a cycle. There can also be several cycles in the same election.

We might attempt to define the problem away by saying all candidates dancing in a Condorcet cycle are tied. But the troubles run deeper. Figure 3.6 shows the results of the pairwise matchups in a Condorcet election.

The graph contains two "touching" Condorcet cycles: A, B, and C make up one, and B, C, and D make up the other. If we try to solve the

---

endorsing an algorithm we now know as *Kemeny's method*. A closely related method was also proposed in 1876 by none other than Charles Dodgson, aka Lewis Carroll, the author of *Alice's Adventures in Wonderland*, who wrote several essays on voting theory.

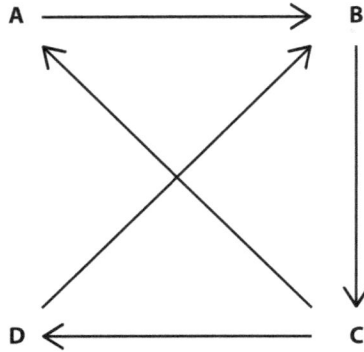

FIGURE 3.6. Two adjacent Condorcet cycles. The missing arrow between A and D is intentional.

cycle problem by declaring a tie among candidates A, B, and C and similarly among B, C, and D, the logical deduction is that A and D should be tied as well. But in practice this doesn't have to be the case— one of A or D might win their pairwise contest, and it would make no sense to declare them tied.

This sort of ambiguity makes mathematicians uneasy. A mathematical process should have a clear outcome. When you open the box containing Schrödinger's cat, you want it to be either dead or alive, not still both (or neither). If you throw milk, cream, sugar, and vanilla extract into an ice cream maker and turn on the timer, you want to know that what will come out at the end is vanilla ice cream. It might not be tasty, but at least you don't have to worry that you'll lift the lid and find beef stew. Or nothing.

Likewise, a voting system should produce a winner. Even more than one winner is acceptable, in which case we could reasonably declare that everyone in the cohort of winners is tied. This is exactly how social choice theory frames the goal: it defines a voting method to be a procedure—a *function*, in mathematical language—that takes in a profile and produces a subset of the candidates called the winners (with a few other bells and whistles that we'll get to later). Everyone in that subset is considered to be tied for first place. If the subset contains one element, then that element is the sole winner.

According to this definition, the Condorcet method is not a voting method. As we just saw, it might fail to produce a subset of candidates who can unambiguously be considered tied. In response, social choice theory has designed other systems that retain Condorcet's commandment—if a candidate wins all head-to-head contests, they should be the winner—while removing the problematic ambiguities.

One popular system is *Copeland's method*, named after A. H. Copeland (1910–1990), a University of Michigan math professor. It awards 1 point to a candidate if they win a pairwise matchup, 0 if they lose it, and 1/2 of a point if there is a tie. In the cycle with three candidates, shown in figure 3.5, each would get 1 point. In the example based on figure 3.6, in which A and D logically ought to have tied but actually, say, A beats D in a head-to-head contest, Copeland's method would give 2 points to A, 1 point to B, 2 points to C, and 1 point to D. Thus A and C would tie for first place.

Most round robin sports tournaments are versions of the Copeland method, possibly with different point distributions. Instead of voters casting ballots, games decide the points. For example, in the first stage of the FIFA World Cup, there are eight groups of four teams. Within each group, every pair of teams play each other (so there are six games in each group). A win is worth 3 points (changed from 2 points in 1994 to encourage a more offensive—and hence more fun—playing strategy), a tie earns 1 point, and losing carries 0 points.* If there are any ties in the total number of points after all six games have been played, the goal differential determines the final ranking.

The top two teams from each group then play in the elimination round, in which the slate of sixteen teams is halved at each stage. This round is represented by the usual "bracket," shown in figure 3.7.

A similar bracket is used for the NCAA basketball tournament, or March Madness. This sort of elimination competition is another—problematic—incarnation of the Condorcet method, called *sequential*

---

* Speaking of encouraging offensive play, my daughter's most recent soccer tournament awarded a colossal 8 points to a win and 3 points to a tie. Points were also given for shutout wins and goals.

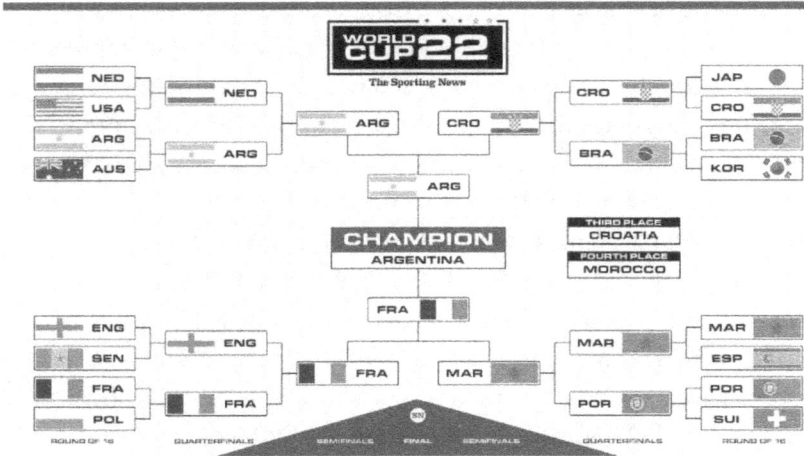

FIGURE 3.7. World Cup final sixteen bracket, 2022.

(*pairwise*) *voting*. In this type of voting, a sequence of head-to-head contests is predetermined. The winners of those initial contests are then matched against each other, also in some preset order. This process continues until the number of candidates standing is whittled down to one.

If we have a ranked profile, we can run a sequential vote for any *agenda*, meaning any predetermined sequence of pairwise competitions. For example, take this profile:

1	1	1
A	B	D
B	C	A
C	D	B
D	A	C

Suppose someone hands us the agenda C-D-B-A. This means we first look at the matchup between C and D, which C wins 2–1. So we pit C against the next candidate in the agenda, B. That matchup is won by B, 3–0. So B goes on to meet A in the final head-to-header, which A wins 2–1, and A is the winner overall.

But now apply the agenda A-B-D-C to the same profile. The first one-on-one contest is between A and B, which A wins 2–1. Then A goes on to meet D, which D wins 2–1. Finally, D and C duke it out and C wins 2–1. The winner changed with the agenda. Moreover, with the second agenda, C won even though *everyone* preferred B to C.

The one case when the agenda does not matter is when there is a Condorcet winner because this candidate will win all pairwise contests and will in particular win against all the candidates they encounter as they progress along any agenda.

That the winner depends on the agenda illustrates the enormous power of those who set it and the danger of manipulation in the sequential method. Having some advance knowledge of the way voters feel about the candidates and placing a preferred option later in the agenda increases its chances of being elected.* Whenever sequential voting is used, the integrity of the process is subject to question.

Returning to the Copeland method to give the fine print: it is not immune to the weirdness we've observed in the other ranked choice methods. Just like instant runoff and the Borda count, it fails independence of irrelevant alternatives and it is susceptible to the spoiler effect.

EVERYBODY (AND NOBODY) IS A WINNER

A natural question arises. We've explored several ranked choice voting methods that, yes, have some flaws but are irrefutably better than plurality. So which one is the best? Which one should we advocate for, write our representatives about, teach our children to love?

Or is it possible we don't have to choose? Maybe they always give the same result, so we can use whichever method we please or whichever is easiest to execute logistically and computationally.

---

* Agendas are closely related to the seeding process in sports tournaments. Typically, however, seeding is determined by an announced algorithm, so it is not as vulnerable to manipulation. Nevertheless, the design of fair seeding algorithms, like the design of fair voting methods, is the subject of active interdisciplinary research.

To investigate, we'll end this chapter like a fireworks finale. We'll hit this profile with all the methods we've encountered so far and see what shakes out:

45	22	15	10	6	2
A	B	C	D	C	E
E	D	D	C	E	C
D	E	E	E	D	D
B	C	B	B	B	B
C	A	A	A	A	A

There are $45 + 22 + 15 + 10 + 6 + 2 = 100$ voters. There is no majority winner because no candidate received more than $100 \div 2 = 50$ votes. Candidate **A wins plurality** with the most first-place votes, 45.

For the runoff method, A and B advance with 45 and 22 votes, respectively. After the other candidates are eliminated, this is what's left:

45	22	15	10	6	2
A	B	B	B	B	B
B	A	A	A	A	A

B wins the runoff, 55–45.

For instant runoff, E is eliminated first with only 2 first-place votes, leaving this profile:

45	22	15	10	6	2
A	B	C	D	C	C
D	D	D	C	D	D
B	C	B	B	B	B
C	A	A	A	A	A

No majority winner yet, so D is now eliminated with the least number of first-place votes, 10. That leaves this:

45	22	15	10	6	2
A	B	C	C	C	C
B	C	B	B	B	B
C	A	A	A	A	A

Still no majority winner, so B, with 22 votes, is eliminated:

45	22	15	10	6	2
A	C	C	C	C	C
C	A	A	A	A	A

Now **C wins the instant runoff** with 55 votes to A's 45.

The Borda count applied to the profile gives this calculation:

A: $45 \times 4 + 22 \times 0 + 15 \times 0 + 10 \times 0 + 6 \times 0 + 2 \times 0 = 180$

B: $45 \times 1 + 22 \times 4 + 15 \times 1 + 10 \times 1 + 6 \times 1 + 2 \times 1 = 166$

C: $45 \times 0 + 22 \times 1 + 15 \times 4 + 10 \times 3 + 6 \times 4 + 2 \times 3 = 142$

D: $45 \times 2 + 22 \times 3 + 15 \times 3 + 10 \times 4 + 6 \times 2 + 2 \times 2 = 257$

E: $45 \times 3 + 22 \times 2 + 15 \times 2 + 10 \times 2 + 6 \times 3 + 2 \times 4 = 255$

Candidate **D wins the Borda count.**

I'll skip the Condorcet calculation of the 10 possible pairs. You can probably guess what happens: candidate **E is the Condorcet winner**, coming out on top of every other candidate.

In summary:

- There is no majority winner.
- A wins plurality.
- B wins runoff.
- C wins instant runoff.
- D wins Borda count.
- E wins Condorcet.

There is no easy way out. Different methods can produce different winners. Each has its own way of deciding who the "best" candidate is. Each regards the election through a different lens and deems different criteria most important: number of first-place votes, overall table placement height, head-to-head comparisons, and so on.

Choosing among these methods is like trying to choose one from a bunch of three-dimensional shapes by focusing on particular features. One shape might have the greatest surface area, so if that's what we care about, this is the best choice. But another might have the largest volume, even if it has smaller surface area.* A third shape might have some other quality—perhaps it's the least curved (it's more cubical than spherical), so it's the best candidate from that point of view. The criteria are different and they describe different features of the shapes. None of the shapes are "best," at least until that becomes the question. Once it does, the shapes enter competition with each other and we suddenly need new tools for comparing them.

This is the same place we found ourselves when we had to figure out which of the two-candidate methods to use. We have a bunch of procedures, each with various advantages and flaws. It is not hard to believe that instant runoff, Borda, and Condorcet are all better than plurality and runoff (and obviously better than other silly things like dictatorship, monarchy, and parity), but which of the three is the best? Echoing the two-candidate discussion, what does "best" even mean? Not only that, but how do we even know there isn't some other method that's even better than these three, but we just haven't found it?

At this point in our discussion of two-candidate elections, we turned to the axiomatic approach, the formalism of mathematics, to help us untangle the Gordian knot of ill-defined vernacular and answer these existential questions. The strategy bore fruit in the form of May's theorem—a definitive, indisputable statement about which voting method we should use when exactly two candidates are running. Let's try it again.

---

* Yes, this is possible; the sphere is the most efficient in this regard because it encloses the most volume for the least amount of surface area. For example, a cube whose sides are 1 foot each has a surface area of 6 square feet and will enclose 1.00 cubic foot of volume. A sphere whose surface area is 6 square feet encloses 1.38 cubic feet of volume.

CHAPTER 4

The Impossible
Democracy

THE DECLARATION of Independence is a mathematical document. It starts by proclaiming certain truths to be self-evident and proceeds to list them. In math, such statements are the *axioms* of a theory. They are things everyone can agree on that constitute the starting point upon which a theory is built. You can't start from nothing, and axioms are the concrete foundation of the house that's being erected. The most famous example is Euclid's axioms.* They say things nobody would dispute, like "given any two points, a line can be drawn through them." And starting from only a few statements, the vast field of geometry arises, full of intricate constructions, algorithms, and theorems.

After the Declaration of Independence establishes its axioms, it also states a theorem: the United States of America is better off without the British Crown. The rest of the document is devoted to the proof of this theorem, establishing its validity by exhibiting how British rule violates the axioms. The framers then continued to build upon those axioms eleven years later as they grew the American democracy further at the Constitutional Convention.

---

* If we allow for broader interpretation, there are plenty of examples even outside the world of math. Any organization's mission statement can be thought of as a collection of axioms upon which the organization is built. Marriage vows are axioms of, well, marriage.

99

Sure, it seems like I've taken some interpretational liberties with one of our nation's most revered documents. But it's not me. The man who drafted it, Thomas Jefferson, was a scholar of Euclid and liked the axiomatic approach. In fact, both he and cowriter John Adams acknowledged the parallels (no pun intended) between the two and were explicit about their intention to model the Declaration of Independence after Euclid.

AXIOMS OF DEMOCRACY

We too have had some success with the axiomatic approach thus far. In the case of two-candidate elections, it produced May's theorem, which settled the question of the voting method we should be using. Let's turn to this strategy again, with more ambition. Let's formalize social choice theory, which, as it's worth repeating, is the study of how individual preferences, opinions, or interests combine to reach a collective decision, or the "will of the people." This theory of course applies to voting and elections, but it also applies to a broader scope of situations, as when legislatures decide policy, courts or committees enact rulings, and so on. The formalism will hopefully give us strong enough wings to develop tools that can answer questions such as: How can a group arrive at an outcome that is most beneficial to its individual members? When is a decision process democratic and representative? How can we rank various alternatives in terms of social well-being?

Before we begin, let's return to something I said before I stated May's theorem. Social choice theory as we know it today did not appear overnight; no consequential branch of academic exploration does. What you're about to see is a lean, academically sanitized final product of a lot of work by a lot of people over a lot of years. Concepts and their definitions have been honed and groomed over time, and discussion about whether they're exactly the right ones continues. Lots of people use alternate but equivalent definitions in a way that suits their particular scholarly line of exploration, while others fiddle with existing definitions or even ignore them. In a word, the subject is still dynamic and vibrant.

Let's first identify the objects of study. Yes, they are voting methods, but can we be precise about that?

It helps to revive a piece of terminology from the previous chapter: a ranking is *transitive* if it does not contain cycles. In other words, a transitive ranking can't place A higher than B, B higher than C, and C higher than A. We'll now say a *voting method* is a function that takes as input a collection of transitive ranked ballots and gives as output a transitive ranking of the candidates.* Transitivity is a natural condition to impose; it avoids paradoxical cyclical preferences that the Condorcet method, for example, might produce. According to this, the Condorcet method is not a voting method (while its modification, Copeland, is). Observe that our definition does not preclude the possibility of ties.

On to the axioms. They will take the form of self-evident criteria that everyone can agree a desirable voting method should satisfy. The four we proposed for two-candidate elections—anonymity, neutrality, monotonicity, and majority—are still relevant because they capture features of voting we continue to find valuable. Here they are, adapted to elections with more than two candidates (we're also keeping in mind that everything we're doing is for ranked voting procedures):

- **Anonymity:** If any two voters exchange their ranked ballots, the election result does not change.
- **Neutrality:** If candidate A is the winner and all voters reverse their preference between A and some other candidate B—that is, if they swap the positions of A and B on their ranked ballots— then B becomes the winner.
- **Monotonicity:** It is impossible for the winning candidate to become a losing candidate by being moved up the ranking on one or more ballots.
- **Majority:** Whenever a majority of voters rank a candidate at the top of their ballots, that candidate is the winner.

---

* Technically, what this definition describes is a *social welfare function*, a more general concept than a voting method, also known as a *social choice function*. The former produces a ranking of the candidates, while the latter just produces the winner(s). I'll blur the difference so that we don't get bogged down in terminology; this won't create problems.

A voting method *satisfies* a particular criterion if the statement of the criterion holds true when that method is applied to any election and any profile. For example, the Borda count satisfies monotonicity because if a candidate is winning, that means they have the most points. If a voter moves this candidate up in their ranking, the candidate will receive even more points and retain first place. This is a general argument that only uses the definition of the Borda count and does not depend on the candidates, voters, or rankings in a particular example.

Extracting the general from the specific does not make a proof. I can't say "the number 4 is even, so therefore all numbers are even." On the other hand, an example is enough to *disprove* a statement. It is sound to say "it is not true that all numbers are even because 5 is not even." In order to say that a method *does not satisfy* a criterion in general, it similarly is enough to produce *one single profile* where the statement of the criterion does not hold. For instance, instant runoff fails monotonicity because we were able to create an example in chapter 3 that violated it—the winner moved higher on some ballots and that caused them to lose the election.

We also showed by example that the Borda count violates the majority criterion. The Copeland method satisfies the majority requirement because if most people rank a candidate at the top of their ballots, that means that they have ranked them above every other candidate. That in turn means that such a candidate will win all pairwise contests and will collect the most Copeland points.

Let's use some of the other strange examples we've seen to formulate new criteria. It seems reasonable, for instance, that if a candidate wins all head-to-head contests—so they are the Condorcet winner—that candidate should be the election winner. This motivates a new criterion.

- **The Condorcet criterion:** When an election has a Condorcet winner, the method selects that candidate as the winner.

We can now draw a freebie observation from the example at the end of the previous chapter, where one election profile produced different winners with different voting methods. Because one of them was the Condorcet winner but instant runoff and Borda each selected another candidate, that example shows that those two methods do not in general satisfy the Condorcet criterion.

Finally, remember the Borda example in which the winner changed because of some movement on the ballots that had nothing to do with the relative position of the winner before and after? To put this behavior formally on our radar, here is another concept, a form of which we've already seen:

- **Independence of irrelevant alternatives (IIA):** Suppose there are two ranked profiles from the same set of voters and the same set of candidates. Suppose the relative positions of candidates A and B are the same for every voter in those two profiles and suppose the method chooses A as the winner in one of the profiles. It follows that the method does not choose B as the winner in the other profile.

One way to think about IIA is that if A is the winner and something changes with the ranked ballots that does not affect where the voters have placed A relative to B (so those who ranked A higher than B continue to do so and vice versa), then B should not win. This does not say that A should continue to win; it only says that B can't win. The new profiles might have put C on top of all lists so that now candidate C is the winner. Whatever the case, the only thing that matters is that B *should not* win, because that candidate's relation to A, the original winner, did not change.*

This definition sounds close to the spoiler effect, which says that a spoiler is a losing candidate who changes the outcome of an election depending on whether or not they are in the race. But there is a key difference: in testing for a spoiler, the two profiles we compare do not have the same candidates because one of them (the potential spoiler) appears in one profile but not the other. By contrast, IIA requires the same

---

* A classic anecdote that is often offered as an illustration of IIA is the story of Sidney Morgenbesser (1921–2004), a Columbia University philosophy professor who was ordering a dessert in a New York diner. The options were apple pie and blueberry pie and Morgenbesser ordered apple. But then the server returned and said that they in fact also had cherry pie. "In that case," Morgenbesser replied, "I'll have the blueberry pie."

TABLE 4.1. Voting methods that satisfy various social choice criteria

	Anonymity	Neutrality	Monotonicity	Majority	Condorcet	IIA
Instant runoff	✓	✓	✗	✓	✗	✗
Borda	✓	✓	✓	✗	✗	✗
Copeland	✓	✓	✓	✓	✓	✗

candidates on both profiles. However, we can adjust our spoiler test. Instead of taking the potential spoiler out of the profile, we move them to the bottom of every ranked ballot. If that new profile has a different winner, then we've confirmed the spoiler and at the same time presented the spoiler effect as a failure of IIA. So our examples where instant runoff and the Copeland method exhibited spoilers can also be interpreted as those methods' failure of IIA.*

In fact, when people in the business talk about IIA, they usually have the spoiler effect in mind. General IIA is hard to check for in actual elections, but the spoiler effect is more easily identified.

Table 4.1 summarizes how our main ranked voting stacks up against the various criteria. One thing is clear right away—no method satisfies all the conditions. Copeland comes close but it fails IIA. Instant runoff additionally fails monotonicity and Condorcet. Borda is good on monotonicity but fails majority. Anonymity and neutrality, on the other hand, don't seem to be the stumbling blocks; all of our ranked methods satisfy it.

Well, maybe we just haven't found the right method yet. There are plenty more we haven't talked about.† What if one of those is the perfect one? Or what if we just haven't discovered (or created) it yet and we need to keep searching?

Kenneth Arrow and his Nobel Prize–winning result says we can call back the search party.

---

* This kind of consolidation is another bread-and-butter operating procedure of mathematics. We might start with a laundry list of observations or criteria, but the goal is always to synthesize and seek coherence.

† Coombs, Kemeny, French, Dodgson, Bucklin, Baldwin, Dowdall, Nanson, Schultze, Minimax, DOC, SET, quadratic . . .

ARROW THROUGH THE HEART OF VOTING

Kenneth Arrow (1921–2017) was a graduate student at Columbia dabbling in mathematical economics when he took a job at the RAND Corporation, a think tank affiliated with the U.S. military. One of the exciting new topics of discussion at RAND was *game theory*, pioneered by mathematician John von Neumann and economist Oskar Morgenstern at the Institute for Advanced Study in Princeton, New Jersey. They laid out the foundations of game theory in their seminal 1944 book *Theory of Games and Economic Behavior*. With the Cold War brewing, game theory was a natural new academic framework for trying to understand the new geopolitical order and America's place in it.

One day, a coffee break conversation turned to the apparent peculiarity that elected officials purportedly speak for everyone, including those who have not voted for them. Was there a way to appraise the degree to which a society has successfully imprinted its preferences on its representatives? Was there a way to maximize this imprinting? Arrow started to think about this and to fiddle around with voting systems, fully determined to show that the perfect system that would optimize the desires of the electorate was out there. But doubt set in after a few days of producing examples that repeatedly shot down his hopes. If a voting system he analyzed was good in some ways, there would be examples that showed it was inevitably bad in others.*

Once Arrow made the mental leap that perhaps he should be trying to argue the opposite, that was it. He quickly managed to prove one of the most consequential theorems in the history of economics: there is no perfect ranked voting system. The result first appeared in 1948 as a RAND research report and then in Arrow's 1951 book *Social Choice and Individual Values:*[†]

---

* William Poundstone's *Gaming the Vote* has a nice account of the history of Arrow's theorem, as does George Szpiro's *Numbers Rule*.

[†] I've stated it a little loosely to avoid technical issues that we don't need to bog ourselves down in.

Arrow's impossibility theorem: The only ranked voting method that satisfies monotonicity and independence of irrelevant alternatives is dictatorship.

Let's break this down for easier digestion: Arrow's impossibility theorem says that if we believe that (a) the winner shouldn't lose by getting more votes, and (b) the winner shouldn't change because something happened in the profile that had nothing to do with them, then the only voting system we can use is the one where a single voter decides what happens no matter how everyone else votes.

So many questions. . . . First, we're talking about ranked systems, so what's dictatorship doing here? Well, dictatorship *is* a ranked system in a kind of a silly way. Everyone still files ranked ballots, including the dictator. Only one ballot matters, though, because the dictator's first choice will be the winner. Technically, this is a ranked method because a winner is produced from a ranked profile.

Also, where's the impossibility? As stated, the theorem doesn't say that something is impossible. But if we agree that dictatorship is bad and we should steer clear of it, then the theorem can be recast as saying that it is impossible for a nondictatorship ranked voting system to satisfy monotonicity and IIA.*

When it was published, Arrow's result was shocking. And spectacular. It still is both. It earned Arrow the Nobel Prize in economics in 1972. It elevated social choice theory to a vigorous multidisciplinary field that ostensibly lives within game theory but is intimately interwoven with political science, history, math, statistics, and so much more. Amartya Sen, who in 1998 won the Nobel Prize in economics for equally important and depressing research that essentially says that individual liberties are incompatible with social needs, called Arrow's theorem the "big

---

* In 1963, Arrow proved a more general version of his theorem. He essentially replaced monotonicity with the *Pareto criterion*: if all voters rank candidate A over candidate B, then B should not be the winner. Instant runoff, Borda, and Copeland all satisfy Pareto. This condition is named after Vilfredo Pareto (1848–1923), an Italian economist and a big name in the field who is known for his foundational work on income distribution and modern welfare economics. He was the first to observe and explain a general principle we are nowadays all too familiar with: very few people control most of the wealth.

bang." William Riker (1920–1993), an influential political scientist and the instigator of *positive political theory*, considered Arrow's result to be the fundamental rationale for the thesis that democracy can never deliver on the populist promise of executing the collective will of the people. He used Arrow's theorem to bring game theory into political science in a transformative way, arguably making it a genuine science.

Possibly the most valid measure of the relevance of Arrow's impossibility theorem is that it is still contested—not on its mathematical merit, because its validity on those grounds is indisputable—but on its interpretation and implications for society. As Riker would argue, Arrow shows that the will of the people is an illusion, but how does that translate into concrete action in the real world? If we don't want dictatorship (and we don't; I hope that's agreed), Arrow says we have to give up either monotonicity or IIA, so how do we decide which is more important to avoid? Do we have to?

LIVING WITH THE IMPOSSIBLE

Impossibility theorems are marvelous and bewildering. They say that no matter how hard we try to establish the validity of something, no matter how much blood and tears we spill doing it, we can never achieve it. Our brains don't naturally twist this way. We like to imagine that even the most outlandish and hypothetical of scenarios might have a chance of coming true. People will never fly? Well, maybe we'll evolve to have wings or we'll be able to hop onto little drones and be on our airborne way. I'll never be rich? Likely true, but there's still an off chance that I'll win the lottery (if I ever buy a ticket). Or maybe I have a rich distant aunt who will leave me all her wealth. The probability is negligible but not zero (I hope). Peace on Earth? Okay, that's a tough one, but sure, it's possible.

An impossibility theorem says there is no hope. Mathematicians are used to this. Early in our careers, we learn that, for example, it's impossible to write the square root of 2 as a fraction, square a circle, or come up with a formula for solving the quintic polynomial. "There does not exist nor will there ever" and "it cannot be done nor will it ever be doable" are phrases woven into the genetic code of mathematics. In fact, one of the most famous impossibility results says that everything we do is built on a shaky foundation. This is *Gödel's incompleteness theorem*, which in one interpretation asserts that there cannot exist a complete and consistent set of axioms upon which to build mathematics. If we can live with this, then we can live with Arrow.

Still, Arrow feels different. Is it because it speaks directly to something we universally appreciate (even if it's denied to us or we're too lazy to participate in it) and deem indispensable—voting? The logical shutdown of the possibility of perfect democracy is unnervingly real. The square root of 2 is a lot easier to ignore than the breakdown of society. Stone-faced mathematics saying "nope, don't even hope for fairness"—how do you move past that?

This is how.

There are ways to live with Arrow's impossibility theorem. One is academic: modify the hypotheses to avoid the negative conclusion. This feels like sleight of hand; we're changing the rules of the game because we don't like losing. But if the changes are informed by experience, then it's legit. And that doesn't invalidate the original theorem; it proves a different one. For example, Arrow assumes "universal domain," meaning that all profiles are possible. In other words, voters are free to rank the candidates in any order. If we are allowed to deny the possibility of some rankings, then there are methods that will satisfy all of our desired criteria. This was observed by the Scottish economist Duncan Black (1908–1991) and follows from his influential 1948 paper titled "On the Rationale of Group Decision-Making." Black's importance extends beyond this; his 1958 book *The Theory of Committees and Elections* laid the foundation for social choice theory.* Presuming that not all rankings are possible is not unrealistic. For example, it is highly unlikely a voter in the 2016 election would have chosen the ranking Hillary Clinton > Donald Trump > Jill Stein.

---

* Black claimed he had proved Arrow's theorem in 1942, before Arrow did. However, the paper containing the result was not submitted for publication until 1949, after Arrow had already announced his version and published it as a RAND Corporation report. Black also advocated for the Borda count, with the caveat that a Condorcet winner, if there is one, should overrule the Borda winner if the two differ.

Other prominent scholars have built on Black's work. They include the aforementioned Amartya Sen and another Nobel Prize laureate, Harvard economics professor and Kenneth Arrow student Eric Maskin. Maskin, along with coauthor Partha Dasgupta, a professor emeritus at Cambridge University, showed that the Condorcet method works well when the universal domain of all possible preferences is restricted. He is also among those who argue that IIA is stronger than necessary, and he shows that the Borda count—and only the Borda count—satisfies all the desired criteria in addition to an appropriate modification of IIA. Donald Saari, a prominent figure in voting theory and a proponent of the Borda count, has proved a similar result and argued that IIA is effectively in conflict with transitivity of preferences and should be weakened to reduce this incompatibility.

But hypothesis-dancing around Arrow doesn't make it go away. Its gloomy shadow hangs over every ranked election. So how about we face this and try to understand its practical implications? Here we're guided by what the theorem *doesn't* say—which is that *every* election we ever run using any of the nondictatorship ranked methods will fail monotonicity or IIA. Arrow himself said: "Most systems are not going to work badly all of the time. All I proved is that all can work badly at times." Maybe the bad behavior—failures of monotonicity or IIA—appears so infrequently in actual voting that we don't really have to worry about it.

The research on this is encouraging. We have data from enough ranked elections that we can examine them empirically and statistically, picking up on idiosyncrasies using powerful analysis methods that were not available until recently. Most of this research is focused on instant runoff because that's the most popular ranked method around the world.

For example, a long-standing concern about ranked voting is that it might not choose the Condorcet winner. Electing such a winner is desirable because it reduces the likelihood the electorate will end up being represented by someone most people dislike. We saw that instant runoff and Borda count do not satisfy the Condorcet criterion, so even when a Condorcet winner is available, these methods might choose someone else. This does not appear to be likely, though. A 2017 FairVote study of

138 instant runoff elections in four Bay Area cities showed that voters elected the Condorcet winner every single time. In forty-six of these elections, there was not an initial majority winner. This is the situation in which the possibility of a non-Condorcet winner arises. Those included seven elections where the candidate who initially had the most first-place votes wasn't ultimately elected (because instant runoff can elect someone other than the plurality winner), but violation of the Condorcet criterion did not occur.

One explanation is that most elections aren't close. FairVote noted that the margin between first and second place was less than 5% in only 11% of the elections they studied. Condorcet is not likely to be violated in wide-margin races because it is not likely that the majority will prefer another candidate over the initial front-runner.

The Condorcet criterion is related to the spoiler effect, which, as we argued, is a manifestation of IIA. According to research by David McCune of William Jewell College and Jennifer Wilson of The New School, when the winner of instant runoff is not the Condorcet winner or when the election doesn't have a Condorcet winner, there is potential for the spoiler effect. Both these situations, however, are rare. (In one study, David McCune and Lori McCune looked at more than 200 elections and found that only one had no Condorcet winner. That election also exhibited Simpson's paradox.) Wilson and McCune found that of the 170 instant runoff elections in California, Minneapolis, and New York that did not have a majority winner and had at least three candidates, only two had a spoiler: mayoral races in Burlington, Vermont, in 2009 and in Oakland, California, in 2010. Adding the 2022 Alaska special election in which Sarah Palin was the spoiler, the rate is 3 of 171, or 1.8%.

Chances of "weak" candidates—ones we regard as having no chance of winning, which is what most people think of as "spoilers"—influencing instant runoff elections are even smaller. For instance, the candidate with the fewest number of first-place votes cannot be a spoiler. Nor can (more generally) any subset of candidates whose total first-place vote count is less than every other single candidate's.* A Con-

---

* This is the same supposition that is made in batch elimination, as I mentioned in a footnote in the previous chapter.

dorcet loser—another type of a "weak" candidate—is highly unlikely to be a spoiler, and that probability decreases as the number of candidates increases.

According to research by Adam Graham-Squire of High Point University, instant runoff also does not exhibit failure of monotonicity when weak candidates are present. One interesting observation is that all three of the spoiler elections mentioned above also exhibited failure of monotonicity. More evidence comes from recent research in which David McCune and Graham-Squire studied 1,079 instant runoff elections for Scottish local government positions and concluded that bad behavior is rare, although close elections tend to have more of it and often bundle together different kinds of anomalies.

But "close" is a tricky notion—for one thing, a researcher must define what it means. Some theoretical research suggests that up to 50% of "close" instant runoff elections show monotonicity failure. The problem is that the 50% incidence rate is an artifact of precisely aligning the definition of "close" with elections that flunk monotonicity. If you assume that the top three candidates receive about 1/3 of the votes, then yes, there's a strong chance of witnessing paradoxes. As Graham-Squire put it to me: "This is akin to saying '40%–50% of people going to bars get into fights' when you've restricted your research to the situation where both people are drunk, have been arguing for the last 15 minutes, and are now both standing up with their fists raised. It is both true and kind of useless. The landscape of U.S. elections is very different from this." Luckily, close elections aren't common, and that explains why violations of monotonicity are rare.

This line of defense against Arrow's impossibility theorem thus says that ranked voting—or at least instant runoff, because that's what we have the most research on—doesn't really show bad behavior in practice. But the nonzero chance of bad things happening might still bother you. In which case, there's one more thing we can do.

We can abandon ranked voting altogether.

CHAPTER 5

To Each Their Own

ARROW'S THEOREM is a bummer: any time we run an election in which voters rank candidates, something could go wrong. A paradoxical outcome is a possibility.

But what if we thought outside the ballot box? These problems are inherent in *ranked* elections. What if we asked voters to evaluate candidates on their own merits without explicitly making comparisons? Voting methods that do this are called *cardinal* (nothing to do with cardinals voting in the conclave or with the St. Louis Cardinals, although St. Louis does use one of these methods). This method contrasts with ranked methods, such as instant runoff, the Borda count, or the Copeland method, which are *ordinal* methods.

You're used to cardinal voting methods. When you choose a few scoops of ice cream, you're using a cardinal method. All you need to know is that you like hazelnut, passion fruit, and cookie dough and don't like any other flavors nearly as much. If you're getting three scoops, you don't need to rank those three or the remaining flavors.

When you rate a product on Amazon or an apartment on Airbnb by giving it anywhere from one to five stars, you're using a cardinal method. In each case, you thought about what you were rating individually without necessarily comparing it to any other alternative. When you give that new TV you bought four stars, you don't first rank all the TVs you've ever owned; you simply judge the new one on its own qualities.

The first way to assess the options, by simply deciding whether you like them or not, is called *approval voting*. The second, in which you choose how many points or stars to give each option, is *range voting*. These are two of the most popular cardinal voting methods, so let's take a look at them.

THEY LOVE YOU, THEY LOVE YOU NOT

Approval voting was first proposed by Robert Weber in his 1971 PhD thesis but was advanced by Steven Brams of New York University and Peter Fishburn of Bell Labs in their 1982 book *Approval Voting*. Brams is still at it forty years later—I was just at a conference where he gave a talk extolling approval's virtues. The method simply requires the voter to indicate which candidates they regard favorably, perhaps by checking a box next to the names they prefer. Checking any number of boxes is fine, including none, only one, or all. A profile for such an election might look like this:

	5	5	3	2	2	1
Candidate A	✓	✓		✓		✓
Candidate B	✓		✓	✓	✓	
Candidate C		✓	✓	✓		

Looking at the columns left to right, we see that five people approve of candidates A and B, five other people approve of A and C, and so on. The tallying is a simple count of check marks:

A has $5 + 5 + 2 + 1 = 13$ approvals
B has $5 + 3 + 2 + 2 = 12$ approvals
C has $5 + 3 + 2 = 10$ approvals

Candidate A is the winner. This is the plurality winner, but in the sense of having the most approvals and not in the customary sense of having the most first-place votes.

This system is familiar. When you talk with your friends about what restaurant to eat at or what movie to see, you might mention a few options you're happy with. In doing so, you're figuratively casting an approval ballot.

The approval procedure is clean and simple. The voter has to think about each candidate separately and make the call on endorsing them or not. No comparison—a potentially daunting task if there are many candidates or if some of them are politically similar—is required.

If all voters decide to check only one box, then approval voting reduces to standard plurality. If a voter approves of all the candidates, their ballot is useless. In the table above, the two people who approved of all three candidates might as well have stayed home because that voting strategy has the same effect as approving of none, or simply not voting. Ties are a possibility with approval voting, and they would have to be resolved somehow.

Approval voting can't cause vote splitting or the spoiler effect. The process doesn't waste any votes because a voter can support more than one candidate with no penalty. In the 2016 Republican primaries, for example, if approval voting had been used, Trump voters could have also indicated that Cruz was an acceptable choice, and that could have resulted in Cruz winning the nomination instead of Trump.

As is the case with other systems that don't waste votes, approval voting encourages the participation of smaller parties and minor candidates. In fact, The Center for Election Science, an organization that advocates approval voting, contends that this method benefits both minor parties and major parties.

Approval voting discourages negative campaigning because it behooves a candidate to appeal to as broad an audience as possible. The electorate often frowns upon negative campaigning, so a better strategy might be to try to get the approval of as many people as possible instead of campaigning against another candidate. This is similar to the incentives associated with ranked methods for which it is in a candidate's interest to climb as high as possible across many ballots rather than attempting to be ranked first on some subset of ballots.

Approval voting is genuinely different from a ranked voting system because the two contain genuinely different information. Given a ranked ballot, there is no way to determine which candidates the voter who filled it out approves of. Maybe the last two candidates ranked by that voter are detestable to them, and they would not under any circumstances want those two to be elected. There's no way to tell this. A 2020 straw poll of Democratic Party primary voters conducted by The Center for Election Science showed that 60% of the voters approved of Bernie Sanders, 55% approved of Elizabeth Warren, 39% approved of Pete Buttigieg, and 36% approved of Joe Biden. This result indicated lack of enthusiasm for Biden, something many voters felt but were not able to express with plurality voting or with ranked ballots. Biden's ultimate nomination may thus be explained as a strategic choice by voters whose overriding goal was to elevate a candidate most likely to beat Trump. Also surprising is that 28% of the voters approved of Amy Klobuchar, who despite this showing did not make it far in the primaries. Approval voting (or approval polling, in this case) gives us access to the pulse of the electorate, providing valuable information about what the voters consider relevant and where they might collectively be headed ideologically.

The discrepancy between rankings and approvals can be dramatic. In 1998, Donald Saari and Jill van Newenhizen showed that preference profiles exist that can produce approval ballots such that *any* of the candidates is the winner, depending on what rank is taken to be the threshold for regarding a candidate as approved by a voter. In particular, the candidate with the majority of first-place votes in a ranked profile might not be the winner in approval voting. (Brams, the champion of approval voting, rebutted by claiming this is in fact one of its advantages.)

Conversely, there is no way to create a ranking from an approval ballot because we do not know how the voter stacks the candidates against each other. If they approve of three options, all we know is that all three are acceptable to them. We know nothing about the relative degree of the acceptability.

Anonymity, neutrality, monotonicity, and IIA can all be defined in the setting of approval voting and that method satisfies all four criteria. The cherry on top is that Arrow's theorem of impossibility no longer applies because we're not in the universe of ranked voting systems!

Approval voting can be found in high-profile settings such as the election of the secretary-general of the United Nations. It is also sometimes used in professional organizations. In fact, a couple of organizations close to my heart, the American Mathematical Society and the Mathematical Association of America, the two largest math organizations in the United States, use approval voting to elect their officers.* If approval voting is the mathematicians' choice, why are we even talking about anything else? Did we find the perfect voting system?

Not so fast. When the Mathematical Association of America adopted approval voting, its president wrote a letter warning that in general, this could lead to the election of a mediocre candidate; we'll see why in a second. But the method nevertheless seems to work in communities such as these two associations because, well, mathematicians are nice people. When we vote for our officers, we are basically not opposed to anyone being elected. Moreover, nominations to officer positions in our societies are regarded as acknowledgments of universally recognized career achievements. For such amicable, uncontentious elections, approval voting works just fine.

Approval voting also works well when there is no limit to the number of candidates to be elected, for example new members of a professional or honor society or a hall of fame. This is why the National Academy of Sciences uses approval voting to elect its new members. These are again elections that are typically at the low end of the contentiousness scale.

Approval voting is not widely used in elections for public office. Fargo, North Dakota, has used it for citywide elections since 2018, and St. Louis, Missouri, has used it since 2020. It has been used in Oregon, Pennsylvania, Texas, and Colorado. It was also used in the Soviet Union in the late 1980s as a test run for democratic elections. (The experiment was interrupted by the country's dissolution in 1991.)

When feelings run high, the strength of approval voting—the relative simplicity of looking at each candidate independently through a binary lens—is also its weakness: there is no way to express the *degree* of approval. A voter might like two candidates but *really* like one of them and really want that candidate to win. As a consequence, they might not

---

* After the Mathematical Association of America adopted approval voting, it elected its first female president in 1987.

indicate approval for the other person, even if they wouldn't mind them winning, in order to avoid hurting their top candidate's chances.

Conversely, voters might check a candidate's name because they want to send a message of encouragement even if at the moment the candidate is far from the voter's most favored choice. But what if many people do this while not enough people indicate approval for any of the top candidates? The result might be that the minor candidate in fact wins—again because degree of approval cannot be expressed. This is similar to the situation with the Borda count, where someone who is nobody's first choice but is highly placed on most people's ballots can rise to the top.

To put it more concretely, suppose that nine of ten voters like candidate A but they also think B is okay, so they approve of them both. The tenth candidate doesn't like A so much but agrees that B is ok, so they approve of B only. In this scenario, A has nine approvals and B has ten. The winner is B, even though 90% percent of the voters think A is a much better candidate. Donald Saari, a critic of approval voting and an advocate of the Borda count, points out that had the 1992 presidential election been conducted by approval voting, a similar scenario could have elected Ross Perot—a decidedly centrist, one might say mediocre, candidate—as the president.

Because the main snag with approval voting seems to be its oversimplicity in the sense that a voter can't say *how much* they approve of a candidate, why don't we try to build this information into the method?

A RANGE OF EMOTIONS

Range (or *score*) *voting* addresses the inexpressiveness of approval voting by asking the voter to give each candidate a score in some range, say 0 to 9 (see figure 5.1).

Sometimes the ballot also includes a "no opinion" option. Selecting it does not hurt or help the candidate. Each candidate's total score is divided by the number of voters who scored that candidate. This gives the average score, scaling each candidate's total to a number between 0 and 9, and prevents candidates from being penalized in case some voters don't rate them at all.

Governor candidates	Score *each* candidate by filling a number (0 is worst; 9 is best)
1: Candidate A →	⓪①②③④⑤⑥⑦⑧⑨
2: Candidate B →	⓪①②③④⑤⑥⑦⑧❾
3: Candidate C →	⓪①②③④⑤⑥❼⑧⑨

FIGURE 5.1. An example of a range ballot.

For example, let's look at this profile:

	Voter 1	Voter 2	Voter 3	Voter 4	Voter 5
Candidate A	5	7	0	6	9
Candidate B	2	1	3	6	0
Candidate C	9	5	8	7	0
Candidate D	0	3		4	1

The range is 0 to 9 and each cell provides the number of points a voter gave to a particular candidate. Add across rows to get the score of each candidate:

A gets $5 + 7 + 0 + 6 + 9 = 27$ points
B gets $2 + 1 + 3 + 6 + 0 = 12$ points
C gets $9 + 5 + 8 + 7 + 0 = 29$ points
D gets $0 + 3 + 4 + 1 = 8$ points

Now divide the totals of A, B, and C by 5 because each of those candidates was scored by five voters. D's score is divided by 4 because one voter did not score them:

A's score is $27 \div 5 = 5.4$
B's score is $12 \div 5 = 2.4$
C's score is $29 \div 5 = 5.8$
D's score is $8 \div 4 = 2.0$
C wins the election.

Range voting is a generalization of approval. If a voter were allowed to put only 0s and 9s in the cells of a range ballot, that would be equivalent to approval voting: the two numbers would be interpreted as either an unchecked box (the voter gives a value of 0 to a candidate) or a checked box (the voter gives a value of 9). But with the full range from 0 to 9, the voter has freedom to express the intensity of (dis)approval. If the voter cares about only one candidate, they can *bullet vote* by putting a 9 next to that candidate's name and 0s everywhere else. Or if all they care about is that one candidate they hate and don't want to see elected, they can put a 0 next to their name and 9s everywhere else. They can signal that they think of candidates as the same by scoring them with the same number. Or they can simulate a ranked ballot by giving their first choice 9 points, their second choice 8 points, and so on. A range voter has lots of options—more than with a ranked ballot—which is arguably a good thing.

The actual range of points is mathematically unimportant; the main consideration is psychological. Greater range allows a more refined expression of preference. But too much of a range might stump the voter. Would anyone be able to rate one candidate as 843 and another as 844 on a scale from 0 to 999? This is why most range voting, including the various rating procedures you take part in regularly, uses something between a 4-point and a 7-point scale.

Any time you express your opinion on a form that says "select 1 for least satisfied, 5 for most satisfied," you're doing range voting. If you ever rated a restaurant on Yelp or a movie on IMDb, you range voted. Various sports, including diving, use (fancier) versions of range voting.

The calculation of grade point average is also a range vote of sorts. The "candidates" are the students, and a teacher "scores" them a certain way (based on the students' performance). The class valedictorian might be considered the "winner" of such an election. We college professors increasingly have to fill out recommendation letter forms for our students who are applying to jobs or graduate schools in which we rate their various attributes and qualities. This can be thought of as providing our degree of approval in each category.

The upvote/downvote system on websites such as Reddit and Quora is range voting. The range can be thought of as −1 (a downvote), 0 (no vote), and +1 (an upvote). The answer to a question with the most points is pushed to the top. The Wikipedia Arbitration Committee elects its members using this scale. There is no mathematical difference between the −1 to +1 spread and, say, a spread of 0–2, but the former has the clear psychological benefit of numerically disapproving of an option.

If you've ever applauded or shouted in someone's support at the end of their performance, the decibels of noise you made provided a range score. In ancient Sparta, this was the method used for electing members to the council of elders: the candidate who received the loudest combination of applause and shouting would become the winner.

We can find range voting in nature. Our close relatives, bonobos, range score their food from bark to grunt on a 5-point scale. Honeybees and ants also use range voting to determine new hive sites and nests.

All the good features of approval voting are retained in range voting. This method is resistant to vote splitting and the spoiler effect and satisfies all the same criteria as approval voting does. Arrow's theorem doesn't apply. Range voting discourages partisanship and encourages honest, positive campaigning.

There is another measure by which range voting does well. It was devised by the mathematician Warren Smith, who is one of the greatest proponents of range voting. In fact, he is the one who coined the term "range voting" in 2000. In the late 1990s, he conducted an extensive simulation to test which of various familiar voting methods generated the "lowest expected avoidable human unhappiness." This is called *Bayesian regret* and is characterized in economics as the discrepancy between the maximum possible utility and the expected utility of a good or a service. More plainly, it is the expected "loss" to an individual or society because some optimal but imperfect procedure was used. According to Smith's analysis, range voting did best in all the simulations, landing closest to an ideal voting system that provides full satisfaction for as many voters as possible. Next in line, in order, were approval, Borda, Condorcet, and instant runoff. Unsurprisingly, plurality is the farthest from the ideal system. The simulations also discovered that Bayesian regret is significantly reduced when voters cast their votes honestly rather than strategically. A user-friendly account of all this can be found in *Gaming the Vote* by William Poundstone, who himself became an adherent to range voting.*

---

* Poundstone later shifted to supporting approval voting, citing the complexity of range voting.

However, range voting has serious flaws. For example, a score of 4 for a candidate might mean different things to different voters. There is no universally accepted notion of what a 4 should mean in terms of the strength of support for or perceived quality of the candidate. Two voters who feel the same way about a candidate might rate them differently, with a 4, a 5, or a 6.

This is further complicated by the fact that the goalposts of 0 and 9 are also moveable. One voter might identify their least and their most favorite among the candidates, assign 0 and 9 to those two, and then compare all other candidates to those extremes. Another voter might have an idealized notion of a candidate who gets a 9 and compare everyone else to that ideal. It's unlikely that anyone will score a 9 on this voter's ballot.

There is also the issue of the distance between scores. A voter might feel like two candidates deserve scores of 4 and 6, while another voter who feels the same way about the candidates might score them 3 and 6 or 4 and 7. Both the score and the numerical range of the scores voters assign to the two candidates are subjective.

All this scoring flab makes computing averages problematic. For example, suppose one voter rated three candidates, A, B, and C, with 2, 4, and 6, respectively, and another voter rated those same candidates with 3, 5, and 7. Candidate C is the favorite for both voters, and that candidate wins with an average score of 6.5. So far, so good.

But now suppose a third voter, who really likes A, decides to give the scores 9, 0, and 0 to A, B, and C. The updated average scores are:

A: $(2 + 3 + 9) \div 3 = 14 \div 3 = 4.66$
B: $(4 + 5 + 0) \div 3 = 9 \div 3 = 3.00$
C: $(6 + 7 + 0) \div 3 = 13 \div 3 = 4.33$

Now A is the winner, even though two of three voters had C as their first choice and A as their last choice! Even if the third voter liked candidates B and C, it was in their interest to give them scores of 0 to lower their averages and give their favorite candidate, A, a better chance of winning. This shows that range voting fails the majority criterion. In addition, it can fall prey to strategic voting—which the third voter used—the topic of the next chapter.

The use of averages also obscures the sample size—that is, the number of voters who scored a candidate. Suppose one voter gave candidate A a score of 9 and candidate B a score of 0. Say there are 999 other voters who did not score A because they never heard of this candidate, but they all love B and uniformly score this candidate with a 9. Now A's average score is 9.000 and B's average score is 8.991. Candidate A wins the race based on a single vote. That's tough for the 999 people who love B; they are about to find out who A is because A will be their elected representative.

Problems with using the average, or mean, are mitigated by systems that use the median instead. In the previous two examples, one with 3 candidates and one with 1,000 voters, the median would have elected the majority winner C in the first example and B in the second. A more elaborate median-based method also tackles the problem of interpreting scores. In *majority judgment* (or *majority grading*), which Michel Balinski and Rida Laraki proposed in 2007, voters rate each candidate on some non-numeric "satisfaction" scale, such as Poor to Excellent. This shift is meant to avoid the issue of numerical scores having different meanings for different people, whereas (the idea goes) qualitative descriptions such as Satisfactory or Very Good hold more universal meaning. Using the median instead of the average is meant to get around the issue of artificial and extremal voting like the 9, 0, 0 wrench described above. Unfortunately, majority judgment is susceptible to the no-show paradox: a candidate might do better if some of their supporters were to make a strategic move and choose not to vote.

———

There are other popular cardinal voting systems, such as *cumulative voting*, in which voters are given a certain number of points or votes they can distribute among the candidates in any way they see fit (usually the number of points equals the number of candidates). Voters can even bullet vote, or give a single candidate all their points. Cumulative voting is commonly used in companies that need to elect members to a board and can be used to fill single or multiple positions. Outside the boardroom, it was briefly at the center of media attention in the early 1990s, when it was promoted by

Lani Guinier, President Clinton's 1993 nominee for assistant attorney general. Guinier advocated cumulative voting and thought that minority voters should bullet vote as a bloc, thereby increasing their chances of securing representation. The backlash against Guinier's views was swift and Clinton withdrew her nomination two months later. To much of the public, an attempt to introduce a method that could give a more equitable voice to the electorate was indistinguishable from manipulation.

Add to the basic options the possibility of *combining* voting methods in the hope of extracting the best they have to offer, and we arrive at dozens of different systems. An example is the STAR (Score-Then-Automatic-Runoff) method, which some political parties in Oregon use. This method conducts a standard range vote but then looks at the top two scorers and asks which of the two is preferred on more ballots. This runoff step then decides the winner.* There are also systems that combine ranked methods and approval voting.

The STAR method was devised to dampen range voting's vulnerability to strategic voting. We saw examples of such voting before with instant runoff and the Borda count, and I kind of brushed it off, hoping you wouldn't say, "Wait, what?!" But I can't downplay it any longer. I'm going to be honest: dishonest voting is a big deal.

---

* Another simulation similar in spirit to Bayesian regret, *voter satisfaction efficiency*, was done by Jameson Quinn, a Harvard statistics PhD student. He found that STAR did the best in this analysis.

CHAPTER 6

Strategy and Manipulation

I'LL SOON BE taking over as the chair of my department. This is not an honor, nothing I've earned; I'm simply the next in line. On the first day, I'll instantly become buried in administrivia and will start the countdown on the 1,094 days remaining in my sentence.

One of the most important jobs of the chair is to conduct job searches for new faculty. The process is meant to be democratic, culminating in a vote by all the faculty, and in my nearly two decades here, it's always succeeded in being thoughtful and amicable. It has always produced a candidate we were delighted to have as a colleague. But suppose I turn into an academic supervillain and decide to mess with the procedure. Maybe that one extra committee meeting that should have been an email nudges me over the edge. So I decide to push for a particular candidate because during the campus interview I find out that they love soccer as much as I do or I'm their father's brother's nephew's cousin's former roommate (yes, I know the reference dates me). I cannot abandon the appearance of democracy, so I turn to the math.

GAME OF CHAIRS

The hiring begins by whittling down the pool of hundreds of candidates to a few who are invited to campus for interviews. After all the candidates have come and gone, I commence my nefarious plan by watercooler chatting with each of my colleagues to find out how they feel about them. My favorite, I learn, is not many people's first choice, but they are highly regarded by most. Based on the intelligence I gather, I predict this potential ranked profile, where V (for Volić/Villain) is my candidate:

5	3	2	2	1
A	C	B	V	V
V	V	C	A	A
C	A	V	C	B
B	B	A	B	C

My own ranking is one of the two in column 3: B > C > V > A. I chose not to put V on top to avoid arousing suspicion.

Then I embark on phase two of my scheme. I suggest we each prepare a ranking of the candidates that will be tallied at the fateful meeting, expecting that everyone will turn in the ballot I gleaned through my covert watercooler work. In addition—and here's the truly villainous part—I say we should use the Borda count because that's generally accepted as a good method. Everyone is used to the idea of assigning points to rankings, and my unsuspecting colleagues agree that this makes sense. At the meeting, their preferences unfold in the way I predicted and the Borda count is deployed:

A: $5 \times 3 + 3 \times 1 + 2 \times 0 + 2 \times 2 + 1 \times 2 = 24$
B: $5 \times 0 + 3 \times 0 + 2 \times 3 + 2 \times 0 + 1 \times 1 = 7$
C: $5 \times 1 + 3 \times 3 + 2 \times 2 + 2 \times 1 + 1 \times 0 = 20$
V: $5 \times 2 + 3 \times 2 + 2 \times 1 + 2 \times 3 + 1 \times 3 = 27$

My candidate wins! Wicked plan executed!

It would not have been smart for me to suggest instant runoff, because in that case, B would have been eliminated in the first round and then my candidate would have been on the chopping block in the next iteration. A would have won and would have got the job.

But let's say one of my colleagues is onto me and argues that we should indeed use instant runoff instead of Borda. No problem—all I need to do is change the order in which I list C and V in my ranked ballot. The new profile would look like this:

5	3	1	1	2	1
A	C	B	B	V	V
V	V	C	V	A	A
C	A	V	C	C	B
B	B	A	A	B	C

The fourth column is my new ranking. Instant runoff then first eliminates B, then C, and V wins! Counterattack to my wicked plan thwarted! Wicked plan B (or V?) executed!

This kind of scheming can be a team effort too. Let's use the Borda count on this profile:

	5	4	4
3 points	D	A	B
2 points	A	B	C
1 point	B	C	D
0 points	C	D	A

The winner is B with 25 points:

A: $5 \times 2 + 4 \times 3 + 4 \times 0 = 22$
B: $5 \times 1 + 4 \times 2 + 4 \times 3 = 25$
C: $5 \times 0 + 4 \times 1 + 4 \times 2 = 12$
D: $5 \times 3 + 4 \times 0 + 4 \times 1 = 19$

But suppose that profile was really based on a watercooler poll and before the actual election four voters in the middle column collude and realize that it would be to their benefit to place C above B. This would be the new profile:

	5	4	4
3 points	D	A	B
2 points	A	C	C
1 point	B	B	D
0 points	C	D	A

A: $5 \times 2 + 4 \times 3 + 4 \times 0 = 22$
B: $5 \times 1 + 4 \times 1 + 4 \times 3 = 21$
C: $5 \times 0 + 4 \times 2 + 4 \times 2 = 16$
D: $5 \times 3 + 4 \times 0 + 4 \times 1 = 19$

Now candidate A, who was the first choice of those four voters, is the winner. Strategic voting by this group paid off. An example involving coalitions of any size can be constructed in a similar vein.

Influencing, rigging, and manipulating elections is as old as elections themselves. As C. L. Dodgson (aka Lewis Carroll) put it, voters are prone to "adopting a principle of voting which makes it more a game of skill than a true test of the wishes of the electors." He followed up with this deep insight: "It would be better if elections were decided according to the wishes of the majority than of those who have the most skill at the game." Thanks for your wisdom, Mr. Dodgson.

The process itself can be controlled or carefully chosen to serve a purpose, as the selection of Borda count did in the Machiavellian scenarios above. Voting theorist Donald Saari once offered the following service: he said he could go to any organization about to have an election where someone would tell him who they wanted the winner to be. After chatting with the voters, he could design a fair-seeming election method that would elect precisely the desired person. Even though this was announced in jest, Saari claims that he was contacted by several members of the House and Senate and by a president of a country. A little less nefariously, Alex Tabarrok of George Mason University has showed that there are voting systems that would have allowed Clinton, Bush, or Perot to win the 1992 presidential election. A similar analysis applied to the 2000 election shows that even Nader could have won with the right method.

Manipulation of the process plays out all over the world, and not even necessarily in secret. Take North Korea, which you might be surprised to hear actually holds elections every five years. The surprise is diminished with the realization that each district has one candidate for the Supreme People's Assembly who is chosen by the Communist Party (Kim Jong Un, really). If a voter wants to express their support for someone else, they must do so publicly. Voting is compulsory, and the penalty for not voting is severe. Turnout is 99.97%.

Another gem of political theater of the absurd is the 1927 presidential election in Liberia, which made it into the *Guinness Book of Records* as the most rigged election ever. The winner received sixteen times as many votes as there were registered voters.

Voting in Eurovision also falls under the heading of "rigged" because many teams of judges rank their top ten countries according to geopolitical considerations, shying away from upsetting their neighbors. Norway and Sweden usually give each other lots of points, as do the UK and Ireland, Ukraine and Belarus, Cyprus and Greece, Serbia and Montenegro, and so on. Cardinals talk to each other and form coalitions before the papal conclave. There is universal awareness that lobbyists are some of the hardest-working people at the United States Capitol and play a huge role in the formulation and passage of legislation.

But the theatrics of pretend democracies or extraneous influences on voters are not where our attention lies. We want to focus on the *mathematical* manipulability of elections, the quantitative exploitation of the methods. As the first order of business, a tallying method has to be chosen. This can be as important a factor in an election as the voting, as evidenced by my rogue chair example. Even for a fixed profile, different methods can produce different winners; remember the example from chapter 3 that had five different winners for five different methods. In the words of playwright Tom Stoppard: "It's not the voting that's democracy; it's the counting."*

Arriving at a decision about a tallying method is usually a political endeavor, achieved by watercooler scheming or Kim Jong Un's scare tactics or, more commonly, some kind of an organizational or legislative consensus. Unfortunately, mathematics is rarely a consideration in this step. This, of course, is why one of the top agenda items of this book is

---

* This is sometimes attributed to Stalin—someone you definitely don't want to quote when talking about fair and democratic elections—although historians agree that he likely did not say it.

to help decision-makers acknowledge the damage inferior mathematics can inflict on democracy.

In any case, let's assume that a tallying method (plurality, ranked, or cardinal) has been selected and cannot be changed. We will also suppose that the counting is legitimate and is not compromised. (Of course, that might not be the case in reality, but that's a topic for another day.) With those two assumptions, let's explore how the methods leave room for mathematical strategy and manipulation.

THE MOST MANIPULABLE METHOD

The mathematics of a voting method can be exploited in two ways: externally, through campaigning or other kinds of strategic machination that steer voters toward or away from certain candidates so the math of the election method works out a particular way; and internally, through voters acting strategically because the math of the method allows them to.

Plurality is extremely susceptible to external manipulation. The simplest example is the exploitation of the spoiler effect. It is now standard practice for Democrat and Republican candidates to run ads for a minor candidate from the other party. The hope is that the smaller candidate will take enough votes away from the bigger one. The GOP group Republican Leadership Council ran ads in support of Nader in several states in 2000 in an apparent attempt to hurt Gore. In his book *Gaming the Vote*, William Poundstone details several examples of this practice from 2006 involving such familiar names from national politics as Rick Santorum, Gabby Giffords, and Rick Perry. One particularly cringy case was the support some GOP agents lent to the 2020 candidacy of Kanye West in the hope of taking Black votes away from Joe Biden.

In the 2016 presidential election, Russian hackers employed a similar scheme. They targeted voters who were displeased enough with Hillary Clinton that they considered voting for third-party candidates, drowning them with negative information about her. This might have played a crucial role in driving those voters to candidates such as Jill Stein and Gary Johnson in swing states and costing Clinton the election.

Tactics for propping up spoilers are getting more and more creative. During the 2018 Senate race, Arizona Republicans sent mailers portraying Angela Green, a candidate from the Green Party, as too liberal. But the wording of the ads could have come straight from the fan fiction of a Bernie Sanders groupie: Green "wanted to cover all pre-existing conditions" and opposed "cutting taxes for businesses." The intended effect was that should this flyer somehow land in a Democratic voter's hands, it might make them think they should be voting for Green. Afraid that she would indeed be a spoiler, Green pulled out of the race and endorsed Democrat Kyrsten Sinema, who ultimately won the election.

Naturally, both sides play the game. In the 2022 midterm elections, Democrats ran ads in key swing districts "criticizing" far-right MAGA* Republican candidates prior to the primary election. However, the ads sounded more like praise of the candidates' character and policies instead of the usual bashing. The strategic objective was for those candidates to win the Republican nomination because they would be easier to beat in the general election.

In multiparty systems, it is standard practice for similar political parties to make arrangements with each other or form coalitions to prevent or minimize the effects of vote splitting and spoilers. If the coalition wins, various cabinet or legislative leadership positions might be split up among the parties according to some prior agreement. In a shadier version of consensus building, parties that seem to be ideologically different will sometimes make an arrangement whereby one of them will try to cause the spoiler effect or vote splitting within its own cohort so its co-conspirator can rise to the top. (This has happened, for example, in Bosnia.) The winning party might compensate by giving its helper party some positions in the government, maybe toss a few ministerial positions its way. Is this legal? Yes. Is it wrong? Absolutely. It's rigged, but nobody cheated.

---

* If you don't know what MAGA is, you've either spent the last seven or so years in a well or you're reading this 150 years from the time it was written. Either way, I envy you. MAGA stands for Make America Great Again, Donald Trump's 2016 presidential campaign slogan.

Then there are the voters. How can they play plurality math to their advantage? First, a bit of terminology. Voting in a way that does not reflect one's true honest preference is called *strategic* (or *tactical* or *insincere*). According to this definition, plurality suffers massively from strategic voting. Remember *favorite betrayal*? A voter who is afraid to "waste" a vote by selecting a minor candidate or worries about contributing to the spoiler effect might choose one of the major candidates they feel best about. But because this is a vote for someone who is not their true first choice, it is a strategic vote. If this sounds like an accusation, it is not. The voter is forced to act in a tactical way because the system does not encourage honesty.

Strategic voting to stave off vote splitting and spoilers happens all over the world. It is estimated that one-third of voters in the 2015 federal elections in Canada voted strategically. The percentage is similar for the 2011 presidential elections in Slovenia. The poll-defying failure of the Podemos party in Spain in 2016 can be attributed to strategic voting driven by a fear of wasted votes. Various surveys found that anywhere from 22% to 32% of voters in the 2019 general elections in England voted strategically, driven by Brexit considerations.*

A voter who wishes to support a small party candidate or a minor candidate from a major party is left in an unenviable place of internal conflict. They want to vote their conscience, but an accusatory inner voice murmurs something like "a vote for Stein is really a vote for Trump." Opponents of strategic voting say that honesty is the best strategy, that this is how democracy should work. But this argument falls flat when the tool that's supposed to implement this democracy does not reward or reflect honesty. A winner is a winner, and everyone else, including the spoiler, is relegated to political oblivion. So a voter casts their vote strategically for one of the major candidates, smaller parties get fewer votes and hence less political clout going forward, it's consequently harder for them to garner support and votes later, and so it goes

---

* While we're on the subject of the UK, a website there (tacticalvote.co.uk) is dedicated solely to telling people how they should vote strategically to block Tories from winning seats in the 650 constituencies (districts).

in a circle. This is the familiar setup of Duverger's law—it is precisely two-party systems that benefit most from plurality. If the danger of the spoiler effect exists as soon as there are three or more candidates, worried voters will strategically gravitate toward selecting two at most to avoid it. Democracy will suffer, and Duverger's casualties, like Angela Green and countless others from all walks of political life and both sides of the aisle, will continue to mount.

Then there is the cunningly tactical flip side—voting in a way that deliberately endeavors to create spoilers. Open primaries, for example, in which voters do not need to be officially affiliated with a party in order to vote in its primary elections, are susceptible to this kind of strategy. Candidates from the other camp might intentionally vote for a potential spoiler of the opposing party with the purpose of depriving a front-runner they dislike of the nomination or forcing a runoff in which the two top candidates will be compelled to dig deeper into their war chests, leaving fewer resources for the general election. Several states, including New Hampshire, Texas, and South Carolina, identified this practice as potentially throwing a monkey wrench into their 2022 midterm elections.

We've beat up on plurality plenty. We already knew it wasn't a great method, so this feels like adding insult to injury. What about the other methods we know? Is the incentive for strategic voting at least diminished, if not eliminated, for any of them?

As I'm sure you can sense, the answer is complicated.

A SCHEME FOR HONEST MEN

Many ranked methods (including instant runoff and Copeland) protect against most vote splitting and spoilers. A voter can vote their conscience regardless of the electability of their favorite candidate or simply decide to support a diversity of political opinions. But as the department chair example illustrates, ranked voting is susceptible to other forms of strategy.

For example, in a practice called *burying*, a voter might deliberately rank someone low if they perceive them as a threat to their top choice.

Here's a tiny example. There are two voters, me and one more person. We are to cast ranked ballots for four candidates. My preference is A > B > C > D and somehow I find out the other person's preference is B > C > A > D. If we are using the Borda count, A gets 4 points, B gets 5, C gets 3, and D gets 0. My favorite loses. So I change my ranking strategically and bury B. My new "preference" is A > D > C > B, and my candidate wins with 4 points to B's 3 points. Everyone else gets fewer points. (I also kept C where they were, because if I moved them up into B's slot, A and C would end up in a tie.)

The first to notice this flaw of the Borda count was the great mathematician and physicist Pierre-Simon Laplace, the inventor of the theory of probability and a buddy of Borda's from the French Academy of Sciences. When Laplace told him about this, Borda offered a precious reply: "My scheme is intended for honest men!" Ah, to be that oblivious! Little did he know that burying would become one of his scheme's most pronounced features.

All of our ranked methods are defenseless against this kind of insincerity. Burying cannot help the favored candidate, but it can hurt the rival. A candidate who is buried will have a harder time advancing up the table in instant runoff, will receive fewer points in the Borda count, and will be at a disadvantage in head-to-head contests in the Copeland method.

Other incarnations of strategic voting are more subtle. Let's revisit the example that introduced us to instant runoff's unpleasant failure of monotonicity. That profile was based on a poll:

6	6	6	4	3	2
A	C	B	B	A	C
B	A	C	A	C	B
C	B	A	C	B	A

Instant runoff winner is A. But then this candidate did something nice before the actual election and some voters decided to elevate them in their rankings: three of four people in the fourth column switched to

A > B > C and both voters from the rightmost column switched to C > A > B. This was the new profile:

9	8	6	1	3
A	C	B	B	A
B	A	C	A	C
C	B	A	C	B

Mind-blowingly, the winner is now C. But this example gives us another form of strategic voting. Suppose we start the second profile, which now plays the role of the poll-based preelection profile. Those same five voters who like A could now decide to *place A lower* in their ballots: three switch from A > B > C to B > A > C and two switch from C > A > B to C > B > A. The new profile is now given by the first table, and their candidate, A, wins. Strategically ranking the favorite candidate lower is our old friend *favorite betrayal*.* All ranked voting systems are susceptible to this strategy, but cardinal methods such as approval voting and range voting are not.

Here is a recent real-life almost example: June 2021 polling in the 2022 Alaska Senate race had Kelly Tshibaka, a Trump Republican, ahead with 39% of voters saying she would be their first choice. Next, with the support of 25% of those surveyed, was Al Gross, an independent who, if elected to the Senate, would likely side with the Democrats. In third place, with 19% of polled first-place votes, was the moderate Republican Lisa Murkowski, who has sometimes sided with Democrats in the Senate. The actual election used instant runoff (in 2020, Alaska became the second state to adopt it).

Democratic voters were in a bit of a pickle. If they voted their political conscience, they would rank Gross first, Murkowski second, and Tshibaka third. According to the percentages reported by the poll, Murkowski would get eliminated and her votes would be redistributed between Tshibaka and Gross. Here's the problem: that same poll

---

* Although this is not exactly a betrayal. In the usual setting of voting betrayal, a voter abandons hope that their candidate will win and throws them under the bus in hope of avoiding a spoiler. But here a voter hopes to help their candidate by ranking them lower.

showed that more of those Murkowski votes would go to Tshibaka than to Gross and Tshibaka would go home with an easy win. A strategic move by Democratic voters would be to rank Murkowski higher than Gross, with Tshibaka again in the last place, in the hope that Murkowski would survive the first round and then win against the likely other survivor, Tshibaka, by collecting all of Gross's votes in the instant runoff. The strategy turns on the expectation that the redistribution of votes for Murkowski would give Tshibaka a bunch of extra votes in the runoff, but the redistribution of votes for Gross wouldn't. Ultimately, Democratic voters breathed a sigh of relief when all this turned out to be hypothetical and they didn't have to betray their favorite: Al Gross dropped out of the race in June 2022.

There is a larger picture here. Instant runoff prevents most spoilers and you are encouraged to rank your top candidate in the top spot whether or not you believe they have a chance of winning. If they don't, your vote will get transferred to the next person on your list, who might be a major candidate. But if that major candidate is in a tight race, the overall redistribution of votes might not give them enough additional votes to survive the second round (or later rounds). Now both of your favorite candidates, a minor one and a major one, are gone. So maybe you shouldn't rank that minor candidate first after all. It walks and talks like the spoiler effect and you want nothing to do with it.

This is the sort of dilemma New York City voters faced in the Democratic primary for the 2021 mayoral election. According to polls, only a few candidates had a real shot at the nomination. If a voter ranked some long shots at the top, it might not have mattered how they ranked the real contenders after that. But if they included the front-runners they liked most near the top, they might have helped these contenders stay in the game, possibly all the way to the end. The New York Democratic primary tallying took eight nail-biting rounds (a counting fiasco that mistakenly added 135,000 test ballots to the count didn't ease the tension). Eric Adams hung on to first place by the skin of his teeth, collecting 50.4% of the votes ahead of Kathryn Garcia's 49.6% and eventually becoming the mayor of New York in a landslide win against the Republican nominee, Curtis Silwa. It is not clear how many people placed

Adams at or near the top because they were afraid of someone else winning, but with such tight margins, strategic voting could have played a decisive role.

And lest you think that Condorcet and his cousin Copeland are getting away with it, consider this profile:

Voter 1	Voter 2	Voter 3	Voter 4	Voter 5
A	A	B	B	C
B	C	A	C	B
C	B	C	A	A

The winner is B with a 3–2 score against both A and C and 2 points total. But suppose Voter 1 buries B by switching B with C. Now it's a three-way tie with 1 point each. This voter has managed to stave off the defeat of their favorite candidate. Perhaps now they can convince Voter 5 to strategically flip their ranking of B and A, which would make A the winner.

There are even more flavors of strategy. For example, Borda count is susceptible to *candidate cloning*, a practice by which a lot of people from the same camp run in the same election, hoping to rack up a lot of points and increase the chances that one of them will be elected. The idea is strength in numbers. What matters is the guarantee that the camp will ultimately be represented; who represents the camp is secondary. We saw in an example in chapter 3 that sequential voting is manipulable because different choices of agenda can yield different results. A voter can also exploit the no-show paradox of instant runoff and Copeland and strategically abstain from voting altogether to benefit their top choice. And all of this is just the tip of the iceberg.

Our favorite ranked methods sure seem to be riddled with opportunities for tactics. An evident question, mirroring the one we asked about ranked methods satisfying certain criteria, is whether a strategy-proof method exists that we haven't discovered yet—one for which it is never in a voter's interest to supply a dishonest ballot. Based on previous experience, you might be bracing yourself for a letdown. Good decision.

Gibbard-Satterthwaite impossibility theorem: For elections of three or more candidates, the only ranked voting system that is strategy proof is dictatorship.*

The theorem was proved independently by Allan Gibbard in 1973 and Mark Satterthwaite in 1975. Both did it as part of their doctoral theses. Gibbard was a graduate student in the philosophy department at Harvard University when Arrow and other heavy hitters such as Amartya Sen and philosopher John Rawls organized a seminar series that was attended by the select few from Harvard and MIT. Gibbard was one of two graduate students who attended, and the seminar is where he first presented his remarkable discovery. He subsequently spent most of his career at the University of Michigan and is the most prominent architect of the *theory of meaning*, or "the meaning of 'meaning.'"

Satterthwaite was an economics graduate student at the University of Wisconsin when he too proved this non-existence theorem. He found out about Gibbard's work when a referee of the paper he submitted pointed it out. Satterthwaite's approach was different than Gibbard's, but the conclusion is the same, so the theorem now bears both their names.† Satterthwaite joined the faculty at Northwestern University straight out of graduate school and has been there ever since.

The Gibbard-Satterthwaite impossibility theorem is of course very much like Arrow's impossibility theorem, but many experts regard it as more important and farther reaching. The language of the theorem can be changed so that it really is an impossibility theorem—it is impossible to have a strategy-proof ranked system that's not a dictatorship. In other words, strategy is an integral part of nondictatorial ranked voting.

---

* One condition is that we should be looking at ranked methods for which it is possible for any candidate to win under some profile. This condition is true for all reasonable methods, including the ones we've studied. Just pick any candidate and imagine that every voter puts them at the top of their ballot. Then, using any of the methods we've seen, this candidate is going to be the winner.

† Gibbard actually proved a more general version of the theorem that holds for all *deterministic* voting systems (randomness in the choice of the winner is not allowed), of which ranked methods are examples.

We could now throw up our hands or we could do what we did after Arrow's impossibility theorem: deal with it. How problematic is Gibbard-Satterthwaite in reality? Can we fix it or ignore it? What about just using cardinal methods instead?

It is problematic and ignore it we cannot. People will be people and they will vote strategically forevermore. Sincerity is not our strong suit. We will continue to betray our favorites and bury our opposition. Allocating time and energy to trying to change human nature would be foolish.

Favorite betrayal and burying are two of the most pronounced ways of strategic voting, but we can at least feel a little better about the perils of some of its more subtle instances. That is because most of our examples required some a priori knowledge of the rankings of all the voters. Only then could a conniving strategy kick in and influence the outcome. But even in small elections on committees or boards, it is unlikely that a voter will have this kind of insight, and this knowledge is essentially impossible to acquire for elections involving thousands or tens of thousands of voters.

DEADLY BULLET POINTS

If you're still worried that Gibbard and Satterthwaite's impossibility might lurk behind every ranked election, let's see how cardinal methods do with strategic voting.

On the face of it, approval voting is strategy proof. From a voter's point of view, it doesn't make sense to disapprove of a candidate they like. If I like pistachio and passion fruit ice cream equally well and my dessert stomach is empty, then I might as well get a scoop of each. It also doesn't make sense for the voter to approve of a candidate they don't like because that would only help that candidate and not increase any of the chances of the voter's favorite. Adding a scoop of dark chocolate, a flavor I don't like, to my ice cream order won't make either of the other two flavors I like taste better.

But we don't live our lives on a binary scale. Our likes and dislikes have gradations. I like pistachio better than passion fruit and I dislike dark chocolate more than mint chocolate chip. It's the same with how we think of candidates. Even if a voter likes one option, they might not approve of them on the ballot because there is another candidate that the voter really, really likes and whose chances of being elected the voter is not willing to risk. This kind of thinking quickly devolves into strategic bullet voting: voters are loath to approve of more than just their favorite candidate. In 2021, St. Louis held seven city council elections using approval voting. The average number of approved candidates across these elections was shockingly low, between 1.1 and 1.4. About one-third of the voters in Fargo, ND, said that they cast their ballots strategically by bullet voting in the 2020 mayoral election that used approval voting.*

Range voting to the rescue? It takes the constricting bifold of thumbs up/thumbs down and spreads it like an accordion into a numerical spectrum of modulated approval. This is exactly what we want, right?

Sort of.

Here is an example. Three voters are range voting on three candidates on a scale from 0 to 9:

	Voter 1	Voter 2	Voter 3
A	8	7	9
B	9	9	7
C	7	8	6

A gets 24 points, B gets 25, and C gets 21. Dividing these scores by 3 gives the averages, which are 8.0, 8.3, and 7.0. So B wins. But Voter 3,

* Proponents of approval voting point to a couple of other elections that appear to provide contrasting evidence. In the 2002 elections for French president, some 2,500 voters were asked to fill out an approval ballot in addition to the usual plurality one. On average, the voters approved of about three candidates. A similar experiment yielding analogous results took place in 1985 with some 6,000 members of the Institute of Management Science, which subsequently adopted approval voting. But the voters in the French election might have voted more honestly because they knew that the experiment had no bearing on the outcome of the election, while the members of the Institute of Management Science might have been the same collegial bunch as the mathematicians of the Mathematical Association of America and the American Mathematical Society who did not feel feverishly passionate about particular candidates.

whose top choice is A and who knows the race is tight, instead decides to bury B and C and give them both a score of 0. Now A has 24 points, B has 18, and C has 15. The new averages are 8.0, 6.0, and 5.0. A landslide win for A, even though two of three voters prefer B to A. Deadly bullet hit its target.

If enough people collude and decide to vote in this type of extreme, bulleted way, the outcome of the election can change. A similar strategy can be employed with cumulative voting, with a voter giving their total allocation of points to a single candidate. The effectiveness of this strategy changes campaigning behavior as well. It is in a candidate's interest to ask their voters to bullet vote, effectively turning cardinal methods into plurality. This was another observed takeaway from the Fargo election.

Burying, bullet voting, and favorite betrayal are strategic behaviors that don't really require much strategizing. A voter can always use them, no matter what the polls say or what the predicted outcome is. Strategy is just another impediment to the conduct of true democracy.

CHAPTER 7

And the Winner Is . . .

ONE OF the most persistent features of my professional life is that I am confused. When I'm trying to figure out some math or prove a theorem or read someone else's paper, the process consists of staring at the wall while my brain is overheating, and then, on a good day, thinking I see something, writing it down, and finally realizing (after minutes, hours, or days) that it was wrong or that I didn't understand what I was trying to do. Rinse and repeat. And repeating is not a guarantee that anything will work out. Sometimes it does, but it might take days, months, or years. Sometimes it doesn't, the fog never lifts, and the proof remains elusive. Or the wannabe theorem isn't true and the entire effort is a foolish pursuit of the impossible. In the end, there can be months of work with nothing to show but unpublishable pages of scribbles. Math is unforgiving that way. It does not give partial credit.

So why do I love my work? The answer is so simple that it feels prosaic and sentimental: every once in a while, the light switches on. Rare as they are, those moments of pure clarity, those epiphanies when things make sense, trigger what must be the most elemental rush of excitement. There is the priceless realization that the mind has triumphed, that I just figured something out that nobody had before me, that I moved the frontier of human knowledge a little bit outward—yes, imperceptibly and probably inconsequentially, but still, I moved it. The feeling might not last—each answer grows tentacles to new questions—but it is as addictive as any drug and well worth the battle.

It is in this context that I have wrestled with social choice theory over the past several years. I've been confused and unconfused countless times and now feel I have a reasonable grasp of the many possible voting methods. If I were operating in my usual mathematical realm with my usual mathematical motivations, this is where I would stop. Voting methods we talked about are simply different, they each have accompanying algorithms and properties that are equally valid and true, and they can coexist peacefully within the domain of mathematics. The functions x^2 and $\sin x$ are different ways of producing an output from an input, and we don't think of them as competing. Each is good for some things and not so good for others. You'd use the first to describe parabolic motion and the second to describe waves, but not vice versa. We'd never ask which is "better" because the question is meaningless. But we're not in the mathematical domain anymore. We are talking about real people, real governments, real choices. Voting is so consequential, so central to our way of life that a comparison among methods is not only desirable, it would be irresponsible if I didn't attempt to supply it.

But I can't appeal to an expert consensus. Each voting method has strengths and weaknesses, and consequently advocates and critics. There's the instant runoff camp, the Borda clan, and the approval club. Each tugs in its own direction, muddying the waters and tripping up the others. Proponents of cardinal methods say that ordinal methods are susceptible to Arrow's theorem of impossibility, that there is a risk of failing monotonicity or IIA. They point to the fact that instant runoff might eliminate a generally well-liked candidate and does not really eliminate spoilers. Moreover, they say, instant runoff's practical complexity for the voter and the necessary centralization of computational labor makes its implementation too burdensome. They say that Borda shouldn't be considered because it might not elect the majority winner or the Condorcet candidate. Copeland, on the other hand, they point out, is problematic because it could result in ties too often.

Instant runoff advocates rebut by saying that real-life failure of monotonicity or instances of the spoiler effect are rare. Borda supporters say that their method maximizes the probability of electing the Condorcet winner. Copeland fans say that their favored method actually elects the

Condorcet winner, which they assert is the true measure of majority support. Taking the fight to the cardinal arena, they all point to approval voting's inability to accommodate a variety of opinions and range voting's failure of the majority criterion. And everyone points at everyone else's vulnerability to strategic and insincere voting and claims that rival methods are more likely to exhibit favorite betrayal, bullet voting, or burying.

So where does that leave us? The best I can do is offer my own moment of clarity that followed many moments of confusion.

I think we should use instant runoff.

Before I present my case, here's something we can agree on: plurality should not be used. Spoilers, vote splitting, Duverger's law, the list goes on. Sure, plurality is simple, but that's about all it has to offer. And there is an urgent need to get rid of it. Our elections are increasingly turning into celebrity events and platforms for extremists. The fields of candidates continue to expand, creating fertile ground for spoilers and vote splitting. These trends make it more likely that the winner will garner support from only a small fraction of the voters, in turn elevating demagoguery, especially through the primary system in which a tiny, gung-ho base steadily demands more radicalism and less openness to conversation. In fact, a recent report from the nonpartisan, nonprofit organization Protect Democracy identifies plurality as one of the main culprits in engendering authoritarianism and extremism.

Mathematically, instant runoff does the job, and it does it well. It works against typical spoilers and vote splitting, avoids wasted votes, elects consensus candidates, and eliminates the dangers of Duverger's law. But also, and importantly—and here is where I step cautiously out of my theoretical cocoon into the real world—we have practical evidence that it works. We can stop worrying that it doesn't satisfy the monotonicity and Condorcet criteria because the empirical evidence shows that those vulnerabilities don't appear in actual elections. This corroborates the confidence of many people who have been promoting instant runoff for decades, such as Rob Richie, the cofounder of Fair-Vote, an organization that has been advocating ranked choice for thirty years. According to Richie, nonmonotonicity is the stuff of academia and has no footing in reality. Mathematicians used to counter that it might simply have gone unnoticed, but with recent research made possible by new data analysis techniques, we know that's not the case. This kind of evidence is currently available only for instant runoff because it's the most commonly used alternative to plurality; other methods are not used often enough to generate the necessary base of data. Some of the most prominent academic leaders in the field, such as the Nobel Prize winner Eric Maskin of Harvard University, agree that we should rally behind instant runoff. He recently told me that even if some of his own papers show that Borda and Condorcet are the best methods under certain circumstances, he feels that instant runoff's theoretical limitations are negligible in comparison to its practical benefits.

We also have encouraging demographic data on instant runoff. For one thing, it helps elect more women and people of color. According to RepresentWomen, an organization dedicated to increasing the participation of women in politics, women won 45% of instant runoff elections in the 2010s, almost double the rate in plurality races. The most recent data point is the New York City Council, which prior to the first instant runoff election in 2021 contained 27% women. This number went up to 61% after the city switched to instant runoff elections. (Other factors contributed to this dramatic change, but instant runoff was one of the main drivers.) Of the thirty-one women elected, twenty-five were women of color. An analysis by FairVote shows that in Bay Area elections, minority candidates in instant runoff elections won 62% of the time, up from 38% in the days before instant runoff. Oakland has elected only female mayors since instant runoff was instituted there in 2010. We also have ample confirmation that instant runoff discourages negative campaigning and promotes political civility.

Voters don't seem to mind instant runoff either. For example, 95% of New York City voters found instant runoff easy to use in 2021 and 77% want to use it in future elections. California voters report similarly high satisfaction rates. Most of us are used to runoff elections, and instant runoff puts them on mathematically prescribed steroids, automatizing them and making them more efficient and complete. Once voters understand this, they recognize instant runoff as an undeniable improvement.

And instant runoff might be the most natural system for us. We're used to ranking stuff. Our brain pits options against each another all the time. Think about ordering food: a clear, uncontested choice rarely presents itself on a menu. It's more likely that we will identify few attractive options and compare them to each other: "I usually like shrimp, but the salmon is almond crusted, which I really like, but wait, the pork is braised and . . . okay, the veggie dish is definitely not my thing, but . . ." There is a ranked ballot brewing somewhere in there, except that in the end you only submit a single choice to the waiter. But if the kitchen has run out of your top choice, you're ready with your next selection.

Instant runoff has momentum. The Pew Research Center reports that some 260 jurisdictions are using a system other than plurality and that instant runoff is by far the most represented. FairVote claims that instant runoff now reaches 11 million voters across the United States. Dozens of editorial boards endorse it and almost one hundred colleges and universities implement it. Jurisdictions in ten states had it on their ballots in 2022 and most approved it. When instant runoff is up for a vote, it tends to pass, as long as the campaign has enough time and resources to educate the electorate.*

Am I taking the easy way out and getting behind a method that's already winning? Is the current head start of instant runoff over other methods simply a reflection of the savviness of organizations such as FairVote that have over the decades become better at fund-raising and moving in the mysterious ways of Washington politics? I'm not, and who cares? My support for instant runoff does not come at the expense of evidence and certainly not at the expense of math; betray it I would never. We've done a lot of work to get here—across six chapters—and we've earned the right to say that endorsing instant runoff is a mathematically sound verdict. That the method is already ahead in real-world implementation is a bonus and a hopeful sign that we might in fact succeed in making elections fairer, more equitable, and more inclusive.

---

* A proposal to implement ranked choice voting in essentially all state and federal congressional elections was on the ballot in Massachusetts in the November 2020 elections. It failed because the campaign could not reach enough people due to the pandemic.

PART II

Representation

DEMOCRACY ON any kind of scale involves decisions about representation—how individuals will be grouped (and divided) into constituencies and how those constituencies will be represented in government. Enfranchisement, strictly speaking, means the right to vote, but an essential test of its functionality happens at the level of representation. Our main objective is to examine the allocation of seats in the U.S. House of Representatives. As different as the many systems around the world might appear, they run on the same mathematical backbone. We'll also look at districting and the rich U.S. tradition of gerrymandering, where the shape of voting districts traces a long history of discrimination and disenfranchisement.

Mathematics provides tools to conduct, understand, and assess the effects of the processes involved in representation. As with voting, the strategy is to identify ideals and construct a system that comes closest to embodying them. The greater complexity of the representation infrastructure gives more scope for political intervention and entails interaction with the courts. Mathematics contributes by constructing objective measures that give legal judgments a firm foundation.

CHAPTER 8

This Old House

I SPENT most of the 1990s in a tuxedo. I was a waiter at a fancy French restaurant in Boston and a tux was the standard garb for a place of that stature. We even sported white gloves for some of the fanciest of the fancy events. I started as a table busser while in high school, right after I moved to the United States from Bosnia, and worked my way up to front waiter, aka captain aka head waiter, via back waiter aka runner, while working my way up through graduate school via college.

Tips, of course, were what paid the bills all those years. Thank you, America, for the 15–20% custom. I don't care if the rest of the world doesn't get it.

A WAITER'S TAKE ON DEMOCRACY

There are two operational models when it comes to tips. One is that each waiter collects the tips from their tables and then pays out to the other staff who worked with them. A few bucks might also go to the bartender who makes the drinks all the waiters deliver to customers.

The other, more socialist practice, and the one used at my fancy French restaurant—France being a fairly socialist country—was to put all the tips in a pile and then distribute equal amounts to all the front waiters, equal but smaller amounts to all the back waiters, and equal but still smaller amounts to all the bussers, with the bartender getting a small taste.

One way to keep track of who gets what is with a point system. At my restaurant, front waiters received 4 points, back waiters 3 points, bussers 2 points, and the bartender 1 point. Because most of the tips were paid on credit cards, the manager would go to the bank once or twice and take out in cash the total amount we received in tips from the previous few days (yes, he would carry piles of cash from the bank to the restaurant). Then, in a ritual that brought joy to all, he (or a designee, which, as a budding mathematician who was automatically assumed to be good at distributing money, became me at one point) would divvy up the cash.

Each of the waitstaff had an old-school eyeglass pouch, the kind that snaps open and shuts on top, with their name written on it in indelible ink. The money would be doled out in rounds for each of the lunch and dinner shifts and stuffed into the pouches for later pickup. The eyeglass cases of the waitstaff who worked a particular shift would be lined up on a table—front waiters in the top row, back waiters in the next row, and table bussers and the bartender in two more rows—and the allocation would begin. The spread might look something like figure 8.1 (except there were no eyeglasses in our eyeglass pouches; their only purpose was to be vessels for cash).

Let's say the tip intake for this shift was $1,500. With 32 points in play, the value of a point is

$$\$1,500 \div 32 = \$46.88$$

Because the manager brought only bills back and no change, the point value would be rounded down to $46. (Rounding up would have meant that we'd run out of cash.) The distribution would commence: $46 for the bartender, 2 × $46 = $92 for the busboys, 3 × $46 = $138 for the back waiters, and 4 × $46 = $184 for the front waiters. (That would have been a good night back in the 1990s on a busy Friday or Saturday.) But if you add up the money that was set on top of the pouches, it actually equals $1,472. There is still $28 to be distributed. This is because we first rounded down and then distributed this lower amount. Without coins, the remaining $28 cannot be split into the required 32 points. The solution would now depend entirely on the distributor, who would decide how to dispense the remaining cash in the way they saw most fit. I would probably give $2 to each of the front and back waiters and

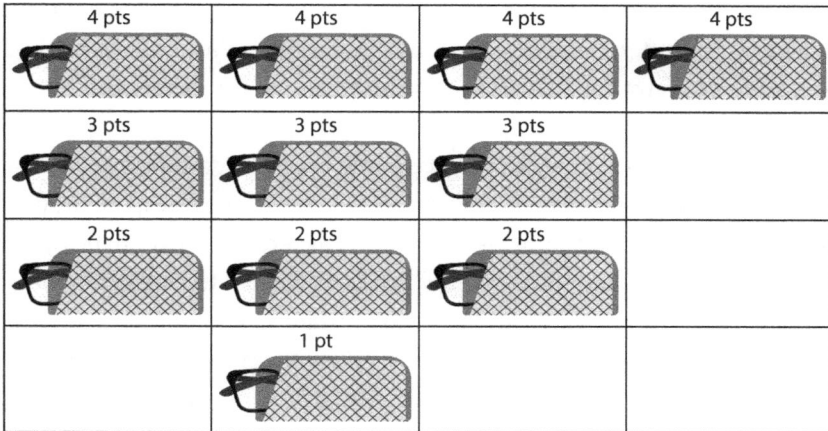

FIGURE 8.1. Arrangement of eyeglass pouches ready
to be stuffed with tips.

the bartender, and $4 to each of the table bussers (because I had been a busser for a long time, I had their back). Someone else might have done this differently. This is when the money would finally go into the pouches and the tip distribution for the next shift would begin.*

The main snag in this allotment process is that we did not have change, so exact amounts could not be given out. Instead, some kind of rounding procedure was required, one that in the end resulted in a subjective distribution.

Nobody ever complained about how the remaining few dollars were distributed because they made up a small fraction of the total intake. But sometimes the remainder in a distribution is more significant. For example, suppose a company has four branches and wants to form a committee of eleven people with each branch represented as equally as possible. It makes sense to take two people from each branch, which

---

* Anyone who wants to hone their mathematical ability should work in a restaurant for a while. Any of us could glance at a signed credit card receipt and figure out instantly what percentage tip the customer left, accompanied by an imperceptible eye roll if the tip was below 15% (European tourists, I'm looking at you) or an even more imperceptible appreciative nod (waiters in French restaurants aren't supposed to show any emotion other than disdain) if the tip was over 18%. Clearly, necessity is the mother of fast calculation.

accounts for eight seats, but who gets the remaining three seats? One branch is bound to be shortchanged. Which should it be? That branch will have 33% fewer people on the committee than the others and hence significantly less power.

The problem is that we can't cut people up. If we could take 3/4 of a person from each branch in addition to the two whole ones, the branch representatives would total eleven people, as required. But instead, we have to round to integers somehow, much as we had to round to whole dollars when divvying up the tips at the restaurant.

This is the *apportionment problem*. It arises any time some amount of something has to be divided but the something comes in indivisible integer values. Sometimes the problem is not really a problem. If fifty stolen diamonds are to be divided among five robbers, there's an obvious way to do it. But if there are four or six robbers, the apportionment problem becomes more serious.

What does this have to do with politics?

There are 435 seats in the House of Representatives. Each state gets some number of them. My home state of Massachusetts has 9. Texas has 38. But as I'll explain in the next chapter, the ideal number of seats is not a whole number; Massachusetts is supposed to get about 9.2 representatives, Texas 38.3. So how were these numbers rounded for all fifty states while sticking to the constraint of 435 total seats? Welcome to the most consequential apportionment conundrum U.S. democracy has ever seen. It has existed for as long as the United States has. It's never been solved.

But first, let's back up to 435. If we were allowed to change this number, then the apportionment problem would have an easy solution. If I'm doling out candy to my kids and their friends and I grab ten pieces for the four of them, that's no good because ten is not nicely divisible by four (and children, as everyone knows, do not believe in dividing candy). But if I can grab two more, then there's no problem; everyone gets exactly three pieces and I don't have to deal with candy fractions.

Cool. We've solved the apportionment problem—just scale the candy example up to the level of the House of Representatives somehow. Well . . . this cannot be done with 435. That number cannot be changed. We're stuck with it.

Why?

No good reason whatsoever. The number 435 is arbitrary, meaningless, and wrong. Like so much of the math in our democracy, it is an artifact of a certain time and its politics. It captures a telling moment in the history of American political innumeracy, so let's talk more about that before we tackle the actual apportionment problem. Let's see why 435 is an artificial restriction, an unjustifiable integer hanging over our legislative system and, through the Electoral College, over our presidential elections.

THE HOUSE HITS THE CEILING

From 1880 to 1920, over 20 million immigrants came to the United States, fleeing famine, persecution, or war or simply looking for a better future. The majority filed through Ellis Island, the scene captured in those iconic images of bedraggled, weary families holding boxy suitcases, weaving through endless lines, waiting for a turn at having their family name butchered by an immigration official. Armed with a clean bill of health and a landing card, they would begin the search for the American Dream.

Most of those immigrants settled in big urban centers in the Midwest and the Northeast, especially in New York City. The Second Industrial Revolution had created low-skilled factory jobs in the cities, and by 1920, over half of manufacturing jobs were held by immigrants. The resulting shift in population was dramatic. In 1880, one-quarter of the U.S. population lived in cities. By 1920, more than half did. As an extreme example, the population of Detroit doubled from 1910 to 1920. In many of those swelling cities, immigrants and their children accounted for over three-quarters of the population.

The Republican Party of the time saw in these changes a problem. The new urban, immigrant masses were a threat, poised to upset the balance of power in the House of Representatives in favor of Democrats. "The New Colossus" at the bottom of the Statue of Liberty might as well have read, "Give me your tired, your poor, Your huddled masses yearning to breathe free . . . and to vote." The soon-to-be-voting masses had to be stopped—and 435 was one way to do it.

To understand, we have to back up some more.

One of the biggest stumbling blocks at the Constitutional Convention of 1787 was the structure of the new country's legislature. Should power be in the hands of the states or in the hands of the people? Smaller states argued for the former through the New Jersey Plan and larger states for the latter through the Virginia Plan. The Connecticut Compromise (or the Great Compromise), written into Article I of the Constitution of the United States, resolved the dispute, establishing the two chambers of Congress. The Senate gave each state two members, which appeased the small states, and the other chamber, the House of Representatives, was supposed to pacify the states with bigger populations by allotting to each state a number of seats proportional to its population.* Proportionality was en vogue in this time of Enlightenment, praised by John Locke (among others), who was a perennial inspiration for the framers of the Constitution. But how exactly was this "proportional allocation" to be achieved? That's precisely the apportionment problem.

To start, how did the framers feel about the size of the House? They couldn't agree. In fact, James Madison complained in Federalist No. 55 that "no political problem is less susceptible of a precise solution than that which relates to the number most convenient for a representative legislature."

The framers faced a classic political optimization problem. On the one hand, the House was supposed to be an extended arm of the people. Ordinary folks were for the most part shut out of the government—at

---

* The notion of "population" has changed over time. The same clause in the Constitution also stipulated that slaves would count for 3/5 of a person, thereby giving southern states an increase in population and consequently more power in the House, although it is impossible to argue that enslaved people were represented by the legislators their fractional body count helped elect. Free African Americans, women, and children were also counted in the population, even though they could not vote. "Indians not taxed" were not counted at all.

the time, they had no say in the elections of the president, the senators (this changed in 1913 with the Seventeenth Amendment), or the justices of the United States Supreme Court—so it was the House of Representatives that would give them a direct voice and a stake in the government. Representatives would need to have a finger on the political pulse of their constituency and be the conduit for their needs and concerns in Congress. This meant that the number of citizens per representative had to be manageable. The representatives should have time to mix with the people, to hear them out, to be one of them. Madison again: the representatives "are to be the great body of the people of the United States." George Washington agreed; the only time he participated in discussion at the convention was to express his preference for smaller districts.

On the other hand, the size of the House had to be reasonable. With a large legislature, reaching agreement becomes more difficult. Too many representatives might spell perpetual gridlock. The public's timeless aversion to too much government was also a factor.

The number that was finally deemed workable was about 30,000 people per representative, with the proviso, true to this day, that each state must have at least one. Yes, 30,000 is yet another arbitrary number, although at least some thought seems to have gone into it. So when the U.S. government began its work in 1789 and the First Congress convened, the House of Representatives had 59 members distributed among the eleven states (North Carolina and Rhode Island hadn't yet ratified the Constitution by that point). Virginia had the most seats, 10, while Delaware had 1, the least possible.

As the framers understood and anticipated, the number of people per representative—let's call it the *representation ratio**—would not make sense forever. It was just an initial value to kick off an iterative process that was supposed to unfold for as long as America existed. The intention was for the House to evolve in size as the population increased and new states joined the Union. After each decennial census—also mandated by the Constitution—the number of representatives was to be

* This quantity has many names; in the next chapter, we'll call it the *standard divisor*.

adjusted. In the same Federalist No. 55, Madison wrote that the process was supposed to "readjust, from time to time, the apportionment of representatives to the number of inhabitants" and "augment the number of representatives."

Even better, Madison had a prescription for doing this. He wanted to make sure that the representation ratio remained reasonable and came up with a graduated algorithm for increasing the number of seats in the House. He proposed the recipe in the First Amendment, one of twelve amendments sent to Congress in 1791. Ten were adopted as the Bill of Rights. One more was ratified as the Twenty-Seventh Amendment in 1992. The one that would ensure that the House grew with the population never made it into law. It was one state short of adoption.

Nevertheless, Congress continued to regard growth as necessary. After the initial census of 1790, the size of the House was set at 105 with the Apportionment Act of 1792, which George Washington signed into law. Over the next 120 years, the number of representatives increased steadily, although not according to any specific formula. After each census, Congress would decide, in mostly an ad hoc way, how large the House should be, adding seats for increased population and new states while making sure that no state lost a seat. The growth over the years is shown in figure 8.2.

Not everyone was happy with the ever-expanding House. But despite occasional legislative attempts to stop it, the only exception to the increase until 1920 was the recalculation after the 1840 census. Like any other census, this one showed an increase in U.S. population, but the Whigs in the Senate were concerned about the unwieldy planned increase in House size that would go along with it. They pushed through the Apportionment Act of 1842, which shrank the size of the House from 240 seats to 223 seats. (The act also mandated that congressional districts as we know them become the standard election units for the House.)

The 1920 elections, the first in which women were allowed to vote, brought Republican Warren Harding to the White House with what remains the greatest popular vote margin of victory in U.S. history. The Republicans retained control of both chambers of Congress. This was also a census year. The size of the House had been at 435 since the 1910

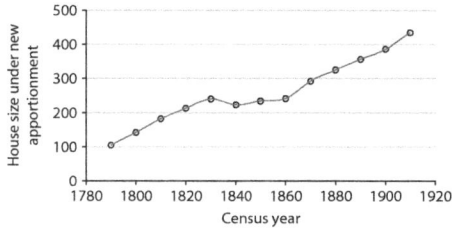

FIGURE 8.2. Size of the U.S. House of
Representatives, 1790–1910.

census (433, really, but 2 seats were added when Arizona and New
Mexico became states in 1912) and it was time to adjust it to reflect the
significant increase in population from the flood of immigrants entering
the country. In time-honored fashion, the Republican chair of the
House census committee proposed in 1921 that the House should increase by 48 seats to 483. That would have been the smallest number
that ensured that no state lost a seat.

But here was the rub: according to the new apportionment, 11 seats
in ten primarily rural states would have shifted from rural to urban areas.
The Republican Congress did not like this. This was a definitive quantification of their fear of losing power to immigrants. Combined with a
growing sense that the House was already too unwieldy and that there
was no more physical space for expansion, this proposal triggered a
stalemate. And indeed, for the first time since 1843, Congress failed to
increase the size of the House after a few unsuccessful attempts at compromise, keeping it at 435.

This crisis, which almost reached constitutional proportions, dragged
on until 1928, when Republican Herbert Hoover, who had been a secretary of commerce with jurisdiction over the Census Bureau, won the
presidency. He called a special session of Congress in April 1929 with
the goal of settling the apportionment issues once and for all.

The result was the Reapportionment Act of 1929. At the time, this
piece of legislation was on brand for Congress; it joined various other
anti-immigration measures it had enacted over just a few years. It froze
the size of the House of Representatives at 435, where it happened to

be when Congress first started bickering about it. Over 110 years later, this monument to racism and anti-immigration sentiment is the sole reason 435 is an untouchable number, carved in stone and enshrined in our politics.*

It should be noted that some scholars disagree with a narrative that foregrounds immigration as the motivation for the 1929 Reapportionment Act. For example, historian Dan Bouk from Colgate University argues that none of the usual reasons that led to the Reapportionment Act—demographic changes, mistrust of the census data verifying the growth of the cities, and academic squabbles over what apportionment method should be used—can be considered precipitators of the act. He instead attributes its passage to a technocratic concern that the government was becoming too large and too unwieldy. What he does not dispute is that capping the House at 435 "shackled democracy."

So why is it so bad the House of Representatives has been stuck at 435 all this time?

ONE REPRESENTATIVE PER SEATTLE

The framers landed on 30,000 people per representative as a good ratio. Madison understood that this number might grow with the country's population but also postulated that 50,000 was a reasonable ceiling for it. That turned out to be unrealistic; by 1840, the larger ratio had already been surpassed. In fact, the ratio grew consistently even as the House expanded, as shown in figure 8.3.

This would have made Madison unhappy, but perhaps he could have come to terms with it as a decent compromise between manageable representation and functioning government—until he saw what happened with the 1929 Reapportionment Act.

Figure 8.4 shows how the representation ratio grew from 1790 to 2020. Before 1910, the line is not so steep because both the denominator (House size) and the numerator (U.S. population size) of the represen-

---

* Mostly. There was a temporary bump in the number in 1959 after Alaska and Hawaii joined the Union, but the number was brought back to 435 three years later.

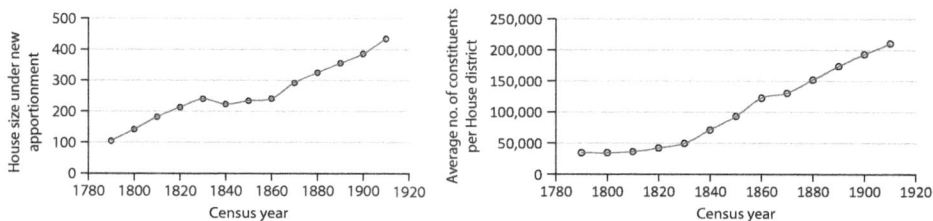

FIGURE 8.3. Size of the U.S. House of Representatives with representation ratios, 1790–1910.

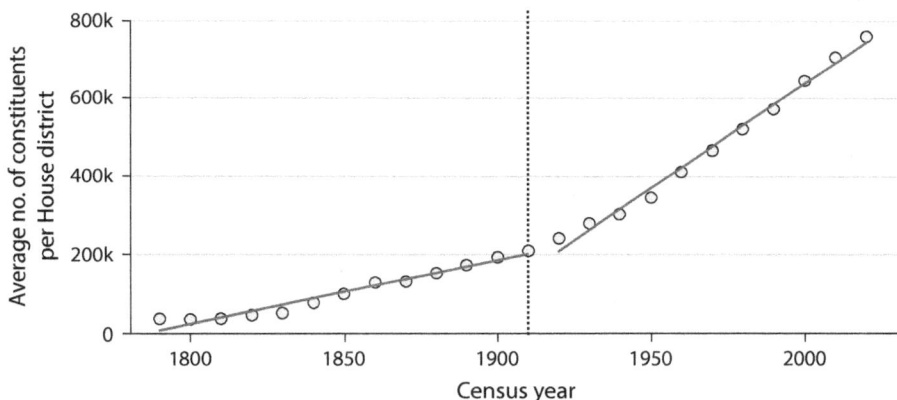

FIGURE 8.4. Population per seat in the U.S. House of Representatives over time.

tation ratio were growing. But with the 1929 Reapportionment Act, the denominator was frozen by law, so the numerator has been running the show since then, making the fraction blow up.

What is the representation ratio now? The U.S. population, according to the 2020 census, is about 331.5 million. The House is still at 435. So the ratio is

$$331,500,000 \div 435 \approx 762,000.$$

This is roughly the population of Seattle. Or Frankfurt, Germany. Or Sheffield, UK (which, by the way, has five representatives in the House of Commons). You can hear Madison shrieking "What?!" from the great beyond. One representative is supposed to advocate for 762,000 constituents, to have a sense of their political preferences and concerns and affirm their needs in Washington?! In 2030, the projected ratio will be 826,000. In 2040, it will be 874,000.

In 2020, the population of the United States was 3.6 times what it was in 1910, when 435 first came into effect. Consequently, on average, a member of the House represents 3.6 times as many people in 2020 as they did in 1910. Research shows that the larger the district, the more likely representatives are to take positions that are at odds with those of their constituents. In addition, they are more susceptible to lobbyists and special interest groups. The gap between a representative and the faceless mass of their district's population creates a vacuum that those entities skillfully fill.

There are scores of other ways the ratio of 762,000 cascades and reverberates across the valleys and troughs of American politics, creating inequities and undermining governance. For example, we know that every state gets a representative. Even states with populations that are less than 762,000 get one, such as Alaska ($\approx$ 736,000), Vermont ($\approx$ 643,000), and Wyoming ($\approx$ 577,000). Then there are states whose populations are not big enough to get two representatives, such as South Dakota ($\approx$ 888,000) and Delaware ($\approx$ 990,000). So Delaware's one representative speaks for almost 1 million people—far over the average 762,000 and almost twice as many as Wyoming's representative must deal with.

Delaware loses by having a population just below the threshold that would bring it an additional representative. States just over the threshold reap the benefits. Rhode Island ($\approx$ 1.1 million) gets two representatives, for example, each of whom represents about 550,000 people—fewer than Wyoming's sole representative. Montana ($\approx$ 1.08 million) received an extra seat after the 2020 census and has a representation ratio similar to Rhode Island's. Delaware, with 94,570 fewer people than Montana, did not get the extra seat. The extra little bit of population Montana has over Delaware was enough to tip the scale of the apportionment formula (which we'll get to in chapter 10) to its favor. This kind of ratio disparity occurs not just with the small states but with states all over the map. The result is a mathematical muddle of representation ratios across

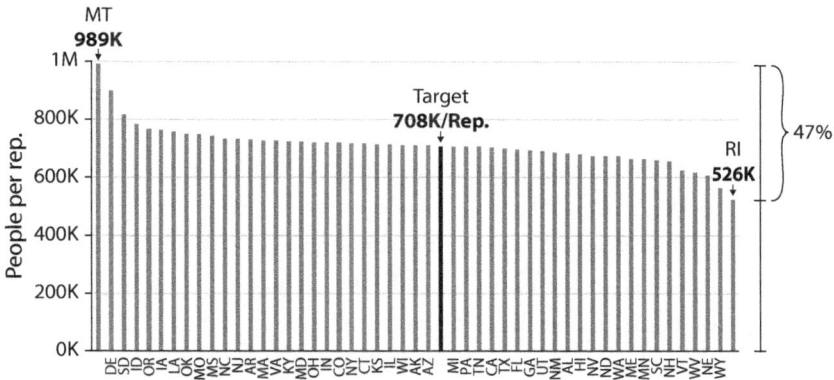

FIGURE 8.5. Current representation ratios for each U.S. state. This visualization previews an analysis by Chris Wilson that we'll discuss in the next section.

the country, all brought to you by the number 435. The intended ratio of 762,000 has been replaced by one that ranges from about 500,000 to almost 1 million (figure 8.5).

The size of the House also affects presidential elections. Remember that each state gets a certain number of votes in the Electoral College. Because this number equals the size of the state's congressional delegation, the Electoral College has been stuck at 435 + 100 + 3 = 538 since 1910 (the 3 additional votes are for the District of Columbia). As we will see, uneven representation ratios translate into unequal numbers of constituents per electoral vote. This inequitable distribution is one of the ways the Electoral College compromises the quality and legitimacy of presidential elections.

And even if the distribution of the 435 seats were fair, the number itself matters. Houses of different sizes would produce different election outcomes because the electoral votes would be distributed differently, even if the distribution were equitable. For example, in the Gore-Bush race in 2020, Bush would have won with a House size of up to 490, but the outcome would flip-flop somewhat regularly among Bush, Gore, and a tie if the number of seats in the House had been in the range of 491 to 655 seats, and Gore would have been the winner for a House that had more than 655 seats.

The legal branch of government has little help to offer. The Supreme Court (in *Wesberry v. Sanders*, 1964) ruled that representation ratios *within* each state should be approximately the same. In other words, it requires that a state's congressional districts be about the same population size. But this ruling does not apply to districts between different states. This is known as the *malapportionment* problem. The highest court thus bans *intra*state but not *inter*state malapportionment. The discrepancy in the representation ratio between states is one of the most glaring violations of the one person, one vote doctrine the Supreme Court upheld in a series of rulings in the 1960s, the gold standard of equity and fairness in the United States.*

A bigger House would go a long way toward correcting the multifaceted mess of the 1929 Reapportionment Act. It would be a release valve for many existing political pressures. The expansion could be done by statute, so Congress can do it at any time. This would not introduce an advantage for either the Democrats or the Republicans, as demonstrated in a report produced by Our Common Purpose, a project by the American Academy of Arts and Sciences.

Of course, more government, especially at the federal level, is infuriating to the average American. One legitimate concern is the additional tax dollars that would have to be spent on more representatives. According to one estimate, the annual cost of a congressperson is about $1 million ($174,000 in salary plus other expenses and office budgets). That sounds like a lot, but it is roughly 0.00000017% of the $6 trillion U.S. federal budget. We could fund 136 representatives for a year by forgoing one F-35B fighter jet (the U.S. military has ordered 2,400, of which

* If we're really looking for legislative malapportionment, we don't need to look farther than the Senate. California and Florida account for over 20% of the U.S. population but are represented by four senators. Meanwhile, the twenty-six least populous states account for less than 18% of the population and hold the majority of the Senate. The Senate, of course, was created like this intentionally, but the House was not.

some 800 are already in use). I concede that the U.S. military budget is the usual target for a cheap shot at the uneven distribution of government expenditure but . . . that's because it's true.*

Another concern is the greater potential for gridlock. More votes would need to be corralled to achieve the majorities required to pass legislation, and more obstinate representatives pandering to a base would surely mean more sniping across the aisle instead of more people crossing it. There are plausible arguments to the contrary, however, such as one put forth in a Fordham University School of Law report that argues that more representatives, and hence smaller districts, would encourage third-party and underrepresented candidates to run because it wouldn't take as much money and political machinery to mount a campaign, even against incumbents with the backing of lobbyists and special interest groups. Such candidates would be more eager to form coalitions and engage in conversations, potentially reanimating the political corpse that is the current Congress. In fact, according to FiveThirtyEight, a statistical political analysis organization, if the size of the House were increased by 158 seats, the number of competitive districts would jump from about 10% to about 25%. Finally, if we look around the world, we see that larger legislative bodies can perform well. The UK, Germany, and France are well-functioning democracies (most of the time) despite their relatively large lower houses. The UK's House of Commons has 650 representatives, Germany's Bundestag has 709, and France's National Assembly has 577.

The workload of Congress in 2020 looks nothing like it did a hundred years ago. A growing population and increasing demographic, cultural, and geopolitical complexities mean that Congress is dealing with a lot more than it did before. It handles a $6 trillion budget and oversees 180 federal agencies with 4 million employees. Each congressperson receives almost 50,000 communications from constituents per year. Meanwhile, the staff that helps legislators do their jobs has shrunk by

---

* There is also the real estate question of whether the Capitol can physically handle more representatives. The answer is that it can, as Danielle Allen explained in her May 2, 2023, piece in the *Washington Post,* "Can the Capitol Hold a Much Bigger House? Yes, Here's How It Would Look."

almost one-third in the past forty years. The lower chamber is overwhelmed and understaffed. It can't keep up. It passes fewer bills, holds fewer hearings, and has lost touch with the people.

THE GOLDEN RATIO

So what should the size of the House be? There seem to be at least 435 answers. Let's consider a few of the strongest proposals for which the math checks out.

One thing we can do is go back to the roots and declare, founders style, that 30,000 is the correct representation ratio. Given the population of the United States now, this means that the House would have 331,500,000 ÷ 30,000 = 11,050 representatives.

Okay, although I think we need a bigger House, it should definitely not be this big. A legislature of that size couldn't function.

What about Madison's ideal cap of 50,000? Then we'd have 331,500,000 ÷ 50,000 = 6,630 representatives.

Nope.

What about going back to 1910, the moment when time stopped as far as the House is concerned? The representation ratio back then was 210,583. If we return to that number, the House would have 331,500,000 ÷ 210,583 ≈ 1,574 representatives.

Getting better, but still a no. Increasing the House almost by a factor of four would be a difficult ask of both the legislators and the American people.

How about returning to *trends* in the data prior to the artificial 435 freeze? For example, from 1870 to 1910, the representation ratio increased by about 20,000 people with each expansion of the House. We could try to extend that trend to the present day, which graphically amounts to extending the left line in figure 8.4 to the present day. If the ratio in 1910 was 210,000, in 1920 it would have been 230,000, in 1930 it would have been 250,000, and so on. In 2030, the next time the apportionment will happen, the ratio would be 450,000, which means that the House would consist of 331,500,000 ÷ 450,000 ≈ 737 representatives.

Now we're getting somewhere. But that's still almost a 70% increase over the current size, an addition of 302 House members. A practical way to achieve this growth might be to catch up to the desired ratio over a few decades. However, the conceptual problem with this model is that the 20,000 figure was not a product of mathematical planning. Congress calculated the changes in the size of the House from 1870 to 1910 in an ad hoc way and it might have been a fluke that the ratio increase was steady for those four decades. After all, prior to 1870, the increase was all over the place; it was as small as 7 from 1850 to 1860. Another more fundamental objection is that in the long run, this approach doesn't solve the problem of large ratios. Having 450,000 constituents in 2030 means having 650,000 in 2130, so we'll eventually be back where we started. We'll have only released some pressure, put a Band-Aid on the crisis, and bought ourselves some time.

Can we look at factors other than the representation ratio? For example, what if we go back to the way Congress used to deal with House size? It would identify states that were slated to lose seats because of shifts in population distribution and then increase the House size so that each of those states maintained the size of its delegation. This is the algorithm that determined the proposed increase from 435 to 483 in the 1920s, an idea the 1929 Reapportionment Act shut down.

For example, the 2020 census showed that seven states had smaller shares of the national population than they had in 2010 and would have to give up seats to other states whose population share had grown. We would start the algorithm by increasing the House size to 436, reapportioning using the formula we'll talk about in the next chapter and seeing how many states lost seats. Then we would increase the House size to 437, 438, and so on. We would stop when none of the seven states loses a seat. It's important to realize that this procedure rarely adds exactly the number of seats lost by states. That is because the reapportionment formula does not necessarily put those states first in line for receiving the next additional seat. In fact, the House would have to increase by 19 seats in order for all seven states to retain their numbers of representatives.

A related proposal, from a report by Our Common Purpose, goes further and recommends an initial jolt of adding 149 seats to the House. This increase compensates for all the seats that have shifted between states in reapportionments since 1920. For example, in 1930, 27 seats were reallocated; in 1940, 9 seats were reallocated; . . . ending in 2020, when 7 seats were reallocated. The total of all such seats is 149. The authors justify their proposal by claiming that it would, to some extent, make up for the representation that was lost over the years by states that were forced to give up seats due to the 435 straitjacket. After this initial increase to 584, the House would continue its no-seats-lost progression.

Although this proposal works as a mathematical algorithm, it's not entirely satisfactory; it is too ad hoc and unpredictable. For one thing, it depends on shifts and relative increases in population between the states. The number of House seats is associated with the size of the federal government and as such it should depend only on the U.S. population. The population sizes of individual states should come into play only in the actual apportionment. Further, there is volatility in where the apportionment formula might place the loser states within the ordered list of states that should receive extra seats. In other words, lots of states might receive extra seats before the losers are compensated—that is why it would take 19 new seats to make up for the 7 taken away from states after the 2020 census. This instability could lead to unpredictable lurches in House size that might inspire an arbitrary cap, as happened in 1910.

A similar state-population-based approach is the proposed *Wyoming Rule*. According to this formula, the representation ratio would be set to the population of the least populous state. Since 1960, this has been Wyoming, hence the name. That state's population is currently about 577,000, which means that the House would be looking at 331,500,000 ÷ 577,000 ≈ 575 representatives.

While adding 140 seats is not outrageous, the method is patently unsound. It defines one state's population to be the requisite ratio and in that way prevents overrepresentation of that state in the House, but it doesn't solve any of the other problems we've enumerated. There is simply no relation between any other state's population and that of the least populous state. With respect to other states, the ratio is unjustifiable.

This recipe can also react wildly to demographic shifts. Imagine, for example, that over the next ten years a bunch of people move from big urban centers to less populated areas because urban living has become more difficult due to—pulling a random thing out of a hat—a pandemic. They swell the populations of states of great natural beauty such as Wyoming, Alaska, the Dakotas, and Vermont to about 1 million each. Suddenly, the least populous state numbers 1 million inhabitants and that's our new ratio. Now we're required to reduce the House size to 331,500,000 ÷ 1,000,000 = 331 representatives. And this is not a fanciful scenario. Had the Wyoming Rule been implemented in 1920, the House would have decreased in size for the next seventy years.

The inverse state of affairs is also easy to imagine. Suppose the population of Wyoming drops by, say, 100,000 people who decamp because Yellowstone and Grand Teton National Parks have been handed over to oil companies. The population of Wyoming now is 477,000 and the House size it determines is 331,500,000 ÷ 477,000 ≈ 695 representatives. We now have an infeasibly large increase.

Such drastic alterations could be regulated by imposing caps on the size of change, both positive and negative, but the caps would again necessarily involve ad hoc decisions. And regardless of any tempering maneuvers, the issue of varying representation would persist. There would always be ratio disparities, as in the Montana-Delaware example from the previous section, where two states with approximately the same population have different numbers of representatives and thus highly unequal representation ratios. In some cases, the unevenness would even be exacerbated by the Wyoming Rule.

Can we tackle the Montana-Delaware problem directly? Remember the bar graph that shows how widely the representation ratio varies across the states? What if we tried to minimize that spread? In technical terms, we'd want to minimize the *variance* that measures the dispersion in representation ratios. This is precisely what Chris Wilson, the director of data journalism at TIME.com, set out to do. He looked at possible House sizes from 435 to 1,305 (3 × 435), apportioned seats to states using the formula currently in use, and then calculated the variance in representation ratios for each of those House sizes. The magic number:

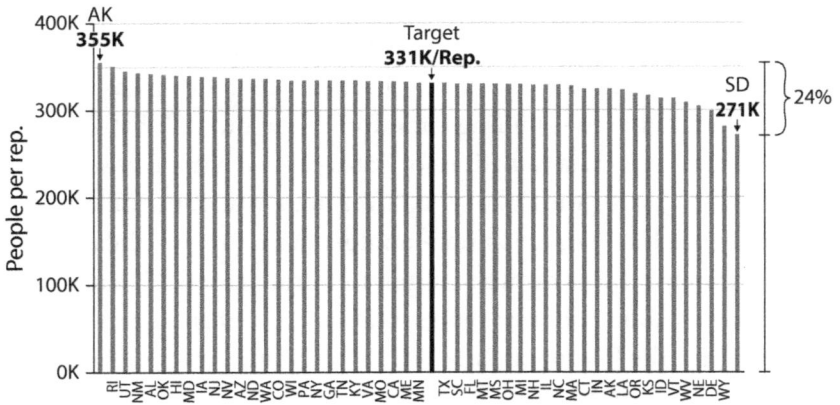

FIGURE 8.6. Minimal representation ratio variance achieved by a 930-seat House, as proposed by Chris Wilson.

930. That's the House size that minimizes the representation ratio variance. The bar graph for that House size appears in figure 8.6.

The heights of bars don't vary as much as they did in figure 8.5. That's the point. The difference between the highest and lowest bars is 24%, as opposed to the 47% difference with the 435-seat House.

This proposal would get us closest to the one person, one vote ideal. But the increase in House size it would require is impractical, as is the case with many of the numbers we've seen. In theory, the increase could be implemented gradually, but it would be hard to find the appetite for such a big change, even if it were spread over a few decades. In addition, this method is also susceptible to demographic shifts. Finally, it is not tied to the actual population of the United States, only to its distribution across states, a shortcoming it shares with the Wyoming Rule.

Instead of continuing to tinker, we can look for clues around the world. What do other democracies do? We saw that the UK, Germany, and France have bigger lower houses. Figure 8.7 casts the net wider and presents a telling picture. The U.S. representation ratio is the highest among all the Organisation for Economic Co-operation and Development (OECD) countries, a far outlier.

Can we extract a rule or formula from the other countries to bring the United States closer to its peers? We can look north to Canada. With

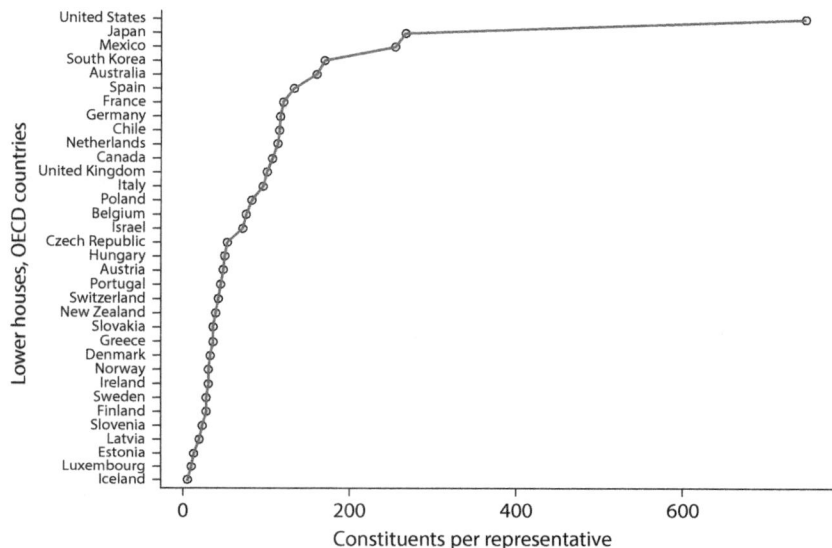

FIGURE 8.7. Representation ratios for OECD countries.

a population of about 38 million, it has 338 seats in its lower chamber, the House of Commons (the equivalent of the House of Representatives). The Canadian representation ratio is about 112,000. Now look south. Mexico has about 129 million inhabitants and its Chamber of Deputies has 500 members. The Mexican representation ratio is 258,000. There's no obvious pattern yet.

But it's too soon to give up. One thing that's often worth doing in math is, well, doing the thing again. To calculate the representation ratio, we divided the size of the population by the size of the legislature. Let's divide that number by the size of the legislature again. For Canada, the first division gave 112,000. Dividing by 338 again gives 331. For Mexico, the first division by 500 yielded 258,000. Dividing again by 500 gives 516. The numbers we're getting are awfully close to the numbers we're dividing by. Something interesting could be going on.

We can formulate what we've done to make it easier to look past the numbers and see the pattern. We took the population, call it P, and divided it by the legislature size, call it L. Then we divided the result by L again and obtained L, approximately. In short, we found that

$(P \div L) \div L \approx L$. Turning this around, multiplying L by itself three times gives something close to P:

$$L{\cdot}L{\cdot}L \approx P, \text{ or, more compactly, } L^3 = P.$$

Finally, because we're looking for a rule for legislature size, we want to rewrite this as a formula for L:

$$L \approx \sqrt[3]{P}$$

The candidate rule is now this: the size of the legislature's lower chamber is roughly the cube root of the country's population. But just because a formula works in a couple of cases doesn't mean it's valid all the time. More testing is needed. And indeed, the graph in figure 8.8 shows that this rule often holds.

The dots compare each country's population to the size of its legislature. For example, South Korea's population is about 52 million and the size of its National Assembly is 300. The curve is the graph of the cube root function. So if the size of a country's legislature is approximately the cube root of its population, its dot falls near the curve. Canada and Mexico are very near the curve, as we would expect. South Korea is below the curve, which means that its legislature is smaller than the cube root of its population.

Many democracies fall on or near the curve. This phenomenon is called the *cube root law*. It's not a law, though. It is an observation that holds often enough to be worth paying attention to. The pattern was first observed in 1972 by Rein Taagepera, a political scientist at University of California, Irvine. The rationale for the law is that the cube root appears to be the number that optimizes the ability of representatives to operate on two communication channels—one with their constituents and one with each other.

Until 1910, the United States followed the cube root law closely. For example, with a population of about 92,000,000 in 1910, the House size of 435 was not a bad approximation to the law, because

$$\sqrt[3]{92{,}000{,}000} \approx 451.$$

2016 population and legislative body* size, select countries

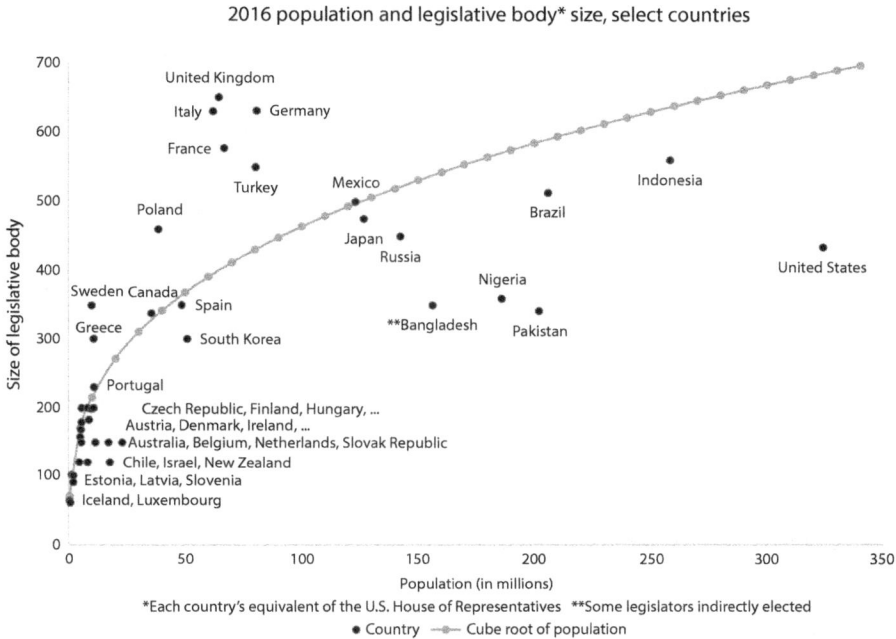

*Each country's equivalent of the U.S. House of Representatives **Some legislators indirectly elected
● Country ━●━ Cube root of population

FIGURE 8.8. Representation ratios across the world compared to the cube root function.

Because the size was frozen at 435, the United States has been steadily dropping below the cube root curve, which means that our representation ratio is unusually high.

Despite having the world's third-largest population, the United States has only the nineteenth-largest lower house. We have kept the size of the lower house static in a way that peer democracies have not. Canada regularly amends the number of representatives in its House of Commons to reflect demographic changes. In 2020, Italy reduced its parliament from 945 to 600, a size more appropriate to its population. When Taagepera's own Estonia became independent from the Soviet Union, it used the cube root law to determine the size of its legislature.

If the United States were to base its House size on the cube root law now, it would have

$$\sqrt[3]{331,500,000} \approx 692 \text{ representatives.}$$

That is about 60% greater than the size of the current House, probably too large for a single increase, but a more phased process could bring it to its cube root value over some period of time.*

The cube root law seems to be the best option at the moment for determining the size of the House on an objective basis. It is tied to the national population and would align the size of the U.S. legislature with the sizes of other democracies. But whatever the method, increasing the size of the House is imperative. It's a straightforward way to alleviate many of the disparities in representation we've enumerated. As we'll soon see, the consequent reduction in district size would also stimulate greater diversity on the political scene and go a long way toward addressing gerrymandering. In short, the compound effect of such a simple change could be transformational.

---

* The cube root law hasn't gone unchallenged. Recent research contends that Taagepera's approach ignored too many other potential sources of interaction among the representatives and used too many simplifying assumptions. The true "law" might be something closer to the square root rather than the cube root, an observation more in line with other statistical work done on this topic. In that case, the United States House would have 18,193 members, each representing 18,193 constituents because the population is in this case the product of 18,193 with itself. As unfeasible as this is, there might be something edifying about each representative looking around the House and always being reminded how many actual people they represent.

CHAPTER 9

Rather Divisive than Indecisive

NOW THAT we know why we're stuck with 435 seats in the House of Representatives, let's return to the apportionment problem. The Constitution requires that after each decennial census, the number of House seats must be distributed among the states according to their population, with each state receiving at least one seat. Figure 9.1 shows the results after the 2020 census.

Seven states lost a seat, six gained one, and one über-winner, Texas, gained two.* How was this redistribution calculated? The Constitution, surprisingly, provides no guidance. We're left to our own mathematical devices.

DIVISORS AND QUOTAS

The most natural way to begin is to mimic the tip distribution example from chapter 8. Allocate the highest possible whole number amount to each state and then deal with the remainder somehow.

---

* According to the Census Bureau, New York was within eighty-nine people of retaining its lost seat. In his fascinating book *Democracy Data: The Hidden Stories in the U.S. Census*, Dan Bouk points out how ridiculous it is to claim such precision. The margin of error of the census in each state is about 1%, which means the New York count is only accurate to within about 200,000 people.

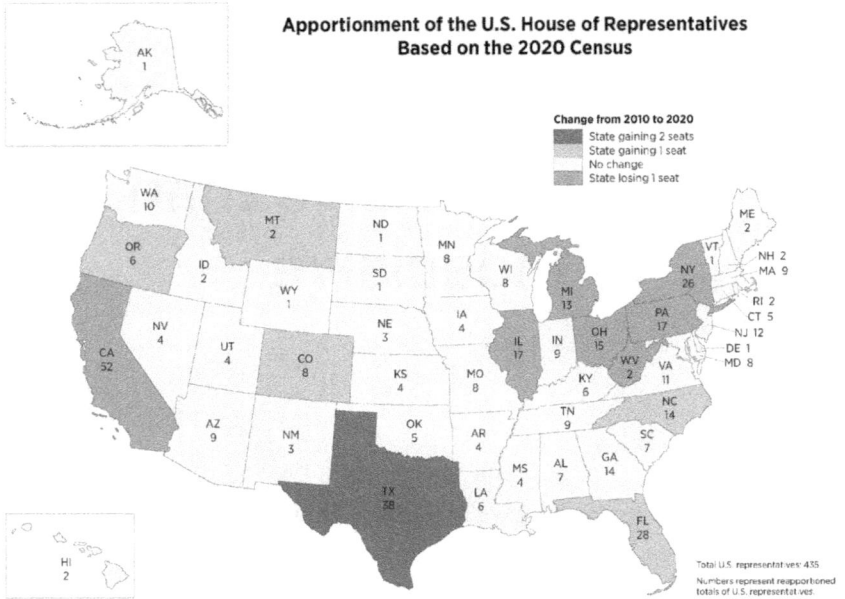

FIGURE 9.1. Reapportionment after the 2020 census.

First, we need to remember the representation ratio, 762,000, obtained by dividing the U.S. population, about 331.5 million, by 435, the size of the House. It can be thought of as the unit of representation that we can now use to measure each state's population. Dividing each state's population by 762,000 tells us how many units of size 762,000 fit into a state's population. That is the ideal number of representatives for that state. Any time a representation ratio is used in this way, it is called the *standard divisor (SD)*; I'll follow the convention and use that terminology and abbreviation.

For example, Maryland should get

population of Maryland ÷ standard divisor = 6,177,224 ÷ 762,000 ≈ 8.11 representatives.

(I'll usually round to two decimal places, sometimes three if necessary.) Similarly, Nevada should get

population of Nevada ÷ standard divisor = 3,104,614 ÷ 762,000 ≈ 4.07 representatives.

Wisconsin should get

population of Wisconsin ÷ standard divisor = 5,893,718 ÷ 762,000 ≈ 7.73 representatives.

And so on, forty-seven more times. The ideal number of representatives is called the *standard quota* for each state.

But we can't allocate 8.11 representatives to Maryland or 7.73 representatives to Wisconsin. Representatives come in integer values. The only time we wouldn't have this problem is if each state's population were exactly an integer multiple of the standard divisor, and the chances of this are practically zero. Solving the apportionment problem means rounding each state's standard quota to an integer in some way that is deemed fair.

One thing we might try is rounding according to the conventional 0.5 cutoff. Namely, we round each state's standard quota down to the nearest integer if the decimal part is less than 0.5 and up to the nearest integer if the decimal part is 0.5 or greater. The problem is that the resulting numbers might not total exactly to the required 435.

A simple example confirms that this happens. Suppose a company has 116 employees in 4 divisions, A, B, C, and D, that have 10, 25, 36, and 45 employees, respectively. A committee of twelve is to be formed and each division is to be represented as proportionally as possible. The standard divisor is the total number of employees (which plays the role of population) divided by the number of seats on the committee (playing the role of the size of the House). So

SD = total employees ÷ committee seats = 116 ÷ 12 ≈ 9.67.

The following table shows the standard quotas. In the last column, those numbers are rounded to the nearest integer according to the 0.5 cutoff.

Division	Employee size	Standard quota = employee size ÷ SD	Rounded standard quota
A	10	1.03	1
B	25	2.59	3
C	36	3.72	4
D	45	4.65	5
Total	116	≈ 12 (not exactly 12, due to rounding to two decimal places)	**13**

The allocation is twelve, but we exceeded that with our rounding, landing on thirteen. The House apportionment would run into a similar problem.

So let's play it safe: round all the standard quotas down to the nearest integer. If we're calculating the House allocation, one exception we'll make is to round up to 1 when the standard quota is less than 1 (each state gets at least one representative). The rounded-down numbers are called *lower quotas* (or *minimum quotas*). In the example above, the lower quotas would be 1, 2, 3, and 4. They add up to 10, two fewer than the number of seats available, but we're not surprised—we lowballed everyone's allocation so it makes sense the total comes out below the target. In the tip distribution example, this is the stage at which everyone had received the maximum possible whole-dollar amount. There was still some money left to be distributed, just as here we have two leftover seats.

We've made some progress. We know how to calculate the minimum number of seats for each state. The apportionment problem is now reduced to figuring out who should get the remaining post-lowball seats.

You might say that at least in the simple example with twelve committee members, the solution seems obvious. The standard quotas for Divisions C and D are closest to the next whole numbers because they have the largest decimal parts. Why don't we just give them the extra seats and start scheduling some meetings? If that is what you'd do, then you're as smart as Alexander Hamilton. But if you're as wise (or politically savvy) as George Washington, you would shut this idea down.

HAMILTON'S PARADOXES

When the Constitution was ratified by eleven of thirteen states,* the initial size of the House was set at 59. The framers computed the first apportionment based on an estimate of each state's population while keeping in mind the proposed representation ratio of about 30,000 people.

---

* In a clever bit of mathematical thinking, the framers decided that the right threshold for ratifying the Constitution was nine states. They had observed that the combined population of any nine states at the time accounted for the majority of Americans, so that threshold ensured that most citizens would be represented in the ratification.

When Rhode Island and North Carolina also ratified the Constitution, the House size was increased to 65. After the first census took place in 1790, giving a head count of about 3.9 million people living in the United States, the number was set at 105. The Constitution mandated that reapportionment take place at most three years after this initial census and every ten years thereafter, ensuring that reapportioning of House seats across the states would occur on a well-defined, regular schedule.

In 1790, for the purpose of apportionment, the U.S. population was 3,615,920.* Alexander Hamilton suggested that the House size should be 120. Dividing the population by this number gives the standard divisor:

$$\text{SD} = \text{population} \div \text{House size} = 3{,}615{,}920 \div 120 = 30{,}132.67.$$

At that point, there were fifteen states in the Union. Each state's standard quota was computed by dividing its population by SD. The lower quota was obtained by rounding those numbers down, except for when the rounding gives 0, in which case the allocation is 1. Table 9.1 keeps track of the calculations.

Each state was certain to receive its lower quota of representatives. As we expect, the total of all the lower quotas, 111, is less than the number of available seats. Nine seats remained to be allocated.

Hamilton proposed a system for distributing the remaining nine seats: give them to the nine states that have the largest fractional remainders, or largest decimal parts. The final apportionment then adds up to the desired 120.

In short, the states with standard quotas closest to the next integer get another representative, in order, until there are no more representatives to give. This is called the *Hamilton apportionment*, or *largest remainders method*. Sounds like a reasonable system, right?

Congress thought so too. But when the bill endorsing Hamilton's proposal was presented to George Washington, he vetoed it. This was

---

* This number is the *representative population*, which counted slaves at 60% and did not count "Indians not taxed." In 1868, the Fourteenth Amendment abolished the 3/5 rule, but it was not until 1940 that all Native Americans were counted in the representative population. Southwest Territory, which became the state of Tennessee in 1796, was also included in the 1790 census.

TABLE 9.1. 1790 Hamilton apportionment

State	1790 population	Standard quota	Lower quota	Extra seat	Hamilton apportionment
Connecticut	236,841	7.860	7	1	8
Delaware	55,540	1.843	1	1	2
Georgia	70,835	2.351	2	0	2
Kentucky	68,705	2.280	2	0	2
Maryland	278,514	9.243	9	0	9
Massachusetts	475,327	15.774	15	1	16
New Hampshire	141,822	4.707	4	1	5
New Jersey	179,570	5.959	5	1	6
New York	331,589	11.004	11	0	11
North Carolina	353,523	11.732	11	1	12
Pennsylvania	432,879	14.366	14	0	14
Rhode Island	68,446	2.271	2	0	2
South Carolina	206,236	6.844	6	1	7
Vermont	85,533	2.839	2	1	3
Virginia	630,560	20.926	20	1	21
Total	3,615,920	≈ 120	111	9	**120**

Note: Maine is not listed because it was still part of Massachusetts.

the first presidential veto in U.S. history and one of only two Washington ever issued.*

Washington appears to have been influenced by his close ally and fellow Virginian Thomas Jefferson. Jefferson argued in a letter to Washington that this apportionment was problematic, because it was based on 120 representatives, a House size that made the representation ratio too small in some states. For example, Massachusetts ended up with 16 seats. Its resulting representation ratio, 475,327 ÷ 16 = 29,707.9, was lower than the threshold of 30,000. This was true for seven other states. Washington sent the veto to Congress the same day with this explanation:

The Constitution has also provided that this number of Representatives shall not exceed one for every thirty thousand; which restriction

---

* The other veto, issued in 1797, prevented a reduction in the number of cavalry units in the army.

is, by the context, and by fair and obvious construction, to be applied to the separate and respective number of the States; and the bill has allotted to eight of the States more than one for every thirty thousand.

The House of Representatives tried to override Washington's veto but could not summon the required two-thirds majority. The next month, Congress drafted another bill that reduced the proposed size of the House to 105 and suggested a different method for allocating the seats. That method was proposed by none other than Jefferson. Washington signed the bill on April 14, 1792, and Jefferson's method, which we'll talk about soon, was used from that point until 1840. If you calculate both the Hamilton and the Jefferson 1790 apportionments using 105 seats, the only differences occur in two states: Delaware lost one seat and—surprise!—Virginia gained one.*

Hamilton's largest-remainder method was resurrected and used from 1850 until 1900 (also known by the name *Vinton's method*, after the representative from Ohio who reintroduced it, apparently without being aware that it had existed before). It shouldn't have been. Whatever politics or home-state camaraderie there might have been between Washington and Jefferson, they were right to shut the Hamilton method down. It was deeply flawed in at least three ways.

The Alabama Paradox

Say a country has fourteen people and three states, A, B, and C, with respective populations of 6, 6, and 2. There are 10 seats in the House of Representatives. (Even with these small numbers, we'll be able to probe the method's weaknesses.) According to the Hamilton method, how should the seats be allocated? First, we calculate the standard divisor:

$$SD = \text{total population} \div \text{number of seats} = 14 \div 10 = 1.4.$$

---

* A detailed account of this drama can be found in M. Balinski and P. Young's book *Fair Representation: Meeting the Ideal of One Man, One Vote.*

Then we compute each state's lower quota by dividing the population by the standard divisor and rounding down:

State	Population	Standard quota (population ÷ SD)	Lower quota	Hamilton apportionment
A	6	4.28	4	4
B	6	4.28	4	4
C	2	1.42	1	2
Total	14	≈ 10	9	10

Because one seat remains to be assigned after the lower quotas have been allocated, Hamilton would identify C as the state with the largest remainder and would give it the extra seat, as shown in the last column.

But now suppose an extra seat is added to the House. Rerunning the calculations with 11 seats gives this result:

State	Population	Standard quota (population ÷ SD)	Lower quota	Hamilton apportionment
A	6	4.71	4	5
B	6	4.71	4	5
C	2	1.57	1	1
Total	14	≈ 11	9	11

Two seats remain to be assigned after the lower quotas, and the two states with the largest remainders, A and B, receive them.

Did you catch that? Even though the total number of seats increased, C lost a seat. Whatever your definition of fair is, I'm sure this isn't it.

This phenomenon, called the *Alabama paradox*, was discovered by the chief clerk of the Census Office as he played around with possible House sizes after the 1880 census. He noticed that as the House increased from 299 to 300, Alabama lost a seat and went from eight to seven representatives. Ultimately, the size was set at 325 and Alabama did not lose a seat, but this freaky side effect of the Hamilton method eroded confidence in the method and led to its replacement in 1900.

The Population Paradox

Even after Congress banished the Hamilton method, people continued to play with it and discovered two additional problematic features. The first is called the *population paradox*. It occurs when an extra seat is awarded to a slower-growing state at the expense of a faster-growing state. It was discovered when a calculation showed the Hamilton method would give an extra seat to Maine after the 1900 census while taking one away from Virginia, even though Virginia's population growth rate was higher than Maine's.

For example, take three states, A, B, and C, with respective populations of 5,300, 9,900, and 22,400. If there are 24 seats to be allotted, Hamilton divides them as 4, 6, and 14, respectively. But if the populations of the states grow to 6,800 (28% growth), 12,500 (26%), and 25,700 (15%), Hamilton would then give 3, 7, and 14 seats. State A has lost a seat to state B even though it had greater population growth.

It gets even worse: the growth of the state that steals a seat could even be *negative*; in other words, an expanding state can lose a seat to a shrinking state.* Again take three states, A, B, and C, with respective populations of 1,450, 3,400, and 5,150 and 10 seats to be allocated. The Hamilton allocation is 2, 3, 5. If over time the state populations change to 1,470, 3,380, and 4,650, Hamilton would then give 1, 4, and 5 seats. State B took a seat from state A, even though its population decreased while the population of A increased.

The New States Paradox

If a new state joins the country, it makes sense to add seats to the House to accommodate the expansion. Redistributing existing seats would be unfair to the rest of the states. This was the case in 1907, when Oklahoma joined the United States. The House size was increased by 5 seats, from 386 to 391, dictated by Oklahoma's population. But had Hamilton apportionment been used to allocate these 391 seats, New York would have

---

* Some people take this to be the definition of the population paradox.

lost a seat to Maine! This is called the *new states paradox*. It is peculiar because the addition of new seats for a new state should not affect the seat allocations for other states.

For example, take the same three states from the second population paradox example above, with the same initial populations totaling 10,000 and 10 seats to be allotted. Now say that state D joins the union with 2,600 inhabitants. It makes sense to add 3 seats to the House to accommodate this new state. But recalculating the Hamilton apportionment now with 12,600 people and 13 representatives results in 1 seat for A, 4 for B, 5 for C, and 3 for D. State A has lost a seat to state B. The addition of a new state should not upset the allocation balance between existing states, but it did.

————————

The source of the paradoxes the Hamilton method creates is the use of the largest remainders to determine which states receive extra seats. The remainder is too sensitive to tiny changes in either the numerator or the denominator. As an extreme example, if one state has a population of 1,499,999 and another has a population of 1,500,000, then with a standard divisor of 1,000,000, Hamilton would regard the second state as worthier of an extra seat. But one single voter moving from the second state to the first would flip the judgment. Or with state populations of 1,000,000 and 1,000,001, a change in the divisor from 1,000,000 to 1,000,001 would flip which state receives an extra seat without any change in the state populations.

Even though the paradoxes of the Hamilton method weren't discovered until long after Washington's intervention, Jefferson was convinced he had come up with a better system. What gave him this confidence? He brilliantly observed that there is no rule that says we have to use the standard divisor.

CHAPTER 10

Divisor and Conquer

PRESIDENT GEORGE Washington vetoed the first apportionment bill Congress sent him, believing it to be unconstitutional. He gave two reasons. The first was straightforward: eight states would receive more representatives than their population justified. The second was deeper:

> The Constitution has prescribed that Representatives shall be apportioned among the several States according to their respective number; and there is no one proportion or divisor which, applied to the respective numbers of the States, will yield the number of allotment of Representatives proposed by the bill.

This was Jefferson speaking through Washington. The quote says that the Hamilton method proposed in the bill requires additional steps beyond the calculation of the lower quotas. That didn't sit well with Jefferson; it seemed nonmathematical, arbitrary. Jefferson sought "an arithmetical operation, about which no two men could possibly differ." He was really asking for it, wasn't he?

JEFFERSON: GOING LOW

Once each state's population is divided by the standard divisor, Jefferson thought, that should be it. This seemed the most objective end to the process, so he proposed that the standard divisor be changed so that all the resulting lower quotas added up to exactly the required number of seats. This is called the *Jefferson apportionment,* or the *greatest divisors method.**

A method of apportionment in which the standard divisor is adjusted to achieve a certain numerical goal is called a *divisor method.* The divisor that achieves the goal is called the *modified divisor.* Because the total of the lower quotas is almost always smaller than the number of available seats, the standard divisor should be decreased to achieve Jefferson's desired effect.† Determining how much the divisor must be reduced is typically a matter of trial and error.

Let's trace the process through an example. Take states A, B, C, and D with respective populations of 8,000, 10,500, 22,000, and 59,500. There are 20 seats to apportion. Because the total population is 100,000, the standard divisor is

$$\text{SD} = \text{population} \div \text{House size} = 100{,}000 \div 20 = 5{,}000.$$

The standard quotas are in table 10.1. The lower quotas allocate 18 seats, so we have 2 more to apportion. Now, giving Jefferson a spin, we decrease the standard divisor. We'll call the modified divisor MD1. Let's set it to, say, 4,700. Take each state's population and divide it by this modified divisor to see what happens. The fifth column of the table shows the result. When we round those numbers down (as shown in the sixth column), they total to 19. So we've managed to allocate one more seat because state D's original standard quota, 11.8, crossed to 12.55, whose integer part is one greater.

But we have one more seat to allocate, so let's decrease the modified divisor even more, say to 4,550. Call this MD2. The next-to-last column shows the new modified quotas and the last column shows those

---

* The Jefferson method is equivalent to the D'Hondt method used in Europe and elsewhere. We'll talk about that in chapter 13.

† Dividing by a smaller number produces a bigger number. For example, $3 \div 5 = 0.60$ and $3 \div 4 = 0.75$. In a particularly striking show of innumeracy, a municipality in Bosnia recently passed a resolution with 3 of 5 of those present voting for it. The bylaws mandate a 2/3 quota for passage, but the legislators said they had enough votes because 3/5 is obviously greater than 2/3 because each number in the first fraction is greater than its counterpart in the second.

TABLE 10.1. Jefferson apportionment example

State	Population	Standard quota[1]	Lower quota for SD	Modified quota[2]	Lower quota for MD1	Modified quota[3]	Lower quota for MD2
A	8,500	1.7	1	1.81	1	1.86	1
B	9,500	2.1	2	2.02	2	2.09	2
C	23,000	4.6	4	4.89	4	5.05	5
D	59,000	11.8	11	12.55	12	12.97	12
Total	100,000	≈ 20	18		19		20

Notes
1. Population ÷ SD.
2. Population ÷ MD1.
3. Population ÷ MD2.

numbers rounded down. We have achieved the desired total of 20, and that's the final Jefferson apportionment.

Think of the divisor as a dial that can be turned to increase or decrease the quotas and thus their roundings. The dial is adjusted until the rounded quotas add up to the right thing. A range of modified divisors will work—for example, we could have used 4,600 instead of 4,550—but if one of them gives the right allocation, all of the divisors in the range will give the same allocation. In my kitchen, I have a radio with a dial and an antenna that stays tuned to a station even if I wiggle the dial a little. This is exactly how the divisor behaves.

The Jefferson method favors larger states; that is, the quotas for states with larger standard quotas increase at a faster rate than the small states as the divisor is decreased. You can think of it this way: if a company says that everyone's salary will increase by 3%, then the employees with the largest salaries will see the largest increase. A big state is thus more likely to be rewarded sooner by the Jefferson method because it will receive a larger quota increase for the same divisor decrease, making it more likely that its quota will cross over to the next integer. This behavior was exhibited the first time the Jefferson method was used in 1792. It allocated Virginia, the largest state, one more seat than it received with the Hamilton method. From 1790 to 1830, New York's standard quotas decade by decade were 9.63, 16.66, 26.20, 32.50, and 38.59. In all those cases, New York's final apportionment was at least the next highest integer (and it was a whopping 40 in the last case). Delaware, a small state, did not seem to get much of an apportionment break; its standard quotas for those years were 1.61, 1.78, 1.95, 1.68, and 1.52, but it received more than one seat only when its quota was 1.95. You can also see this bias in Table 10.1. The two extra seats in the end went to the two largest states. If we had used Hamilton's method in this example, it would have awarded to A, the smallest state, the extra seat that went to C with Jefferson's method.

Dropping the fractional part of the quota also has disproportionate effects on smaller states. If state A's standard quota is 1.9 and state B's standard quota is 1000.9, then dropping the 0.9 accounts for 47% of state A's quota and only 0.09% of state B's quota. The extra seat has a much greater impact for state A than for state B. If the Jefferson method gives neither A nor B an extra seat or the large state gets an extra seat, that can have a huge effect on the disparity between representation ratios for the two states.

On the positive side, Jefferson's method always avoids the Alabama, population, and new states paradoxes. In fact, any divisor method does. But it runs into a different issue—it violates the *quota rule*, which means that it might allocate a number of seats that is neither of the integers closest to the standard quota. If the standard quota is, say, 3.45, we'd expect the allocation to be 3 or 4, not 2 or 5. But Jefferson's method might actually assign 5 or even more.

In the apportionment following the 1820 census, for example, New York's standard quota was 32.5 but Jefferson gave it 35. In 1832, when New York's standard quota was 38.59, it received 40 seats. In fact, if the Jefferson method had continued to be used, it would have violated the quota rule for some state in every apportionment to date. Hamilton's method satisfies the quota rule by design because it rounds up or down from the standard quota.

The Jefferson method's violation of the quota rule undermined confidence in it. Political storms were brewing and the distribution of power between the North and the South hung in the balance. The mathematics of apportionment was once again front and center.

WEBSTER: TAKING THE MIDDLE ROAD

James Polk of Tennessee, a rising star who became the eleventh president of the United States, was good at math. After the 1830 census, he calculated that, if Congress stuck to the rule that no state should lose a seat and set the divisor at the precise value of 47,700, the House would increase from 213 to 240 and the Jefferson method would give extra seats to Georgia, Kentucky, and New York, states that were crucial to his later ascent to the presidency. It would also take a seat away from Massachusetts, which of course delighted the southern states. Many other divisors were at play, each packing its own political agenda, but Polk's divisor had an unfair advantage: he was the chair of the House reapportionment committee.

To blunt the effects of Polk's proposal, John Quincy Adams, a former president and a representative from Massachusetts, proposed a solution in a letter to Massachusetts senator and future secretary of state Daniel Webster, who was the chair of the Senate reapportionment committee. Jefferson's method ought to be "inverted," Adams said, so that the quotas are *rounded up* rather than rounded down. In other words, the Adams method would modify the divisor so that the rounded-up values add up to the right thing. This method would have given Massachusetts and New Hampshire additional seats, so its appeal to Adams is understandable.

Adams's method suffers from the opposite problem of Jefferson's: it favors small states. This makes heuristic sense. If rounding down favors large states, then rounding up should favor small states. For example, if Jefferson and Adams were to apportion seats after the 1970 census, the former would give 41 seats to New York while the latter would give it only 37 and would distribute the remaining 4 to smaller states.

Webster also received a letter from James Dean, a mathematics professor at the University of Vermont, who claimed that the best apportionment formula would be one that produces the least deviation from the divisor.

Here is what Dean meant. In 1830, the population of Illinois was 157,147. If the House was set at 240 seats and a divisor of 47,700 was used, then the standard quota for Illinois would have been $157{,}147 \div 47{,}700 = 3.29$.

So Illinois could have received 3 or 4 seats. If it had received 3, then its representation ratio would have been 157,147 ÷ 3 ≈ 52,382. If it had received 4, the ratio would have been 157,147 ÷ 4 ≈ 39,287. The difference between the first ratio and 47,700 is 4,682, so the first allocation would have produced district sizes that deviated from the ideal by 4,682 people. The difference between the second ratio and 47,700 is 8,413. Because the difference is smaller for the first allocation, Illinois should have gotten 3 representatives with this method.

Now compute the same two differences for each state, take the smaller difference in each pair, and then average these chosen numbers across all the states. Go back and change the divisor to something else and recalculate everything. Do this for lots and lots of divisors. Finally, choose the divisor for which the average is the smallest. The apportionment is then done using this divisor. This is *Dean's method*.

This procedure works if we allow the size of the House to vary so it can accommodate the best allocation for each state. But when the House size is fixed, it has to be thought of a little differently. The thing to do is look at the pairwise differences in the representation ratios between states and try to minimize them in the context of the constraint of a fixed House size. From this angle, one way to rephrase the procedure is this: Dean's method allocates seats in such a way that the difference in representation ratio between any two states cannot be made smaller by a transfer of a seat from one to the other. It's a mouthful, but if our notion of fairness of an apportionment method is to minimize the representation ratio differences, then Dean's method is the way to go. Because the representation ratio is the same as the size of a district in each state, this method aims to make the district sizes across states as uniform as possible by choosing the right divisor.

If you look back at the bar graph of all the representation ratios in figure 8.5, Dean's method is this: think of the divisor as a tunable knob that, as you turn it, has the effect of increasing or decreasing the height of each of the bars in that graph. Turn the knob to the place where the bars are as equal in height as possible, the place where the variance is minimized.*

---

* Reducing this variance was also the goal of one of the methods we saw for determining the correct size of the House.

Webster did not act on Adams's and Dean's propositions, but he nevertheless continued to think about the apportionment formula. In 1832, he proposed a method he deemed superior. It was a simple compromise between Jefferson and Adams: the rounding shouldn't always go up or always go down but should be determined by the usual 0.5 cutoff. This is the *Webster apportionment,* or *major fractions method.*[*]

In the Webster method, almost everything works the same way as in the Jefferson method. Compute the standard divisor and every state's standard quota. If the standard quota has a decimal part that is 0.5 or greater, round it up. If the decimal part is below 0.5, round the quota down. Add up these rounded numbers. If the sum is exactly the number of available seats, great, that's the apportionment. Otherwise, modify the divisor until the procedure gives the right sum. One new feature is that when the roundings are added, the total might be smaller or greater than the given number of seats. This means that the Webster method might require us to increase the divisor in some cases, unlike the Jefferson method, in which the divisor is always decreased.

In the example from table 10.1, the Webster method would round the initial standard divisors to 2, 2, 5, and 12 seats. These add up to 21, more than the number of seats, so we'd need to increase the divisor. A divisor that works is 5,120, with which Webster allocates 2, 2, 4, and 12 seats. The allocation is different from Jefferson's, which means these two methods are genuinely different.

Because Webster's method is a compromise between Jefferson's method, which favors large states, and Adams's method, which favors small states, it would make sense if it favored neither. That is true, but it's not obvious and remains to be qualified, which I will do shortly.

Just as Dean's method can be regarded as minimizing a certain kind of representational disparity, so can Webster's. While Dean looks at representation ratio (aka district size), Webster considers the reciprocal of that, namely the share of a House seat that "belongs" to each voter in a state. For example, the size of my congressional district is about 781,000, so "my share" of the House seat for my district is 1 ÷ 718,000, or

---

[*] In a different guise, this is the Sainte-Lägue method, which is used in Europe and around the world. We'll see it in chapter 13.

0.00014%. Increasing this ratio is in a state's interest because it means its voters in some intuitive sense have more power. Webster's method minimizes the differences between the shares across the states. It allocates seats so that the difference in seat share between any two states cannot be made smaller by a transfer of a seat from one to the other. If that is what we mean by fair, namely making sure the influence of each voter on the selection of representatives is as equitable as possible across the country, then Webster's method is what we should use.*

The 1840 census sparked new debate about the size of the House, the divisor, and the apportionment method. On one day in April 1842, fifty-nine different divisors were proposed in the House. The Senate did its own divisor dance with twenty-seven proposals, finally settling on 70,680 (proposed by James Buchanan, another future president) and *reducing* the House size to 223—the only time in history that happened. But most significantly, Congress adopted the Webster method (favored by Adams, who was at the same time fervently opposed to the reduction in House size because that meant even less representation for New England).

Webster's method didn't survive long. Following the apportionment of 1850, Ohio representative Samuel Vinton became the loudest proponent of two changes, hoping they would end the decennial squabblefest: the House size should be fixed and *his* apportionment method should be used. The Vinton Act of 1850 passed, establishing the House size of 233 and mandating the use of Vinton's method—which, it turns out, was none other than the Hamilton method.

The Vinton Act was in force until 1910, but it was never strictly followed. The House kept growing, and the Vinton/Hamilton method was adapted and refined after almost every census to accommodate the political currents of the time, especially when the Alabama, population,

---

* It might seem that Dean and Webster are saying the same thing, only with reciprocals. But that is not the case, essentially because the difference of the reciprocals is not the same as the reciprocal of the difference.

and new states paradoxes were discovered. Deviation from the method even caused controversy around the election of Rutherford Hayes over Samuel Tilden in 1876. Even though Tilden won the popular vote, Hayes won the presidency with 185 electoral votes over Tilden's 184. Had the Hamilton method been followed to a tee, Tilden would have won.

By 1880, trust in this method had dissolved. Jefferson's method was not an option because it favored big states and violated the quota rule. Apportionment posed such a conundrum that the 1882 method wasn't even specified. The House size was set at 325 because the Hamilton and the Webster methods gave the same allocation for this number. A similar work-around was used after the 1890 census, with the magic number of 356. The apportionment following the 1900 census was a combination of the Hamilton and Webster methods, with the latter finally deciding the calculation. The writing was on the wall for the Hamilton method.

The 1910 apportionment was the final nail in the coffin. When allotments were calculated for House sizes from 350 to 400 using the Hamilton method, all kinds of fluctuations were observed, such as Maine flip-flopping between 3 and 4 seats. Congress decided to use the Webster method and set the House size at 433, again the number at which no state lost a seat. When Arizona and New Mexico joined the United States in 1912, one seat was added for each. That is how we arrived at the infamous 435. And that is where math and politics finally came head to head as never before.*

HUNTINGTON-HILL: A MEAN BATTLE

The 1910 switch to Webster's method is especially noteworthy because it marks the first time a serious scholar of apportionment had a say in the problem. That scholar was Walter Willcox (1861–1964), a philosophy professor at Cornell, who studied apportionment methods carefully. Based on his recommendation, the House Committee on the Census

* The only notable remaining use of the Hamilton method is in Switzerland, one of the world's oldest democracies and a place the founders looked to as a touchstone. There it goes by the name Hare-Niemeyer method and is used to allocate the 200 seats of its National Council among the twenty cantons.

adopted the Webster method. Willcox, along with other scholars, also argued for the necessity of a permanent agency that would do the nation's counting and statistical analysis while simultaneously being independent and removed from politics. In a commendable show of forward thinking, Congress accepted the recommendation of the experts and created the Census Bureau as a permanent scientific institution in 1902. The bureau attracted top statisticians and evolved into a training ground for applied social scientists. It was an academic organization, working closely with universities and keeping up with the latest advances in research.

But then came the 1920 census. That was the moment when the urban-rural shift in population due to immigration, industrialization, and World War I was put on full display. Agricultural states began to worry about losing seats and used their strong presence in Congress to block the 1921 reapportionment. For the first time, apportionment did not happen—in direct violation of the Constitution. No progress was made for several years.

In order to avoid further calamity with the 1930 census, Senator Arthur Vandenberg of Michigan proposed legislation that would put the House size in suspended animation at 435 seats and would use the previous method, which would be Webster's in this case, if Congress couldn't figure out the apportionment within some specified period after the census.

That was the gist of the 1929 Reapportionment Act. But Vandenberg's version wasn't quite its final form. Rural states fought against Webster's formula because a competing method that emerged as a serious challenger in the early 1920s allocated seats in a way that was kinder to small states.

That was the *Huntington-Hill,* or *equal proportions method.* It was devised by Joseph Hill (1860–1938), who taught at the University of Pennsylvania and Harvard and worked as the chief statistician at the Census Bureau. Hill accepted the reasoning of Dean's method but decided that it is the *quotient* of the representation ratios, and not their difference, that matters most.

Here is an example. Suppose you pay $180 for a plane ticket from Boston to Cleveland. The airline then says you need to pay an extra $50 to check your luggage. Meanwhile, I buy a ticket from Boston to Sarajevo for $1,200 and the airline quotes me the same luggage fee. If neither of us can get out of checking a suitcase, then the price is the same for us both, $50. But this absolute difference doesn't paint the whole picture because it doesn't take the ticket price into consideration. I might be less reluctant to pay the fee because I'm already dishing out a lot of money for my ticket; the extra $50 doesn't seem like a big deal. For you, however, $50 feels like robbery. One way to capture this contrast is to consider the quotients of the new cost and the old cost. For you, this is $(180 + 50) \div 180 = 1.28$, and for me it's $(1,200 + 50) \div 1,200 = 1.04$. In other words, your price with the luggage fee is 128% of the old price, and for me it's 104%. These are the increases in price relative to the baseline ticket price. Unlike the absolute difference, they're not the same for you and me. Because my quotient is smaller, my luggage fee is better than yours according to this measure.

Hill's method applies this reasoning to the quotients of representation ratios. Say state A has 45,000 people, state B has 25,000, and the standard divisor is 10,000. Suppose state A gets 5 seats and state B gets 2 seats. The representation ratio for state A is $45,000 \div 5 = 9,000$ and for state B, it's $25,000 \div 2 = 12,500$. The quotient of these two numbers, where the larger number goes in the numerator, is $12,500 \div 9,000 = 1.39$. But what if A gets 4 seats and B gets 3 seats? Then the ratios are $45,000 \div 4 = 11,250$ and $25,000 \div 3 = 8,333$, and the quotient is $11,250 \div 8,333 = 1.35$. A quotient that is closer to 1 is better because that means the representation ratios are closer to each other. In this case, Hill would consider A getting 4 and B getting 3 seats the better allocation.

As with the other methods, Hill would now do this for all pairs of states. The final allocation would be one that minimizes the quotients of representation ratios across the board. More concisely, Hill's method apportions seats in such a way that the quotients of representation ratios between any two states cannot be made smaller by a transfer of a seat from one to the other. If the relative representation ratio is the criterion we deem to be the best indicator of fair apportionment, then Hill's method is what we should use.*

I'm using the word "quotient" because "ratio" is already reserved for representation ratio. But if we remember that a state's representation ratio is just the size of its districts, then we can free up "ratio" and say succinctly that Hill's method minimizes the ratios of district sizes across states.

Hill's method is secretly a divisor method, just like Jefferson's and Webster's. The difference is in the rounding. Webster's conventional rounding at the 0.5 cutoff is familiar and probably feels natural. But this rounding is just that, conventional, and it is not the only possibility.

Declaring 0.5 to be the cutoff for rounding implies we are using the *arithmetic mean* to find the splitting point. The arithmetic mean of two numbers x and y is the usual halfway point between them, calculated as

$$(x + y) \div 2.$$

If x and y are adjacent integers, namely x and $x + 1$, the arithmetic mean is

$$(x + x + 1) \div 2 = (2x + 1) \div 2 = (2x \div 2) + (1 \div 2) = x + 0.5 = x.5.$$

So we split up or down at $x.5$. Nothing unusual there.

But there is a different notion of a mean. What if instead of adding the two numbers, we multiply them? The role of division by 2 is now played by the square root. The result is the *geometric mean*, computed as

$$\sqrt{x \times y}.$$

This mean is called geometric because it gives the side of a square with the same area as the rectangle with sides x and y. If x and y are

---

* Remember how Webster's method minimizes the *differences* between the individual shares of House seats across the states? If the *quotient*, instead of difference, is used, then Hill's method actually minimizes these quotients across the states. In short, this is because the quotient of two fractions and the quotient of their reciprocal can be computed from each other; the final numbers are reciprocals of each other.

adjacent integers, then the formula is $\sqrt{x \times (x+1)}$.* This is a natural—and sometimes better—notion to use when, for example, the numbers exhibit correlation or when it is desirable to dampen the effect of extreme values. Finance and other disciplines that model growth often use the geometric mean.

Armed with the geometric mean, it's now straightforward to express Hill's method as a divisor method. Follow the same procedure as in Webster's method, but round according to the geometric instead of arithmetic mean.

Let's look at this example:

State	Population	Standard quota[1]	Geometric mean	Initial Hill apportionment	Modified quota[2]	Final Hill apportionment
A	2,560	4.267	$\sqrt{(4 \times 5)} = 4.472$	4	4.335	4
B	3,233	5.388	$\sqrt{(5 \times 6)} = 5.477$	5	5.475	5
C	1,190	1.983	$\sqrt{(1 \times 2)} = 1.414$	2	2.015	2
D	5,012	8.353	$\sqrt{(8 \times 9)} = 8.485$	8	8.488	9
Total	12,000	20		19		20

Notes
1. Population ÷ SD.
2. Population ÷ MD.

The standard quota column is the same as always, obtained by dividing each state's population by the standard divisor, which is 600. The next column shows the geometric mean of the two integers closest to the standard quota. (For the Webster method, everything in this column would have the form $x.5$.)

Next compare the standard quota to the geometric mean. If the standard quota is greater, we round up. If it is lower, we round down. So, for example, state A gets 4 seats because 4.267 is less than 4.472. State C gets 2 seats because 1.983 is greater than 1.414. The fifth column shows that the initial Hill allocation adds up to 19. So there is one more seat to be allocated, which means the divisor has to be modified.

---

* The geometric mean, like the arithmetic mean, can be computed for more than two numbers.

By diligent trial and error, I found that the modified divisor (MD) of 590.5 works. The modified quotas appear in the next column. Comparison with the initial allocation reveals that D is getting the extra seat: its modified quota passed the geometric mean cutoff.*

The geometric mean is always less than the arithmetic mean. Here are the comparisons for the first 10 consecutive positive integer pairs $(x, x + 1)$:

	$(1, 2)$	$(2, 3)$	$(3, 4)$	$(4, 5)$	$(5, 6)$	$(6, 7)$	$(7, 8)$	$(8, 9)$	$(9, 10)$
Arithmetic mean	1.5	2.5	3.5	4.5	5.5	6.5	7.5	8.5	9.5
Geometric mean	1.414	2.449	3.464	4.472	5.477	6.481	7.483	8.485	9.489

I've laid this out because something interesting is happening: the geometric mean inches toward the $x.5$ value as the pairs of numbers increase. It slowly closes the gap, with the difference between the arithmetic mean and the geometric mean diminishing. Which brings us back to the 1929 Reapportionment Act.

Rural states opposed Vandenberg's proposal to lock in Webster's method because they saw it as solidifying urban advantage. But they liked what they saw in Hill's method: it favors small states, because the geometric mean is significantly less than the arithmetic mean for small numbers, which means that it is easier for small states to clear it. With advocacy from these states, the bill that finally passed decreed that in order to apportion House seats after a census, the following would happen: the president would send allocations to Congress based on both Webster's method and Hill's method, as well as the allocation based on the method used in the previous apportionment, whatever that might be (Webster's, Hill's, or something entirely different). If Congress couldn't agree on one of these apportionments, the allocation produced by the previous method would take effect (which would have been Webster at the time, because that was the method used in 1910).

---

* There is a small wrinkle here: because state C's modified quota crossed into the next integer value, namely 2, it should now be compared to the geometric mean of 2 and 3, which is 2.449. Because 2.015 is less than 2.449, the allocation for C stays at 2.

The real reason Hill's method became recognized is due to the work of one of the most eminent mathematicians of his time, Edward Huntington (1874–1952), a professor at Harvard famous for his research in the field of abstract algebra who was well liked in the mathematical community. His renown was such that he was frequently called to testify in Congress. In 1921, Huntington came across a letter written by Hill, whom he knew from his student days and from some work Huntington had done with the Census Bureau. The letter explained Hill's equal proportions method. Huntington liked it. He reworded it, made some corrections and modifications, and gave it a more rigorous math coating. This is why the procedure is now known as the *Huntington-Hill method.*

Huntington went on to become a staunch advocate of equal proportions. In a January 16, 1921, letter to the *New York Times,* he called it a "mathematical principle which admits no gradation between truth and falsity." He argued that with the proposed House size of 483 (the result of the time-honored calculation according to which no state would lose a seat), the Webster method is flawed because it would give Missouri eight times the seats of Montana even though it had only 6.29 times the population. He sent a letter to the chair of the House Committee on the Census in an attempt to persuade him to use equal proportions rather than Webster's method. The letter was used as a pretext for the Senate to kill the House proposals to set the size of the House first at 483 and then at 453 with apportionment done by the Webster method. The rural states, which would have lost 11 seats if the Webster apportionment had been used, seized upon Huntington's letter as evidence that more investigation into the right method was necessary. This foot-dragging led to a failure to reapportion and set the stage for a showdown that would not be resolved for years.

What followed was billed as the "battle of the methods" or the "war of the quotients." Huntington became increasingly strident about the superiority of his method, engaging in a two-decade-long struggle with Willcox.* But the battle was mathematical, not political, and the arena

---

* This is reminiscent of the current debate among supporters of instant runoff, approval, and range voting.

was often reputable publications such as the journal *Science*. Huntington showed mathematically that the only divisor methods that optimize certain kinds of fairness criteria are Jefferson, Adams, Dean,* Webster, and Huntington-Hill. Naturally, he thought that the criterion his and Hill's method optimizes—quotients of representation ratios should be as close to 1 as possible—was the most important. Willcox could not be persuaded.

In 1929, fearing a replay of the non-apportionment mess after the 1930 census, the Speaker of the House asked the National Academy of Sciences to study the proposed methods. A panel of four noted mathematicians was assembled. Their conclusion: the Huntington-Hill method is the best.

As much as I'd like to point to this as the perfect example of the capacity for mathematics to triumph over politics with its objectivity and rationality, I can't. This decision was not without controversy. There are claims that the panel, in its deference to the looming presence of Huntington and his mathematical illustriousness, simply went along with the method Huntington supported. The verdict might have been more an expression of loyalty than mathematics. The report itself confirms this to some extent; the panelists began by asserting the relative difference of representation ratios as the quantity to be minimized and then, lo and behold, they endorsed Huntington-Hill as the method that does precisely this.

That is how Huntington-Hill made its way into the 1929 Reapportionment Act as a method, along with Webster's, that must be used to compute an apportionment for Congress to consider. Congress could thank its lucky stars after the 1930 census, because the Webster and Huntington-Hill methods gave the same apportionments (figure 10.1). Disaster was averted.

---

* Dean's method seems to stand out here because all the others can be regarded as divisor methods. But so can Dean's: it compares the standard quota to the *harmonic mean*, yet another version of computing the average of two numbers. It is calculated as the reciprocal of the arithmetic mean of the reciprocals. The harmonic mean is the right choice for many physics applications.

State	Population as enumerated Apr. 1, 1930	Indians not taxed	Population basis of apportionment	Apportionment of 435 representatives by method of—	
				Major fractions used in last preceding apportionment	Equal proportions
Total	122,288,177	194,722	122,093,455	435	435
Alabama	2,646,248	6	2,646,242	9	9
Arizona	435,573	46,198	389,375	1	1
Arkansas	1,854,482	38	1,854,444	7	7
California	5,677,251	9,010	5,668,241	20	20
Colorado	1,035,791	942	1,034,849	4	4
Connecticut	1,606,903	6	1,606,897	6	6
Delaware	238,380		238,380	1	1
Florida	1,468,211	20	1,468,191	5	5
Georgia	2,908,506	60	2,908,446	10	10
Idaho	445,032	3,496	441,536	2	2
Illinois	7,630,654	266	7,630,388	27	27
Indians	3,238,503	23	3,238,480	12	12
Iowa	2,470,939	519	2,470,420	9	9
Kansas	1,880,999	1,501	1,879,498	7	7
Kentucky	2,614,589	14	2,814,575	9	9
Louisiana	2,101,593		2,101,593	8	8
Maine	797,423	5	797,418	3	3
Maryland	1,631,526	4	1,631,522	6	6
Massachusetts	4,249,614	16	4,249,598	15	15
Michigan	4,842,325	273	4,842,052	17	17
Minnesota	2,563,953	12,370	2,551,583	9	9
Mississippi	2,009,821	1,667	2,008,154	7	7
Missouri	3,629,367	257	3,629,110	13	13
Montana	537,606	12,877	524,729	2	2
Nebraska	1,377,963	2,840	1,375,123	6	6
Nevada	91,058	4,668	86,390	1	1
New Hampshire	465,293	1	465,292	2	2
New Jersey	4,041,334	15	4,041,319	14	14
New Mexico	423,317	27,335	395,982	1	1
New York	12,588,066	99	12,587,967	45	45
North Carolina	3,170,276	3,002	3,167,274	11	11
North Dakota	680,845	7,505	673,340	2	2
Ohio	6,646,697	64	6,646,633	24	24
Oklahoma	2,396,040	13,818	2,382,222	9	9
Oregon	953,786	3,407	950,379	3	3
Pennsylvania	9,631,350	51	9,631,299	34	34
Rhode Island	687,497		687,497	2	2
South Carolina	1,738,765	5	1,738,760	6	6
South Dakota	892,849	19,844	673,005	2	2
Tennessee	2,616,556	59	2,616,497	9	9
Texas	5,824,715	114	5,824,601	21	21
Utah	507,847	2,106	505,741	2	2
Vermont	359,611		359,611	1	1
Virginia	2,421,851	22	2,421,829	9	9
Washington	1,563,396	10,973	1,552,423	6	6
West Virginia	1,729,205	6	1,729,199	6	6
Wisconsin	2,939,006	7,285	2,931,721	10	10
Wyoming	225,565	1,935	223,630	1	1

FIGURE 10.1. Apportionment after the 1930 census calculated using the Webster method and the Huntington-Hill method.

But not for long. After the 1940 census and in accordance with the 1929 Reapportionment Act, President Franklin D. Roosevelt sent Congress two apportionment calculations. They had come out different. Webster gave 18 seats to Michigan and 6 to Arkansas. Huntington-Hill gave 17 to Michigan and 7 to Arkansas. Because Michigan leaned Republican at the time and Arkansas was solidly Democratic, the Democrats resurrected the issue of the method. Their goal was to adopt Huntington-Hill permanently, to bake it into the legislature as the one and only method to be used thereafter.

Roosevelt signed the Apportionment Act of 1941 into law. It passed because the Democrats controlled both chambers of Congress. No Republican voted for it. It solidified the size of the House at 435 and made Huntington-Hill the permanent apportionment method. Both these things are true to this day.

Uneasiness surrounding the politically charged circumstances of the 1941 Apportionment Act prompted the Speaker of the House in 1948 to ask the National Academy of Sciences for help yet again. The academy assembled another team of mathematicians that included John von Neumann from the Institute of Advanced Study, the father of game theory we met back in the tale of Arrow's theorem of impossibility, and Princeton mathematician Luther Eisenhart, who had been a member of the first panel nineteen years earlier.* That one-third of the new panel had served on the old panel did not bode well for the independence of the review. The new members were likely loath to go against the recommendation of their esteemed colleagues, one of whom was sitting right there with them. Sure enough, after a few calculations, the committee reaffirmed that Huntington-Hill was the best method. But the theoretical question of which method is superior had not been answered to any degree of satisfaction.

Is Huntington-Hill really the right choice?

* The third member of the 1948 panel was Marston Morse, another Institute of Advanced Science mathematician and a giant figure in topology, my own field of research.

THE FINAL IMPOSSIBILITY

Except for the Hamilton method, all the methods we've seen are divisor methods. They all follow the same procedure and differ only in the way the rounding cutoff is decided. But cutoff can be set in infinitely many ways. Pick your favorite decimal between 0 and 1 and call that the cutoff point. Or sprinkle cutoffs randomly between pairs of consecutive integers.

This leads us to a now-familiar question: How can we tell which method is the "best"? The "fairest"?

We've encountered several such ill-defined questions with elusive answers and found that stepping back to a formal, axiomatic approach can clarify the picture. As before, the first order of business is to formulate desirable criteria and then hope to extract an apportionment method that satisfies them all. (Can you tell where this is going?)

It was in this Arrow-like spirit that mathematicians Michel Balinski (1933–2019), one of the creators of the majority judgment voting system, and Peyton Young (1945–) approached the problem. In their seminal 1982 book, *Fair Representation: Meeting the Ideal of One Man, One Vote*, they identified and analyzed three criteria that rule out failings observed in various apportionment methods:

- **House monotonicity:** If the number of seats increases, no state should lose a seat.
- **Population monotonicity:** If the population growth of state A is proportionally greater than the population growth of state B, then A should not give up a seat to B.
- **Quota rule:** Each state's apportionment should be one of the integers obtained by rounding the state's standard quota up or down.

An apportionment method that satisfies house monotonicity avoids the Alabama paradox. If it satisfies population monotonicity, it avoids the population paradox. You'll notice that these criteria do not address the new states paradox explicitly, but that's because the new states paradox can be expressed as a special case of the population paradox. Think of the new state as always having been there, but with population 0. The moment it joins the union is equivalent to pretending its population went from 0 to whatever it really is.

Balinski and Young prove that the Hamilton method does not satisfy population monotonicity but that all the divisor methods do (and hence also avoid the new states paradox). Moreover, divisor methods are the only methods that do so. Further, any method that satisfies population monotonicity also satisfies house monotonicity. Therefore, divisor methods also satisfy house monotonicity.

In a nutshell, we should use a divisor method because unlike the Hamilton method, they all avoid the paradoxes. So far so good.

But what about the quota rule? I promise, this is the last time I'm going to do this:

Balinski-Young impossibility theorem: There is no apportionment method that satisfies the quota rule and population monotonicity.

So whatever method we have in mind, it will either be vulnerable to the population paradox or it can fail the quota rule, or it can fail both. The theorem says that if we stick with divisor methods, which we know are immune to the population paradox, they will necessarily be susceptible to failing the quota rule. We already know that this is the case with the Jefferson method because it happened in the 1820 and 1830 apportionments. Balinski-Young says this can happen with *any* divisor method, regardless of which one we pick from the infinity of them.

But things are not so bad.

In practice, Webster and Huntington-Hill appear to be extremely unlikely to fail the quota rule. Huntington-Hill has never violated it and it would not have done so had it been used prior to 1940. The same is true for Webster. In fact, simulations show that these methods would on average fail the quota rule only about once every few hundred thousand apportionments, with Webster doing slightly better than Huntington-Hill.

So we're on the same three-loop roller coaster with Balinski-Young as we were with Arrow and Gibbard-Satterthwaite. First comes a stunning result that we find simultaneously impressive and disorienting. Then the brows furrow as our thoughts turn to the implications for real-world democracy. And then, finally, we gather enough evidence to show that actual use is probably okay and we have little to worry about.

Should we then just go with the mathematical elite of the first half of the twentieth century and say that Huntington-Hill is the winner? That's not so clear. The issue of whether one should look at the absolute or relative discrepancy of representation ratios—the distinction between the Webster and Huntington-Hill methods—remains unresolved. The decision by the panel assembled in 1929 to back Huntington-Hill might have reflected deference to Huntington and his larger-than-life presence in the mathematical community. It didn't help that the way Huntington-Hill was ultimately baked into the apportionment legislature in 1941 had everything to do with politics and the fact that the Democrats stood to gain one seat by using this method instead of Webster's.

There is one more scale we might use: the way the various methods treat small and large states. We know that Jefferson's method favors large states and Adams's method favors small states, simply because the rounding they employ is always one sided—down for Jefferson, up for Adams. We also know that Huntington-Hill favors small states because geometric means are lower than arithmetic means and more so for smaller numbers. Both Huntington and the 1929 panel were conveniently silent about this bias in their analyses and arguments.

That leaves Webster's method.

It turns out that overall, Webster is not biased in favor of large or small states. Well, it is, but in both ways equally, so not overall. If the initial apportionment is less than the size of the House, it is slightly favorable to large states. When the initial apportionment overshoots, Webster favors small states. Because these situations are equally likely to occur, the net effect over many apportionments is zero bias. These are the conclusions of Balinski and Young themselves, who crunched the numbers for twenty-two apportionments, from 1790 to 2000. The results are summarized in figure 10.2. They went even further and proved that the Webster method is the unique unbiased method. That means

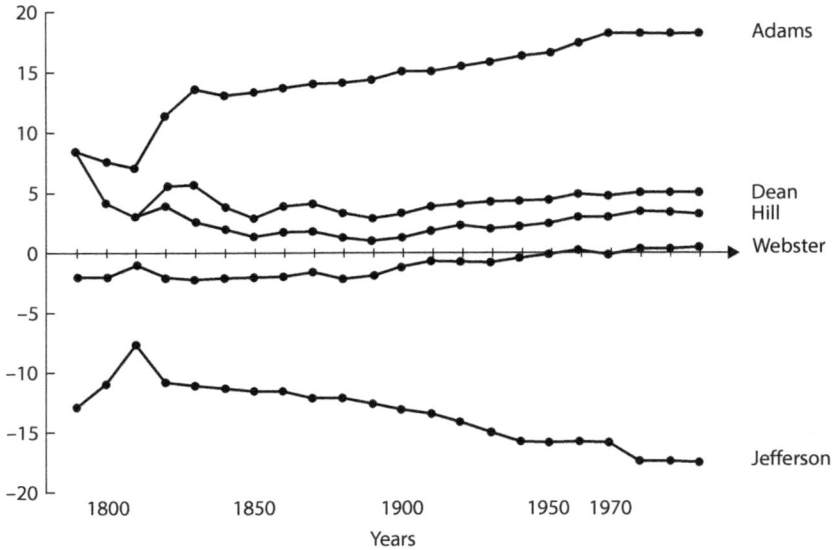

FIGURE 10.2. Average bias of the five main apportionment methods across twenty-two censuses.

that if we went digging among the infinitude of divisor methods for a less biased one, we'd never find it.

The Webster method is the least biased and the least likely to violate the quota rule, so that is what we should be using.* We should not be using Huntington-Hill.

There is an important contrast between how measured I am about the Huntington-Hill mistake and how impassioned I am about the mistake we're making with voting. Plurality is so manifestly bad, so detrimental to democracy that my palms start sweating as soon as I start typing something about it into these pages. But I don't feel the same way about the Huntington-Hill method of apportionment. I'm mildly annoyed that we're using it. Mildly, because the question of apportion-

---

* Some scholars argue that we should use the Jefferson method, which favors large states, to partially mitigate the effect of the Senate, which is all about giving power to the small states. But using one flawed mechanism to compensate for another is always worse than improving both.

ment method is the least contentious of all the mathematical issues that afflict our democracy. Most of the time, Huntington-Hill and Webster will agree and give the same apportionment—of the 4,350 seats that were apportioned between 1910 and 2000, only 10 would have been allocated differently had the other method been used.

Unlike with voting, the reason we are not using a terrible method for apportionment is that experts were involved in the discussion and decision-making—ranging from Jefferson the citizen-mathematician to the academic mathematicians serving on the two National Academy of Sciences panels to the public mathematicians associated with the census. Even though their participation did not remove politics completely, mathematics remained the primary guiding principle.

But the decades-long labor of the mathematics community around apportionment and its ultimate fate also illustrates the limitations of mathematics in the political arena. The final decision to enshrine Huntington-Hill into law in 1941 was so petty, so shortsightedly political, that the mathematics ended up being simply a convenient justification for its implementation. Alma Steingart, a Columbia University professor and a scholar of the interaction of mathematics and politics, puts it this way in her essay "Democracy by Numbers": "The perennial challenge is that quantification and its opaque rigors can all too easily be strategically deployed and conflated with the interpretive work of politics. Numbers and quantification—so often taken to be objective, unbiased, and merely descriptive—actually can end up formalizing political arguments." While I disagree that the rigors of mathematics are opaque—unless Steingart means they are perplexing to most people and require specialized training to understand—this statement perfectly captures the unrelenting tautology of politics: politicians are politicians. No matter what mathematics says, they will ultimately act in their own interest.

All of this is a reminder that when math operates within human systems, it's subject to a web of constraints—political, of course, but also legal, physical, and social, among others. The key is to keep a grip on the mathematics without ignoring the context in which it's embedded.

Apportionment has given us an instructive illustration of how to manage that stance when the mathematical goal is clear. In the next chapter, we'll see a greater challenge—raging right now—in which the mathematical goal is still being formulated, often in interaction with the courts. We're about to meet gerrymandering.

CHAPTER 11

A Country Divided

BRAINTEASER TIME: a man goes to sleep in one district and wakes up in another. He didn't move. How did this happen?

That is exactly the predicament Hakeem Jeffries found himself in. It was 2002 and Jeffries, a Democrat, was running for New York State Assembly against the Democratic incumbent, Roger Green. The party establishment didn't like the potential shakeup in its ranks, so it conveniently redrew state district lines in a way that placed the city block where Jeffries lived in a different district. This did not make him ineligible to run against Green in that election cycle (although it did in the next), but it made it more difficult for Jeffries to win because a number of his supporters were also carved out of the district. Jeffries: "Brooklyn politics can be pretty rough, but that move was gangster."

In politics, this gangster move—redrawing district lines so that a candidate's home now belongs to another district—is one of the bread-and-butter techniques of *gerrymandering,* the practice of drawing local, state, or congressional voting district lines to achieve a certain political effect.*

---

* Jeffries was later elected to the United States House, representing District 8 in New York. He is an outspoken critic of gerrymandering and an advocate of redistricting by nonpartisan, independent commissions.

BECAUSE AMERICA CAN HACK
ITS OWN ELECTIONS

The drawing of district lines—districting for short—is where the rubber meets the road in representative democracy. Apportionment gets us part of the way, allocating some number of representatives to a population. Districting determines which voters have the right to elect which representatives. Because it's almost always tied to where voters live, districting adds another layer of political and mathematical complexity, bringing geography and geometry into the mix and with them, rich opportunities for manipulation such as gerrymandering.

We can already see the complexities in a small example with manageable numbers, and your job is to imagine this scaled up to hundreds of thousands or millions of voters. (For simplicity, in these examples I'll conflate population and voting population—the electorate—but this won't create a problem.)

Suppose there are fifty voters in a state, pictured in figure 11.1. In a previous election, the Triangle Party got 30 votes and the Circle Party got 20. Let's assume that this is how voters will vote in the next election too. The state has been apportioned five legislative seats and must be divided into five districts, each one of which will elect one representative. Each district is required to have the same number of people (ten) and must be *contiguous*, which means connected, so it shouldn't consist of two or more separated chunks.

Panel A of figure 11.1, which has districts that look like vertical strips, seems sensibly divided. Each of the districts is composed only of Triangle or Circle voters and each elects a representative unanimously. Of the five representatives of this district, three are from the Triangle Party and two are from the Circle Party. This result achieves perfect proportional representation; both the population and its delegates are divided in the 3–2, or 60%–40%, ratio. This is not the only way to draw districts to achieve this representation, but it's the most obvious.

Panel B is where things get weird. By drawing districts horizontally, we put six Triangle and four Circle voters in each district. Because voting is by plurality, the Triangle candidates win in all five districts. Even

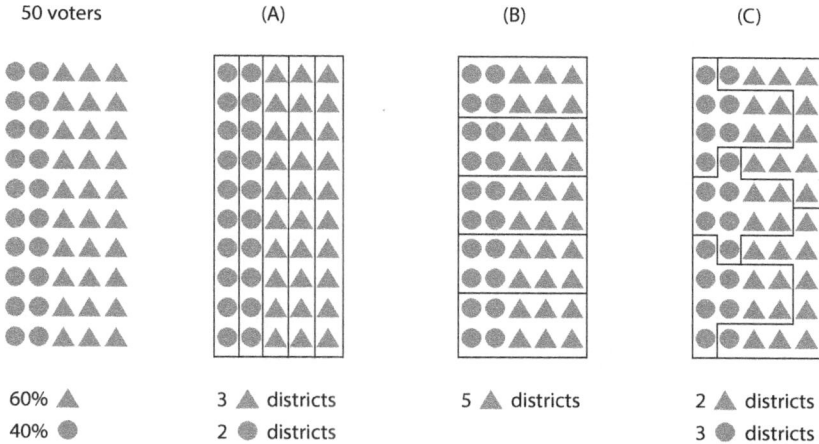

50 voters   (A)   (B)   (C)

60% ▲   3 ▲ districts   5 ▲ districts   2 ▲ districts
40% ●   2 ● districts   3 ● districts

FIGURE 11.1. Three ways of dividing fifty people into five districts.

though Circle voters account for 40% of the population, they get no representation. If I were a Triangle partisan and controlled districting in this grid state, this is how I'd carve it up. My party would get all the power this way.

Panel C is even more interesting. The state has been dissected so that Circle wins three seats (three districts have six Circle voters and four Triangle voters) and Triangle wins only two seats (the two remaining districts have nine Triangle voters and only one Circle voter). So Circle has more overall power in the legislature, even though it has the support of only a minority of the voters. Clearly, this is how a Circle partisan would want to carve up the state.

This example makes plain why gerrymandering inspired the adage "politicians choose the voters, not the other way around." It is the politicians who place voters in particular districts to make the math work out in a favorable way. As Trevor Noah explained it on *The Daily Show* episode titled "Gerrymandering: Because America Can Hack Its Own Elections": "It's back to front. It's like getting to your Uber and your driver's in the back seat and he's like 'start driving, asshole, I'm late.'"

Two standard gerrymandering techniques are at play in panel C. One is *packing*, which fills one or more districts with voters of one party, as happens with the two majority Triangle districts. These are "throwaway districts"; the party doing the gerrymandering (in this case, the Circle Party) knows it will lose in those districts, so it might as well lose big. The complementary technique is *cracking*, whereby the remaining voters of the opposing party (in this case, the Triangle Party) are split across other districts in numbers that deny them wins. Cracking is what happened in the three districts that elected Circle representatives. None of them had enough Triangle voters to counter the Circle vote. The outcome is that the Circle gerrymanderers successfully diminished the voting power of the Triangle population by letting them win a couple of districts where they are consolidated but left enough districts where the Circle Party holds a majority, giving it overall control. In Panel B, where the Triangle Party won all five seats, the Circle population has been cracked across the five districts.

John Oliver explained packing and cracking on his show *Last Week Tonight with John Oliver* in terms of table assignments at a wedding: "You have the option of cracking your eight awful relatives across eight tables in order to water down their awfulness, hoping their voices will be drowned out by the other people at those tables. Or you can cut your losses by packing them all together in 'one insufferable table of the damned' that you try to ignore for the duration of the reception." I'd add that if your tables seat only, say, six people, you can pack *and* crack, packing one table with six of them and cracking the remaining two across two other tables.

Now imagine that the awful relatives are members of the opposing party or some group whose voice you're trying to dilute and that the tables are voting districts. Also imagine that you have access to data clouds filled with demographic and prior voting data. You know which party each precinct went for historically, you can see the party registrations of people who live in those precincts, and you have all kinds of precinct-level census data about variables such as income and ethnicity. You can use sophisticated, fast software and geographic information systems that allow you to run the digital scalpel down any block or street with surgical precision. All of these tools are readily and publicly available. That's all you need. You're ready to craft a map that sends the most members of your party to Congress or the state legislature and guarantees a sustained reign in each of the districts your party has seized.

The number of *competitive* congressional districts—those that aren't preordained by their demographic makeup to elect a member of one or the other party with certainty—has plummeted over the years, in large part due to gerrymandering. According to the Cook Political Report, in the mid-1990s there were 164 competitive districts among the total of 435. Twenty years later, there were 72. After the 2021 redistricting, there are only 35. This means that more than 90% of races happen in *safe* districts where incumbents are protected and need to do little to get re-elected. In 2022, the average margin of victory for a contested House of Representatives race was 28%. After the 2016 elections, in which the average victory margin was a whopping 37%, former California governor Arnold Schwarzenegger, an outspoken critic of gerrymandering, tweeted "There are dictators who win by less."*

Less competitiveness means more disenfranchisement. In his book *One Person, One Vote: A Surprising History of Gerrymandering in America*, Nick Seabrook, a professor of political science at the University of North Florida, puts it this way: "To gerrymander is to distort, to corrupt, to turn the institutions that should be working on behalf of the people's interests into perversions that serve only the powerful, the moneyed, or the politically connected. Whatever political machination over gerrymandering is afoot, no matter who wins, it's always the voter who loses."

The big parties love noncompetitive districts. They don't have to spend campaign money on those races. All a candidate needs to do is toe the party line and satisfy the base, which, by design, makes up the majority of that candidate's voting constituents. There is no incentive to reach across the aisle or to think about how policy might affect populations beyond those that reliably vote for the candidate. There is no

* Definitely a fun jab at gerrymandering, but it's even more fun to try to find some dictators who might have won by less than 28 percentage points. Here are some attempts: Kim Jong Un—100% in 2014, Kim Jong-il—99.9% in 2009, Saddam Hussein—100% in 2002, Raúl Castro—99.4% in 2008, Bashar al-Assad—97.6% in 2007, Gurbanguly Berdimuhamedov—97% in 2012, Mikheil Saakashvili—96% in 2004. . . . Okay, I give up.

incentive to compromise or move forward in a substantive way. Safe passage is guaranteed as long as the boat is not being rocked. Gerrymandering has made being in Congress a cushy, highly partisan job that can last for decades—at the expense of the millions whose voices have been fragmented into silence.

MANGLING DEMOCRACY SINCE 1788

How did we get to a place where gerrymandering is endemic? The practice seems to have originated in England in the eighteenth century with the evolution of "rotten boroughs" that contained small numbers of eligible voters who nevertheless got to elect two Members of Parliament. The votes in such sparsely populated districts would then be easy for resident grandees to buy. There was even a borough, Old Sarum, that had eleven inhabitants on the books, but none of those people lived in the borough. Populated only by sheep, this borough still elected two MPs to every parliament from 1708 to 1832.

In the United States, gerrymandering is as old as the republic; states such as Virginia, North Carolina, and South Carolina practiced it in the late eighteenth century. In fact, the gangster move that shut Jeffries out of the New York Assembly could have come out of the playbook of Patrick Henry, a governor of Virginia who in 1788 created a new congressional district that forced his political enemy James Madison to run for the House of Representatives against his friend and ally James Monroe. Madison managed to win, but only after reluctantly endorsing the Bill of Rights, which he had previously opposed. Redrawing district lines to force two (usually like-minded) candidates to run against each other, thereby guaranteeing that one of them won't make it, is called *double bunking* or *incumbent pairing*. It's kind of shocking it happens often enough to have not one, but *two* names.*

---

* To round out the terminology, there is also *stacking*, which is the practice of drawing district lines such that there is a large number of minority voters but an even larger number of white voters. A variant is to draw a district that has many low-income, less educated minority voters, who typically vote at lower rates, such that they might even make up the majority—that is,

Gerrymandering was first put on the map in a big way by Elbridge Gerry, a founding father who later served as the fifth vice-president. Gerry called for "less democracy" at the Constitutional Convention and fought against direct election of representatives to the House. He successfully argued for the power of Congress to override a presidential veto. Oh, and he refused to sign the Constitution.

But Gerry remains most famous for something he did late in his career as the Democratic-Republican governor of Massachusetts. In 1812, the Democratic-Republican Party legislature sought to disempower the Federalist Party members in the state senate and redrew the districts into shapes the Federalists called "carvings and manglings." Apparently acting in accord with his proclamation at the Constitutional Convention that "the evils we experience flow from the excess of democracy," Gerry signed the redistricting into law.

One of the districts the Democratic-Republicans created was depicted as a ghoulish salamander in the *Boston Gazette* (figure 11.2).

The derogatory portmanteau "gerrymander," blending "Gerry" and "salamander," was coined not too long after that at a Federalist gathering.* But it was the Democratic-Republicans who were laughing, all the way to the state House. Even though they and the Federalists both received about 50,000 votes in the 1812 elections, Democratic-Republicans captured 29 seats in the Massachusetts senate but the Federalists received only 11.

After an "era of good feelings" from 1815 to 1825, when the two parties did not compete as viciously and so did not feel the need to mutilate the districts as much, gerrymandering became more common, beginning in the 1830s with the creation of the Whig Party.

District butchering took off after the Fifteenth Amendment granting Black men the right to vote was passed in 1870. Southern states intensified their gerrymandering to dampen or dilute the Black voice. Nothing fancier than what we saw in the Triangle-Circle example was needed,

---

create a *majority-minority* district—along with wealthier, more educated white voters who then turn out to vote in greater numbers and elect their candidate.

  * Even though Elbridge Gerry's last name is pronounced like "Gary," for some reason we now pronounce the name like "Jerry" in the word "gerrymandering."

FIGURE 11.2. Elbridge Gerry's salamander, the original gerrymander.

just an understanding of demographics and ingrained racism with a side of packing and cracking. For example, South Carolina created a "boa constrictor" district in 1882 that consolidated Black voters (who at the time made up the majority of the state's population) so other districts could comfortably elect white representatives. An 1852 redistricting of Indiana gave Democrats ten of eleven districts even though they won only 53% of the popular vote (the Democratic Party was the dominant party in the South at the time).

The first half of the twentieth century was marked by the disenfranchising exploits of Jim Crow in the southern states, which used such familiar weapons as poll taxes, literacy tests, and intimidation. But gerrymandering was also an effective tool in this bigoted plan, often counterintuitively exhibiting itself in the *refusal* to redistrict. Back then, states redistricted however often they wanted to (or didn't want to). The motive behind choosing not to redistrict was to contain the increasingly minority and immigrant urban populations in as few districts as possible. Alabama, for example, redrew its districts in 1901 and then left them unchanged for sixty years.* By the 1960s, one of Alabama's rural state districts contained about 15,000 people while the district containing Birmingham, which had a large Black population, had more than 600,000. This is essentially packing, but with no limit on the size of the packed population. The result is a grossly disproportionate distribution of power between urban and rural areas. Some states were controlled by legislators who represented less than 20% of the population.

Gerrymandering inventiveness in the South wasn't limited to inaction. For example, the Alabama legislature, fearing an increase in registered Black voters in the wake of the Civil Rights Act of 1957, redrew the city boundaries of Tuskegee, changing the city from a rectangle into the twenty-eight-sided polygon shown in figure 11.3. Why the geometric gymnastics? The veering city limits carefully and methodically excluded almost all Black voters from the city while keeping the white voting population intact.

The 1960s brought long overdue changes meant to put a stop to racial gerrymandering (at least in theory). The Supreme Court curtailed the practice of circumscribing a population defined by a racial or economic status in a number of decisions. In the 1960 case *Gomillion v. Lightfoot*, the Court ruled that the Tuskegee polygonal gerrymander violated the Fifteenth Amendment. In *Baker v. Carr*, the Court ruled that Tennessee had to update its district lines, which had been unchanged for sixty years, causing, as in Alabama's case, an imbalance of power between expanding urban and shrinking rural areas. Redistricting could now be litigated in federal courts, paving the way for the 1964 cases *Wesberry v. Sanders* and *Reynolds v. Sims*, brought by voters in Georgia and Alabama. Those decisions mandated that congressional districts across the United States and state districts within each state must have approximately the same populations. This is known as the one person, one vote doctrine. It marks the entrance of a quantitative, computable criterion onto the districting scene. In addition, all district borders had to be revisited after

* At the other extreme is Ohio, which redistricted seven times in the period 1878–1892.

FIGURE 11.3. Redrawn boundaries
of Tuskegee, Alabama, in 1957.

each census to keep the district populations about equal.*

These decisions complemented the Civil Rights Act of 1964 and the Voting Rights Act of 1965, which aspired to give minority populations the opportunities to elect candidates of their choice and defend them against discriminatory voting practices. Among other things, Section 2 of the Voting Rights Act mandates the creation of majority-minority districts under certain circumstances. The ruling in the 1986 Supreme Court case *Thornburg v. Gingles* operationalized some aspects of the Voting Rights Act by identifying three criteria that determine racial gerrymandering, thereby laying further legal foundation for challenging and litigating redistricting practices that hinder the representation of minorities. *Thornburg* was partially rolled back in the 1990s with the *Shaw v. Reno* and *Miller v. Johnson* rulings, which said that race cannot be the only or the primary reason for creating a district, but racially concentrated districts can be acceptable if the main impetus in drawing district lines is partisanship.

At the state level, the notion of *communities of interest* also has legal districting implications. How such a community is defined varies from state to state, but the general idea is that a group of people might have shared traits or interests and that districting should attempt to preserve them. About half the states explicitly require districting authorities to pay attention to such communities. The requirements are sometimes baked into constitutions or statutes and sometimes simply appear as recommendations. Colorado, for example, says that "communities of interest, including ethnic, cultural, economic, trade area, geographic,

---

* The 1929 Reapportionment Act removed the requirement of uniform size and requirements about the geometry of the district shapes. This permissiveness allowed significant malapportionment until the uniform size requirement was reestablished in 1964.

and demographic factors, shall be preserved within a single district wherever possible." Urban, rural, and industrial areas; media markets; occupation and lifestyle patterns; reservations; ski areas; and service delivery areas are just some of the categories that define communities of interest across the states. Requirements of compactness and contiguity, two spatial criteria we'll talk about in the next chapter, are sometimes conflated with the preservation of communities of interest. Identifying such communities is generally difficult, as recent attempts to collect input from the public in some states indicate. The self-sorting can produce many and varied categories, bringing into question the feasibility of building the concept of communities of interest into districting at all.

———

Despite the best efforts of the 1960s courts, gerrymandering has thrived, especially in the past twenty years. Mapping technology, the availability of professional redistricting software such as Maptitude,* and the speed of data processing have given gerrymandering new wings. There is no more need to pore over giant paper maps spread on the floor, pencil and eraser expertly guided by the rare wizard who understands the politics and idiosyncrasies of every voting district. A computer can generate tens of thousands of possible maps in seconds, and anyone with some understanding of local politics and some dexterity with spreadsheets can play the gerrymandering game, securing their own party's election success.

Around 2010, the Republicans became the undisputed champions of this game.

HE WHO CONTROLS REDISTRICTING

One of the Republican Party's shrewdest moves was REDMAP, short for Redistricting Majority Project, a plan hatched by party operative Chris Jankowski after the Republicans suffered massive losses in the 2008 elections. The details of the project can be found in the excellent

* You can also try your hand at districting: visit districtr.org or districtbuilder.org.

documentary *Slay the Dragon* and in David Daley's exhaustively re-searched page-turner *Ratf**ked: Why Your Vote Doesn't Count.* Here is a plot summary.

In the hierarchy of election cycles, one reigns supreme: the years divisible by ten. This is because the census happens every ten years, in a year that ends with a zero. With each new census, the Huntington-Hill method spits out the reapportionment of House seats and it's go time for redistricting each of the fifty states into however many seats it has been allocated. Whoever takes the reins of a state's legislature in the year ending in zero also gets to (in most states) take their scissors to the maps and essentially determine the distribution of power for the following decade. In the Republican analysis, all they needed to do to tip the balance of power was to win a certain number of seats in certain state legislatures.

Phase I of REDMAP dumped money into local races in a number of key states—places such as West Lafayette, Indiana, and Round Rock, Texas, that did not look relevant at first glance—to secure exactly the seats the party needed to flip their legislatures in Republicans' favor.

Phase II would then fall into place by itself; those same legislatures would control the decennial redistricting process and would gerrymander those sometimes-competitive states into Republican bedrock. This state-level effort would ensure that Republicans secured and maintained more than their fair share of seats for at least a decade.

The first phase worked amazingly well. In 2010, Republicans gained over 700 state-level seats and flipped twenty state chambers. All it took was $30 million in campaign investments. The redistricting stage was set. Republicans had won the opportunity to draw the borders in 219 of 435 districts.* Wasting not a moment, they took to the map of the United States with the fervor of Edward Scissorhands going to town on a topiary.

The second phase of the plan also worked beautifully. In 2012, Democratic candidates for Congress won 1.4 million votes (about 0.9%) more

---

* By contrast, Democrats controlled the districting in only forty-four districts. To round out the count, eighty-eight districts were controlled by independent or bipartisan commissions, seventy-seven were drawn jointly by the two parties, and seven were at-large districts, which means that those seven representatives were chosen by statewide election.

than their Republican counterparts, but the Republicans won 33 more seats (about 16.4%) than the Democrats, solidifying their grip on the House. Why the discrepancy? Let's look at some states. In Pennsylvania, Democrats won 2.79 million votes while Republicans won 2.71 million, but the Democrats sent five people to the House while the Republicans sent thirteen. In other words, Democrats won over 50% of the votes but only 28% of the seats. In Michigan, Democrats received 2.33 million votes and won 5 House seats, while Republicans received 2.09 million votes and won 9 seats. In North Carolina, Democrats had 2.22 million votes, which earned them 4 House seats, and Republicans got 2.14 million votes and 9 seats.

A similar story unfolded in state-level elections. In Wisconsin's state legislature, Democrats got 1.42 million votes compared to the Republicans' 1.25 million, but the legislature seat spread was 39–60 in favor of the Republicans. In Michigan, the Democrat-Republican vote split was 54%–46% while the spread of the state seats was exactly flipped, 46%–54%.

Republican good fortune continued in subsequent elections as gerrymandering kept bearing fruit. In 2014, the party picked up 13 more seats in the House. In 2016, Republicans won 49% of the vote while Democrats won 48%, a winning margin of 1%, but Republicans took 55% of the House seats, a 10% margin over Democrats. Ohio sent twelve Republicans and four Democrats to the House in 2016, but the true shocker is that this spread would have occurred if the Republicans had won anywhere from 49% to 79% of the vote, a thick cushion carefully fashioned into the maps. In 2018, Republicans won about half of the votes in North Carolina but took 10 of 13 House seats. You get the picture.

Figure 11.4 shows a few of the REDMAP state outcomes. It is difficult to tell simply from the district shapes whether and how much gerrymandering they reflect. Are they the result of the kind of maneuvering we saw in figure 11.1? Or do the shapes have reasonable explanations? Chapter 12 will give us the mathematical tools we need to make headway on these questions.

Over a period of just a few years and a couple of election cycles, Republicans took control of thirty-two states, double what they had before

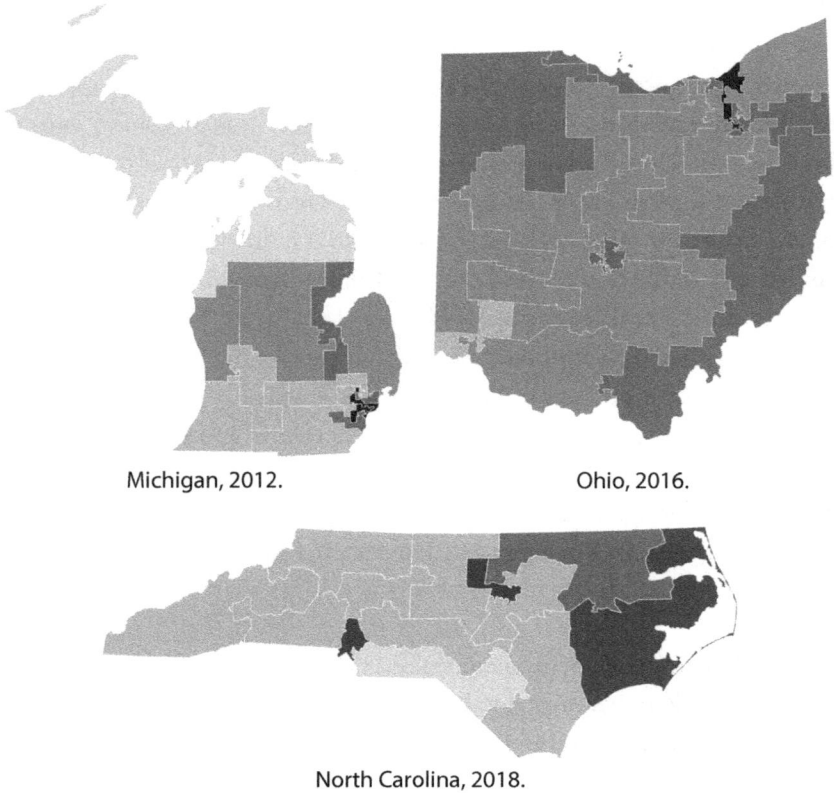

Michigan, 2012.   Ohio, 2016.

North Carolina, 2018.

FIGURE 11.4. Congressional district maps.

REDMAP. And with that came bulletproof control of the House of Representatives, at least until Democratic voters surged against Donald Trump and turned out in droves in 2018 at the polls to take it back. But in the 2020 elections, gerrymandering was in evidence again. Democrats lost 13 seats, even though their presidential candidate, Joe Biden, won 4.5% more votes across the country than Trump.

The Democrats couldn't say they weren't prepared. Surely some of them had read Karl Rove's foreboding proclamation in the *Wall Street Journal* in 2010: "He who controls redistricting can control Congress." In fact, Democrats are pretty good at the game themselves and have answered in kind. In Maryland, sometimes dubbed "the most gerrymandered state," Democrats received 62% of the vote in 2012 and won 88% of House seats. In Illinois the same year, they received 54% of the vote and 67% of House seats thanks to gerrymandering. Oregon and Washington are heavily gerrymandered as well. As a matter of fact, Democrats were the first to harness the power of computing in districting when they locked in their power in the California General Assembly in the 1980s, courtesy of a savvy operative, Representative Phil Burton. He described the map he created as "his contribution to modern art." California finally established an independent districting commission in 2008.

If you're of the opinion that this is all par for the course because politics isn't about playing nice, we'll see soon that the courts would agree with you. But this raises a question: Why is the process usually so shrouded in secrecy? It's because extreme gerrymandering that locks in an uneven distribution of power for a long time is illegal, violates the equal protection clause of the Fourteenth Amendment, or denies minorities representation. State legislature czars know this and conceal the process with an invisibility cloak. Any parent whose children suddenly fall silent knows something fishy is going on. The same is true for the process of gerrymandering, whose hiddenness screams volumes about its guilt. In 2011, the Wisconsin Republican Party outsourced the post-census districting (Phase II of REDMAP) to the law firm Michael Best & Friedrich. Only a select few had access to the "map room," where the law firm carved out state districts. One by one, Republican lawmakers were taken to the room to see their new district (and only their district), and each had to sign a nondisclosure agreement before they left the room. No Democrat saw the new state map until a public hearing that was held one week before the redistricting was voted into law. Every Republican voted for it; not a single Democrat did.

Secrecy is at the center of the still-unfolding drama precipitated by the death in 2018 of Thomas Hofeller, a Republican gerrymandering mastermind. When the North Carolina Supreme Court ordered that some of over 75,000 files on Hofeller's four external hard drives and eighteen thumb drives that contained information on gerrymandering in a bunch of states be turned over, the documents revealed a pattern of racial gerrymandering in the creation of 2011 and 2017 maps. One of Hofeller's presentation slides said, "Treat every statement and document as if it was going to appear on the FRONT PAGE of your local newspaper. Emails are the tool of the devil."

As REDMAP got under way, gerrymandering battles began to rage in the courts. Tribunals up and down the judicial hierarchy have had their hands full with redistricting cases. The poster child is North Carolina, whose 2011 map was struck down in 2017 as racially gerrymandered.* Joining North Carolina in the infamous circle of perpetual offenders are Illinois, Oregon, Pennsylvania, Texas, Wisconsin, Maryland, Florida, New York, Ohio, and several others.

A particularly relevant case, poignantly documented in the film *Slay the Dragon*, was *Gill v. Whitford*. The case was brought by a group of citizens from Wisconsin who challenged that state's 2011 redistricting (conducted in the secretive "map room"). In a 2016 ruling that struck down a map for partisan bias for the first time in over thirty years, a federal court sided with the plaintiffs. Wisconsin appealed, and the case came before the Supreme Court in 2017. However, the court punted and remanded the case to the district court for new arguments on the plantiffs' standing to bring suit.

Then came the 2019 case *Rucho v. Common Cause*. In a 5–4 decision on Maryland's and North Carolina's maps, the Supreme Court took the wind out of the sails of opponents of gerrymandering: "Partisan gerrymandering claims present political questions beyond the reach of the federal courts." In other words, partisanship in redistricting is not a matter for the courts at the federal level. Yes, the court said, the map is clearly gerrymandered, but unless somebody can establish a direct line to an infraction of the Constitution exhibiting a racial gerrymander that violates the Voting Rights Act, we're going to ignore this. The court summarily dismissed *Gill v. Whitford* a few months later.

---

* The state's Republican legislature redrew a temporary remedial map and went to great pains to explain how the districting intentions were not racial but rather were aimed at disabling Democrats in the state. The redrawn map, which packed Democrats into three districts, still did the job; in the 2016 elections, Republicans in North Carolina won 53% of the vote but took 77% of the House seats.

The deciding vote in *Rucho* was cast by Justice Brett Kavanaugh, who had recently replaced Justice Anthony Kennedy. Some fifteen years earlier, Kennedy had expressed an openness to the idea that federal courts might have a role to play in gerrymandering cases. But after *Rucho*, state courts became the last line of defense. And with state legislatures aiming to weaken state courts (according to one count in May 2021, twenty-six states introduced ninety-three bills that would diminish the power of state courts), as well as the increasing politicization of the courts themselves, the message of the *Rucho* ruling was that unbridled partisan gerrymandering is acceptable.* In her dissent, Justice Elena Kagan wrote: "Election Day . . . is the foundation of democratic governance. And partisan gerrymandering can make it meaningless."

A CIVIC SOLUTION

After the 2020 census, most districting was done by partisan actors. In nineteen states, Republican-controlled legislatures drew the maps (41% of congressional districts). Democrats ran the districting in seven states (11%). Courts were involved in drawing of maps in eight states (21%) as a result of political stalemate (such as the governor and the legislature disagreeing on the map), usually appointing experts (special masters) to do the job. Politically appointed bipartisan commissions decided the maps in five states (6%). Finally, independent commissions produced the maps in four states (19%): Arizona, California, Colorado, and Michigan.†

And this is where we arrive at an effective civic solution to gerrymandering: independent commissions. These are composed of people who are not elected officials or legislators and are screened for biases before they take on the role. As the name suggests, they are tasked with drawing district lines in a nonpartisan way that respects the guidelines and

* North Carolina's courts show strong susceptibility to politics. Its Democratic majority declared in 2022 that the state's maps were illegal gerrymanders, but in 2023, the Republican-dominated court reversed the decision.

† This adds up to forty-three states, but seven more have only one seat and therefore do not district for congressional seats.

rules of geography, fairness, and opportunity. Many countries, including the UK, Australia, and Canada, use independent commissions.

Independent is not the same as bipartisan. A commission whose members are appointed by politicians—even if they are from different parties—isn't as effective at achieving fairness. In New York in 2021, for example, members of the bipartisan commission couldn't agree on a map, so the process went back to the Democrat-controlled legislature, which unleashed a full gerrymander on the map. (That map ultimately ended up in the courts.) Virginia's advisory commission could not agree on a map and the state supreme court had to step in. Bickering and political favoritism also complicated the work of partisan commissions in Montana and New Jersey.

It is additionally important that independent commissions have the final say. In Ohio, the Republican legislature simply ignored the commission's map and drew its own. A similar story played out in Utah, costing the Democrats a potentially competitive district. Democrats in New Mexico also disregarded the recommendation of an advisory commission.

Some barriers to fair representation will persist. Factors such as self-sorting and geography get in the way of even the most independent of commissions. For example, Democrats naturally pack into cities. Take Missouri, where they are clustered around Kansas City and St. Louis. Of the eight House seats that state gets, there's little the Democrats can do to win more than two (or maybe three) seats. The story is similar in Milwaukee and Madison in Wisconsin. This talks and walks like gerrymandering, but it's not. In fact, a standard Republican defense against accusations of gerrymandering is that the Democratic population clusters in metropolitan areas, giving the appearance of packing. Although this is true to an extent (I'll elaborate in the next chapter), it alone cannot explain the wide prevalence of districts with a strong and lasting Republican grip.

In the UK, which also has single-member districts like the United States and does its districting with independent boundary commissions, elections can still exhibit bias toward certain parties, especially the Labour Party. This can be attributed to districts not having the same size (Scotland, Northern Ireland, and Wales have fewer people per district than England), Labour tending to win in districts with lower turnout, and voters being distributed more efficiently.

Nevertheless, there is strong evidence that independent commissions do reduce gerrymandering overall. The Center for American Progress concluded in 2016 that legislature-run commissions are twice as likely to produce districts that "deny voters choice in the general elections." Mathematical measures such as *partisan bias* and the *efficiency gap*, which we'll see in chapter 12, corroborate this.

Maps drawn by independent commissions also have almost twice as many competitive districts as their legislature-drawn counterparts. This was substantiated, for example, in California, where the average percentage of competitive districts went from 5.6% in the decade preceding the 2010 introduction of an independent commission to 14.6% since then.*

If you're convinced that independent commissions are the way to go but the prospect of advocating for one seems daunting, consider how Michigan voters forced one into existence. The grassroots movement started in the wake of the 2016 elections on the Facebook page of Katie Fahey, a young activist who was fed up with the broken system and simply posted: "I'd like to take on gerrymandering in Michigan. If you're interested in doing this as well, please let me know." Her organization Voters Not Politicians soon had thousands of volunteers across the state who gathered 400,000 signatures to put Proposal 2 on the ballot. Despite lawsuits and attempts by conservative groups to quash the initiative, it passed by over a 20-percentage-point margin. Michigan now has a districting commission whose thirteen members are prohibited from having ties to politics. The map they produced is one of the most balanced and competitive in the nation.

---

* California's initiative to enact an independent commission had support from across the political spectrum. People were sick of the decades-long bickering over district borders. The strongest opposition came from the predictable places, the cushy chairs of the Democratic incumbents, including Speaker of the House Nancy Pelosi. Just like Republicans, Democrats are in favor of ending gerrymandering when they're on the losing end of districting but very much for it when they're in charge. California now uses a lottery system to choose a panel that then chooses the mapmakers; thousands of people typically apply. (Chapter 18 of the book *Political Geometry*, edited by Moon Duchin and Olivia Walch, details the process.)

CARVING UP PRISONS AND SCHOOLS

Although the courts work to refine what a gerrymander is in a narrow electoral sense, the lay concept has expanded to encompass the one-two punch of manipulation of boundaries and undermining of civic goals. It's worth digressing for a moment to mention other host institutions this parasite has invaded.

Prison gerrymandering is the variant in which districts containing prisons count inmates as residents. This practice gives the districts greater electoral power and increased resources—even if the inmates can't vote (prisoners can vote only in Maine and Vermont) or benefit from the resources. Inmate numbers don't typically make a big difference in congressional races, but they do at the state and local levels, where the prison population might even constitute the majority of the residents. Seven New York state districts would not even be districts if they didn't contain prisons. The population of one of the city districts in Anamosa, Iowa, is 96% inmates. Of the 1,400 official residents, only about 100 are eligible to vote. In 2008, Danny Young, a candidate from that district, won a city council seat with just two write-in votes, one from his wife and one from his neighbor.

Rural areas benefit more from prison gerrymandering because they are better able to attract prisons due to lower building costs. The practice has escalated in the era of mass incarceration that began in the 1970s; the number of people in U.S. prisons is now more than 2 million. Because the inmate population is disproportionately Black and Brown while the rural population is mostly white, the implications for racial inequality are evident.

Thirteen states have banned prison gerrymandering and will record each prisoner's pre-incarceration address in the 2030 census. The swiftest way to enact such a change nationally would be for the Census Bureau to change its residence rule and prohibit the use of a prisoner's temporary address—the prison—in the census.

———

Then there is *educational gerrymandering*, which usually takes the form of *school secession*. This happens when affluent, usually white communities carve themselves away from integrated school districts to form their own district. Such resegregation often comes with financial deprivation for the abandoned district because the seceding communities take their property taxes with them. For example, in 2014, six Memphis suburban towns separated from the consolidated Memphis school district, wreaking havoc on the remaining schools, which had to cut spending drastically and lay off teachers.

According to EdBuild, an organization dedicated to fair school funding, over seventy successful secessions were enacted in the period 2000–2017. In twenty states, there are no provisions that discourage secession. Only eight of the remaining thirty states pay attention to the social, economic, and racial impact of this practice. Most secessions happen in the South. Erika Wilson, a law professor at the University of North Carolina who has studied the problem, describes this as a "form of destructive localism. People who are fortunate enough to form utopias do so on the backs of other folks who have been excluded."

This derivation of gerrymandering feels particularly deviant. Maybe that's because it's directed at that most fundamental opportunity for equity and upward mobility—education. Plenty of examples show how education can be integrated with positive outcomes for all; Montgomery County, Maryland, and Cambridge, Massachusetts, are among them. Open access and equal housing opportunities are known to work and to promote economic growth for everyone. Gerrymandering does not.

CHAPTER 12

Math v. Gerrymandering

GERRYMANDERING UNDERMINES democracy—by design. It's also much easier to do than it is to detect and it's easier to suspect than it is to prove. Fortunately, as with other parts of our electoral infrastructure, mathematics offers an objective and clarifying viewpoint on a tangle of geographic, demographic, legal, and partisan forces.

The study of gerrymandering is one of the most exciting areas of applied mathematical work. The pace at which new methods are emerging is staggering, so this chapter will take us into the messiness of live research. I don't have the luxury of hindsight to give the orderly picture that I presented for voting theory and apportionment. Nothing is cut and dried. Instead, in this chapter, I show you a selection of methods that I believe best represents the direction of progress.* Before we unleash the math, however, we first need to examine the rules of the game.

AXIOMS OF DISTRICTING

Federal districting rules mandate that states must respect the 1965 Voting Rights Act, must create minority opportunity districts, and each district within a state must have the same number of people. Most states also try to preserve communities of interest. Some states also require that whenever possible, districts should not split towns, cities, or coun-

---

* With sincere apologies to partisan Gini score, declination, cut edges, I-cut-you-choose, moment of inertia, and many other methods.

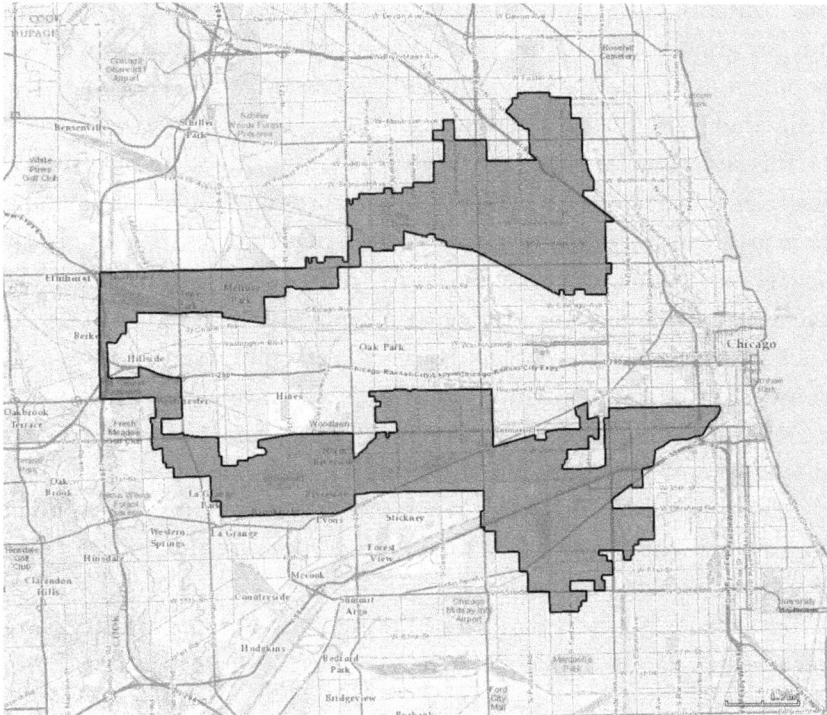

FIGURE 12.1.   Chicago's "earmuff" district.

ties. Fifteen states place this condition on congressional districts and thirty-four require it for state districts. These no-splitting rules are bendable, especially because equal-population or Voting Rights Act considerations must take precedence.

But what about mathematical attributes such as the shape of a district? Let's look at figure 12.1, the 4th Congressional District of Illinois, which is located in Chicago, dubbed "earmuffs" (tip your head sideways).

The tiny strip joining the two muffs runs along a highway. Nobody lives there. But the two big blobs of this district have something in common—they contain large Latino populations, so creating a district like this ensures that this constituency can elect a candidate to best represent their interests. The space inside the earmuffs is another district that contains a large Black population. Is this map gerrymandered? Yes, but for good reasons. These two districts give the Latino and Black populations a voice that they would not necessarily have if districts were defined in some gridlike manner.

This situation illustrates the difficulty of imposing strict rules on the shape of a district. The 4th district of Illinois is ugly, but it has a strong civic justification. Plopping down a bunch of rectangles (or triangles or hexagons or some other regular shape) onto a state map to please the eye and prevent gerrymandering with rigid uniformity would be silly. Such borders would ignore geography (rivers, mountains, coastlines), demographics, and population densities as well as communities of interest and opportunities for minorities to secure representation. Figure 12.2 shows four more greatest hits that drive home the challenge.*

Are these borders forced by geography? Do they gather like-minded communities or aim to dilute their voice? These are the million-voter questions with no easy answers.

Because it's not necessarily a bad thing if the borders meander and because there are many ways to meander, the rules have to be few. There are only two.

The first, and more nebulous, requirement is that districts should be *compact*. This is generally taken to mean that a district shouldn't have a strange shape. None of the districts we've looked at so far is especially compact. Shapes that are pleasing to the eye, like squares and circles, are.

Some thirty-two states have rules about reasonable compactness of their state districts and fifteen states require that their federal districts be compact. But what does that mean? California says that a compact district should not detour from a population in order to include another population farther away. In Arizona and Colorado, a district is more compact if its boundary doesn't "meander too much." In Michigan, Missouri, and Montana, a compact district is not "dispersed," meaning it shouldn't have tentacles emanating from some central blob. Idaho just

---

* In fact, there are so many wonky-looking districts that an entire font, called Ugly Gerry, has been created out of them. Visit fontsarena.com/ugly-gerry/ and use the font to write to your representative.

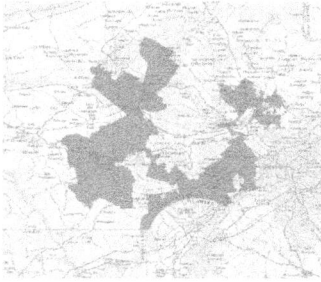

Goofy kicking Donald Duck:
Pennsylvania's 7th Congressional District.

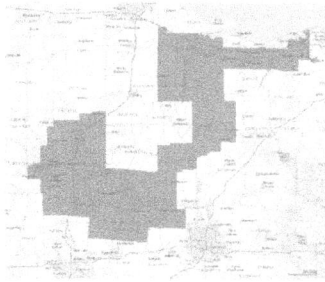

Duck:
Ohio's 4th Congressional District.

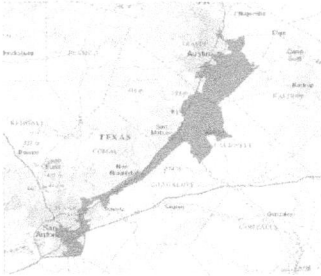

Upside-down elephant:
Texas's 35th Congressional District.

Praying mantis:
Maryland's 3rd Congressional District.

FIGURE 12.2. Some strange-looking congressional districts.

says districts that are "oddly shaped" are less compact. The mathematician's favorite is Iowa, which says compact districts "are those that are square, rectangular, or hexagonal in shape, and not irregularly shaped, to the extent permitted by natural or political boundaries." I can't even. . . . Why only those three shapes? What about pentagons and octagons? And somebody please tell them that a square is just a special case of a rectangle.

As you can see, there is no real definition of compactness to be found here, and consequently over thirty ways to measure it have been proposed. Compactness is one of the primary tests for detecting gerrymandering (and was in fact established as an instrument of prevention), but its indeterminacy is a terrible itch to a logic-prone mind. Having no definition to ground your theory is a mathematician's nightmare (but also a tantalizing challenge).

The second requirement is easier to state. It's also archaic and un-necessary. This is *contiguity*, a stipulation that a district should be con-nected. More formally, you should be able to travel from any point in the district to any other point in the district without leaving it. Some-times this requirement is artificially fulfilled, as with the strip of highway that connects the two earmuffs in Illinois District 4.

Most states follow the contiguity rule, even if not all of them require it formally. Forty-five states mandate it for local districts and eighteen mandate it for congressional districts. Exceptions exist where islands are adjoined to mainland districts, in which case a ferry might play the role of a connector. Bridges also count. Leaning the other way, Hawaii doesn't allow "canoe districts" where a canoe would be required to get from one part to another. There are many other special cases.*

The contiguity requirement was one of the earliest standards (it first appeared in the Apportionment Act of 1842), but it makes less sense these days. It was meant to preserve communities and avoid splintering them. In the meantime, the notion of community has changed. With increased mobility and ease of virtual communication, our lives do not necessarily happen where we live. In my slice of suburban America, many of my neighbors just sleep here. They work in Boston and prob-ably party there on the weekends too because, well, my town is pretty sedate. Should these people be made to vote for a candidate who repre-sents suburban interests (whatever those might be)? They have little in common with me and in all likelihood do not have the same vision of how our representatives should speak for us. It would likewise be best for me not to vote together with my neighbors but with my own demo-graphic of middle-aged, middle-class professionals who are increasingly worried about how they'll pay for their kids' college education. In many ways, that's my community of interest.

Canceling geography from the districting equation would likely bring together slices of the population that are truly politically alike. This is not a new idea and is in fact mandated in many countries. Armenians, Assyrians, Jews, and Zoroastrians have their seats in the Iranian Parlia-

* See, for example, Wisconsin's 47th or 61st state district.

ment. Italian and Hungarian populations have their seats in the Slovenian Parliament. More than thirty countries reserve such seats for minorities. Croatia reserves three seats for its diaspora. Hong Kong lets people of a certain profession, such as teachers, vote for one of their own to represent them.

However, in a place as diverse as the United States, identifying populations that would constitute voting districts independent of their physical location within a state might be a doomed proposition. I wouldn't recommend it. But there is another solution, namely creating *multimember districts*. Such districts would still be contiguous but would now be electing more than one candidate and would consequently be bigger. This means that a subset of the population in such a large district could rally behind one of the candidates and that subset would not have to consists of physically proximate voters. An entire state could even be a single district that elected all of its representatives in at-large elections, or it could be divided into a few large districts, each of which would elect some number of representatives. This is one of the topics of the next chapter, so I'll postpone the details, but suffice it to say that the benefits of multimember districts would be manifold. FairVote, an organization that advocates ranked voting, has modeled such districts in all fifty states to demonstrate the benefits for equitable representation all over the country. Third parties would get a shot at representation and noncompetitive "safe" districts would be eliminated. And, more pertinent for this chapter, gerrymandering would be significantly diminished even in multimember districts that elect as few as two or three candidates.

SYMMETRY AND BIAS

In the previous chapter, I repeatedly identified elections in which the proportion of votes a party garnered didn't correspond to the proportion of seats won. So, for example, in the 2012 elections in Pennsylvania, Democrats won 50% of the vote but only 28% of that state's congressional seats. Does that indicate foul play? Gerrymandering?

Pump the brakes.

Let's go back to Massachusetts. Its Republican voters, which constitute some 30% of the population, can never be represented in the House in the current system because they are spread across nine congressional districts with no majority in any of them. A group of researchers proved that a division of Massachusetts that would elect a Republican candidate to the House *cannot exist*.* So of the essentially infinitely many district maps that could be overlaid onto the map of Massachusetts, not a single one will provide representation for 30% of Massachusetts voters. Gerrymandering? No, it's just how things are.

And what about all the single-member states? One party will always take 100% of the seats without winning 100% of the votes, so there's no chance for proportionality.

Gerrymandering always produces a deviation from proportional representation—that's essentially its purpose—but deviations from proportionality are not always evidence of gerrymandering. They're often forced by unalterable circumstances.

Consider this more concrete example: a state has 3 million voters, 1 million of whom vote Orange; the rest vote Green. All of the Orange voters are concentrated in an urban area; Green voters are spread around the rest of the state. This state is to be divided into four districts, each with 750,000 people (roughly the size of a U.S. congressional district).

We can make one district in the urban area. This makes sense because it gathers a community of interest, it will be contiguous, and it will be reasonably compact. It might look small on the map, but that's because it's densely populated. The remaining 250,000 Orange voters will have to be dispersed among other districts because there's no room for them in this urban district. No matter how they're split up into the remaining three districts, they will never have the majority and Green will carry all three.†

---

* This is true even without imposing the contiguity requirement: no collection of Massachusetts voting precincts could be combined into a district that would contain majority Republican voters.

† This is an example along the lines of the *Merrill v. Milligan* case that went before the Supreme Court in the fall of 2022, except in that situation, two majority-Black districts could have been created in Alabama but were not.

So 33% of the state will vote Orange and 66% will vote Green, but Orange will win only 25% of the seats while Green will win 75%. Green will have 9% more seats—the difference between its seats share and its votes share—than it deserved. That is not proportional representation, right?

But there is no better way to divide this state. Try it. One option is to split Orange down the middle and place 500,000 voters into each of two separate districts. Because they would be the majority in both those districts, Orange would now win two seats. The remaining two districts, populated entirely by Green voters, would go to Green. In short, both Orange and Green will win 50% of the seats. This seems even less fair! Now Orange will get 17% more seats than it deserved.

You're seeing here something called the *seats bonus* (or *winner's bonus*), a common outcome in plurality, single-member district elections whereby the winning party gets a larger percentage of seats than the percentage of votes won. This makes sense because all it takes to win a seat is to win more votes; the margin of victory makes no difference. Just think of the extreme case in which a party wins 51% of the votes in every district, thus garnering 100% of the seats. That's quite a bonus! Every congressional election since 1946 has exhibited the seats bonus phenomenon to some extent.

The seats bonus doesn't care which party won; it's simply an artifact of the election method. Another way to think about this is in terms of switching roles—if two parties were reversed and the elections were otherwise fair, the other party should get the same bonus. The results of the election should be identical, only with the parties flipped. This should sound familiar—it's the neutrality condition we evaluated our voting systems against.

This observation leads to the concept called *partisan symmetry*, which allows us to distinguish between seats bonus and possible gerrymandering. It asks this question: If the vote share were to flip between two parties, would the seat share also flip? If it does, that's an indication that the seats bonus is at play. If it doesn't, it might be time to think about filing a lawsuit.

Let's look at another example. Suppose a state has 400,000 voters and four districts and the total statewide Orange-Green spread is 56%

Orange and 44% Green voters, so 12% overall in favor of Orange. By districts, the distribution is this:

District 1: 60,000 Green, 40,000 Orange
District 2: 40,000 Green, 60,000 Orange
District 3: 45,000 Green, 55,000 Orange
District 4: 30,000 Green, 70,000 Orange

Orange wins three of four seats, or 75% of them. This seems like a solid seats bonus, 19% more than the share of votes. Now let's flip 12% of the voters from Orange to Green in each district. This is called the *uniform partisan swing* because we're doing the same percentage flip in each district.* Each district has 100,000 voters, so 12% means a 12,000-voter flip, giving new vote distributions:

District 1: 72,000 Green, 28,000 Orange
District 2: 52,000 Green, 48,000 Orange
District 3: 57,000 Green, 43,000 Orange
District 4: 42,000 Green, 58,000 Orange

Now Green wins in three districts and Orange wins in one. The roles have switched exactly. According to the partisan symmetry test, all is good.

But now suppose the same population is divided among the four districts like this:

District 1: 85,000 Green, 15,000 Orange
District 2: 35,000 Green, 65,000 Orange
District 3: 35,000 Green, 65,000 Orange
District 4: 20,000 Green, 80,000 Orange

Districts still have 100,000 voters each, the statewide vote spread is still 56%–44% in favor of Orange, and Orange still wins three seats. Let's flip 12%, or 12,000, of the voters from Orange to Green in each district:

---

* This is a useful simplifying assumption, but not a realistic one. There are more sophisticated ways to model the districtwise vote swing.

District 1: 97,000 Green, 3,000 Orange
District 2: 47,000 Green, 53,000 Orange
District 3: 47,000 Green, 53,000 Orange
District 4: 32,000 Green, 68,000 Orange

Orange still has three of four seats! Something is surely afoot; the partisan symmetry test tells us so. Indeed, what might be happening is that the Green voters have been packed into District 1 and cracked across Districts 2, 3, and 4.

There is a souped-up version of partisan symmetry that goes like this: instead of flipping the ratio of the votes and checking what happens with the seats won, what if we make the change gradually, percentage point by percentage point, and watch how that affects the outcome? So if we start with the actual election results in the previous example—56% Orange and 44% Green—we don't immediately go to 44% Orange and 56% Green but do it incrementally: 55% Orange, 45% Green; 54% Orange, 46% Green, and so on. At each step, we check the seat spread and we graph the results. What we get is the *seats-votes curve*. Each point on the graph gives the share of seats a party would receive if they won a particular share of votes.

This curve can be generated for state or national elections. For example, figure 12.3 shows the seats-votes curve for the Democrats in the 2018 House elections.

Plotting graphs like this does two (mathematical) things for us. First, we can think about the effect of the districting in general, not just for the results of one or two elections. And second, we can think about what characteristics the curves should have if the districts are fair and we can develop measures that tell us how close or far from fair they are.

The dot that represents "Actual 2018 outcome" on figure 12.3 is what happened: Democrats won 55% of the votes and received 54% of the House seats. Not too bad at first glance. But looking at just one point doesn't tell the whole story of the districting. (Would you buy a car that had been tested at only one speed?) Maybe the election results landed by chance on a point where the vote share produced approximately the same seats share. The curve also tells us what would have happened if the Democrats had won, say, 48% of the votes: that would have earned

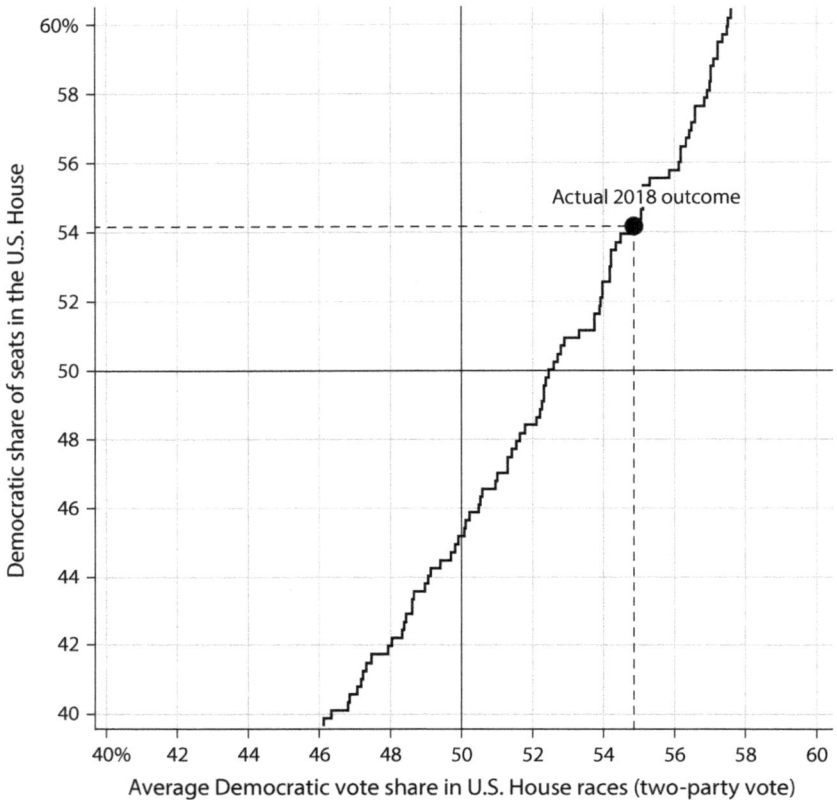

FIGURE 12.3. Democrat seats-votes curve for the 2018 U.S. House elections.

them about 42% of the seats because the point $(48, 42)$ lies on the curve. What if we look at other points? The discrepancy between the percentage of votes won and seats received varies depending on the location. What's the right place to look for a measure of gerrymandering?

The convention is to look at the 50% vote mark. If the Democrats had won half the votes, what share of the seats would they have received? The answer in this example is about 45%. The difference $(45\% - 50\% = -5\%$ in this example$)$ is called the *partisan bias* (figure 12.4).*

---

* Partisan bias is an example of a *symmetry score*; other types of symmetry scores can be extracted from a seats-votes curve; see chapter 2 of *Political Geometry*, edited by Moon Duchin and Olivia Walch.

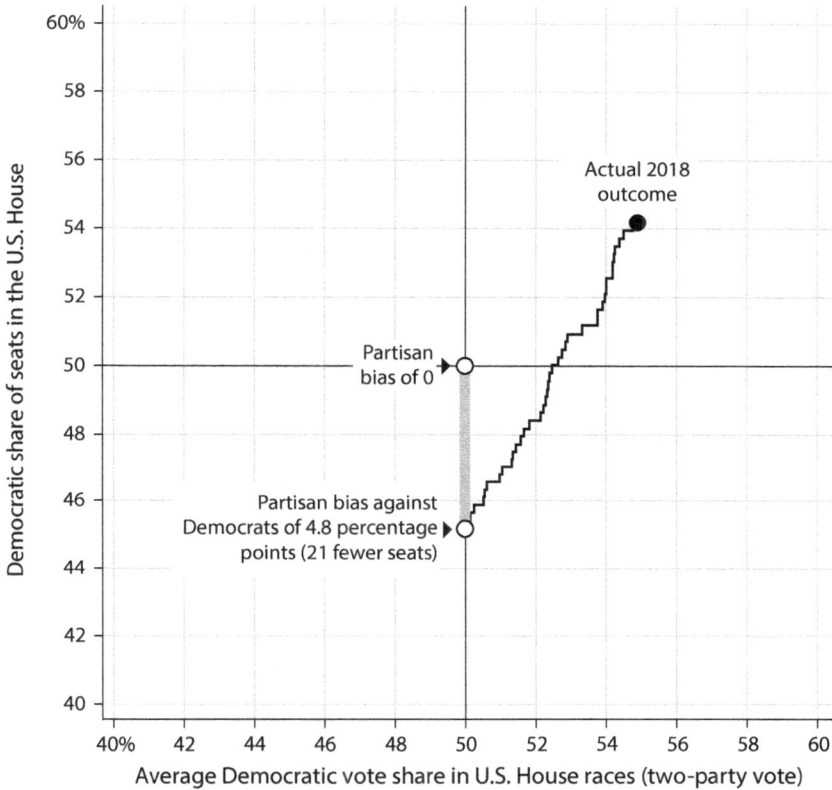

FIGURE 12.4. Partisan bias against the Democrats in the 2018 House elections.

This is the bias *against* Democrats because the number is negative; they receive fewer seats than they should for winning 50% of the votes. So the bias is the difference between seats received and the number of seats that *should* be received, expressed as the percentage of the total number of seats in the hypothetical situation when each party wins exactly half the votes.

If we were to plot the Republican seats-votes curve, we would see that at 50% of seats won, the curve would pass through the point $(50, 55)$, a place that's the vertical reflection, across the center point $(50, 50)$, of where the Democrats' curve passes the 50% vote mark. The partisan bias would now be 55% − 50% = 5%. Because this is a positive number, the bias would be in the Republicans' favor; they would receive more seats than they "deserve."

The entire Republican curve would be a flip of the Democrat curve across that center point: take a point on the Democrat curve, draw a line going through it and the center, and then mark the place on the line that's the same distance to the center as the original point. Doing this for all the points on the Democrat curve generates the Republican curve.*

The ideal districting has a curve that is a straight line going diagonally upward from left to right at 45 degrees through the center point (50, 50). On this line, no matter what share of votes the Republicans win, they'd receive the same proportion of seats. And the Democrat curve would look exactly the same. This is perfect proportional representation (figure 12.5).

In real life, things aren't ideal (shocker!). Consider the curves for the 2006 Texas congressional elections in figure 12.6. The 11.93% partisan bias means that at 50% of votes won, both curves are 11.93% away from 50% seats received, one in deficit and the other in surplus.

Figure 12.6 also reports two other numbers, *symmetry* and *responsiveness*. For symmetry, look at the difference for each percentage of the votes on the x axis. This basically means that we would draw lots of vertical line segments connecting the two curves and measure them, with plus or minus signs indicating whether one curve is on top of the other or vice versa. The average of all these differences, or the lengths of the segments counted with positive or negative signs, is *partisan symmetry*. If the districting map treats both parties the same, then this quantity will be zero, as in the case of the two lines coinciding in figure 12.5. If symmetry is to be achieved, then the partisan bias must be zero: each party will receive precisely half the seats when it wins half the votes.

Responsiveness is the measure of how readily gains in votes translate into gains in seats. If a curve has a horizontal, flat section, responsive-

---

* Put precisely, if (x, y) is a point on one curve, then $(100 - x, 100 - y)$ is a point on the other curve. For example, if $(47, 51)$ is a point on the Democrats' curve, that means that when the Republicans win $100 - 47 = 53\%$ of the vote, they receive $100 - 51 = 49\%$ of the seats, so a point on their curve is $(53, 49)$. All of this happens under the simplifying assumption that only two parties are running.

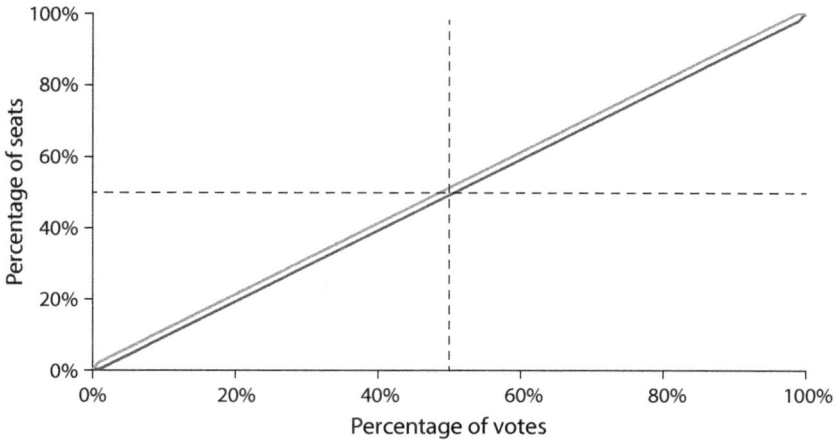

FIGURE 12.5. A symmetric seats-votes curve.

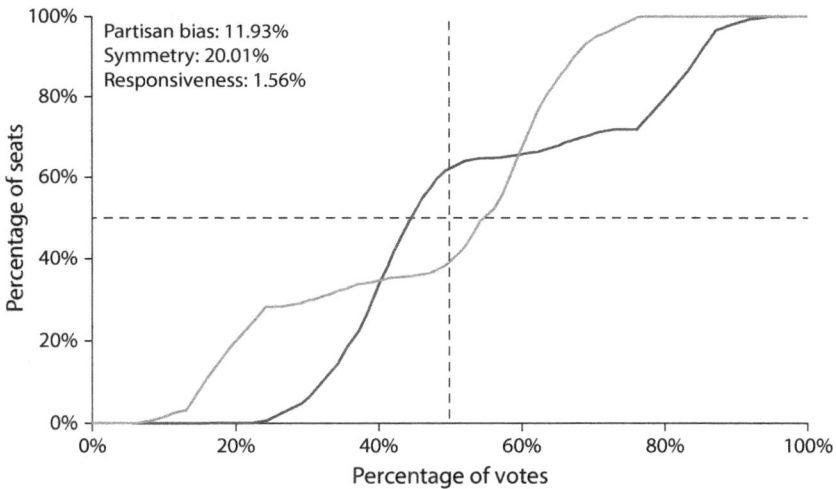

Partisan bias: 11.93%
Symmetry: 20.01%
Responsiveness: 1.56%

FIGURE 12.6. Seats-votes curve for the 2006 U.S. House elections in Texas.

ness over that interval is low because winning more votes isn't producing more seats.* For example, in 2016, Ohio's Republicans received 12
House seats to the Democrats' 4, but that would have been the result if
the Democrats had won anywhere from 21% to 51% of the votes. A simi-

---

* Technically, responsiveness is the average *derivative* (the slope of the tangent line) of the
seats-votes curve over some range of values, usually 45% to 55%.

lar thing happened for the 2012 Pennsylvania congressional election, where Republicans won 13 of 18 seats: that would have been the result if the Democrats had won anywhere from 40% to 60% of the vote. The seats-votes curves are flat for the vote ranges for those two elections. High responsiveness is desirable because it indicates competitiveness: small changes in the distribution of votes are reflected in changes in the number of seats each party receives. Put bluntly, votes matter.

Even when partisan symmetry and partisan bias appear to have detected foul play by being nonzero, the analysis is not conclusive without considering other factors such as geography and demographics.* Look back at the second version of the Orange-Green example, the one that suggests the Green vote might have been cracked across Districts 2, 3, and 4. How do we know this isn't just demographic self-sorting, namely that the Greens didn't pack themselves into District 1 and the rest of that population is scattered across the other three districts? We don't. More exploration is needed.

CRACKING DOWN ON PACKING

In their 2015 article "Partisan Gerrymandering and the Efficiency Gap," law professor Nicholas Stephanopoulos and political scientist Eric Mc-Ghee introduced a way to quantify how "efficiently" the winner of an election won by measuring "wasted votes," the votes that did not contribute decisively to the victory. Any votes cast for the winner over the requisite 50% are wasted because the winner didn't need them. All the votes cast for the loser are also wasted because that candidate wasn't elected.[†]

Here's the math: look at each of the districts and figure out the number of wasted votes for both parties. Add up the wasted votes for each party over all the districts. Then subtract the smaller number from the

---

*There are also mathematical problems with symmetry scores; see, for example, the paper "Implementing Partisan Symmetry: Problems and Paradoxes" by Daryl DeFord and coauthors.

[†] We're as usual assuming there are two candidates from two parties in the race, although everything extends to more candidates.

bigger number and divide the result by the total number of voters. This is the *efficiency gap*, the difference in total wasted votes between the two parties expressed as a fraction of total votes cast.

The idea behind this definition is that a gerrymandering party will produce a lot of wasted votes for the other camp. If Orange packs Green into a few districts, then Green will win big in those districts, but the efficiency gap flags winning big as overkill and identifies lots of Green wasted votes. If Orange cracks Green across a bunch of districts where Green can't win, then all those Green votes will also be recorded as wasted.

Let's look at an example:

	District 1	District 2	District 3	District 4
Orange Party	32,000	30,000	28,000	10,000
Green Party	18,000	20,000	22,000	40,000

In each of the four districts, there are 50,000 voters, so their total number is 200,000. To win in any of the districts, 25,000 votes are required (well, that plus one, but let's ignore the extra vote so the numbers will shake out nicely). In this example Orange won in District 1. Because they have 32,000 − 25,000 = 7,000 votes they didn't need, that is their wasted vote count in that district. Because Green lost, all of their 18,000 votes in that district are wasted. Similar calculations work in Districts 2 and 3. District 4 is different because Green won there, so their waste is 40,000 − 25,000 = 15,000 votes and all 10,000 cast for Orange are wasted. The totals are as follows:

Wasted Orange votes: 7,000 + 5,000 + 3,000 + 10,000 = 25,000

Wasted Green votes: 18,000 + 20,000 + 22,000 + 15,000 = 75,000

To calculate the efficiency gap (EG), subtract the smaller number from the larger number and divide the result by the total number of voters:

$$EG = (\text{Wasted Green} - \text{Wasted Orange}) \div \text{Voters} =$$
$$(75,000 - 25,000) \div 200,000 = 50,000 \div 200,000 = 0.25.$$

This number can also be expressed as 25%, interpreted as the percentage of total votes the Green Party wasted beyond what the Orange Party wasted.

Calculating the efficiency gap produces a number, but what does it tell us? Is a 25% efficiency gap worrisome or unsuspicious? Researchers stipulate, based on analysis of prior elections, that an efficiency gap greater than 7% is problematic and points to foul play.* Our efficiency gap of 25% thus flags a problem, warranting an investigation of potential Orange gerrymandering.†

Table 12.1 shows actual numbers from the 2016 elections in Maryland. The total waste for Democrats is 510,000; for Republicans, it is 789,000. The difference in wasted votes is 279,000. The total number of voters is 1,636,0 00 + 962,000 = 2,598,000. That's all we need to compute the efficiency gap:

$$EG = (789,000 - 510,000) \div 2,598,000 = 279,000 \div 2,598,000 = 0.107.$$

The efficiency gap is thus 10.7% in favor of Democrats and against Republicans. This is no surprise because we already know Maryland is heavily gerrymandered by the Democrats.

After the 2020 census and subsequent redistricting, all the maps but one drawn by independent commissions have had efficiency gaps of 5% or less. In contrast, the gap is at least 13% for Democrat-drawn maps and at least 7% for Republican-drawn maps (in favor of the party that drew them, of course). Assuming that independent commissions draw fairer maps (and don't just try to rig them so the efficiency gap comes out small), this data lends the efficiency gap some credibility.

But here comes the requisite downer: the efficiency gap's simplicity is also its doom. We know by now that there's no way such a simple concept and a single number can capture all the complexities of districting. As with partisan bias, other factors might be at play that the measure can't account for.

* This number is based on the calculation of the gap for various elections and the observation that the outliers had an efficiency gap at least this large.

† The efficiency gap is rarely this high in practice. I say that with a straight face as I write this in a state with a gap of 24% in favor of Democrats. Check out your state's EG at FiveThirtyEight's "The Atlas of Redistricting" page at https://projects.fivethirtyeight.com/redistricting-maps/.

TABLE 12.1. Wasted votes in the 2016 congressional elections in Maryland

| District | Dem. votes | Rep. votes | Votes to win | Wasted Votes | |
				For Dem.	For Rep.
01	104,000	243,000	173,000	104,000	69,000
02	192,000	103,000	147,000	45,000	103,000
03	215,000	115,000	165,000	50,000	115,000
04	238,000	69,000	153,000	84,000	69,000
05	243,000	106,000	174,000	69,000	106,000
06	186,000	133,000	159,000	26,000	133,000
07	239,000	70,000	154,000	85,000	70,000
08	221,000	125,000	173,000	48,000	125,000
Total	1,636,000	962,000	1,299,000	510,000	789,000
				Net: 279,000 Rep.	

Note: All numbers are rounded to the nearest thousand. The Republican net waste is 279,000 votes greater than that of the Democrats.

One deficiency of the efficiency gap is that it does not adjust for natural voter packing and self-sorting. Look back at District 4 in the Orange-Green example. This district composition could just be natural packing of the Green voters into an urban area. Green winning there by a landslide is not evidence of a wicked packing-cracking plan by the Oranges. And maybe Orange won the other three districts because those are rural areas where Oranges are more likely to live than Greens.

The efficiency gap also doesn't like competitive races. Suppose there is a district with 100,000 voters and Orange wins with 50,001 votes (now I temporarily care about the +1) and Green loses with 49,999. Orange has wasted 0 votes and Green has wasted 49,999. Those wasted vote numbers could contribute to the calculation of the statewide efficiency gap in a way that points an accusatory finger at the Oranges. But did Orange gerrymander the district? Definitely not. Nobody would gerrymander in a way that produces a competitive race. And this race was *really* competitive—if only two voters switch their votes from Orange to Green, Green would have won and the wasted vote tally for this district would have flipped. In mathematics, this kind of volatility is bad. If tweaking the input a tiny bit results in huge changes in the output, we're violating the principle of *continuity*, and things that are discontinuous are notoriously unreliable and hard to work with.*

The efficiency gap has some pleasant mathematical properties that make it fun to play with. But as neat as they are, they lead to further erosion of confidence. For example, the total number of wasted votes in each district is always half the voter turnout. Look at the Orange Party and Green Party table above again. In the first district, the total waste was 7,000 + 18,000 = 25,000. In the second district, it was 5,000 + 20,000 = 25,000. Do the math for the other two districts and you'll get 25,000 both times, half the number of voters. Take your calculator to the 2016 Maryland numbers in table 12.1 to see this in a real-life example.

This means that the efficiency gap is not a district-by-district measure. It is a global quantity. It's not designed to be a local quantity because in each district it's a predictable number (half the votes) regardless of whether the district was gerrymandered. The efficiency gap is designed to say something in the aggregate, and the more districts there are, the better it is at that. Judicially, this property of the efficiency gap is problematic because many gerrymandering cases hinge on claims that a particular district or a small subset of districts exhibits extreme partisan or racial gerrymandering. Those cases rely on proving that an individual's right to equal representation has been violated. The argument for that claim cannot be that the vote of a district or a subset of districts has been wasted because that's nothing special; it's always true for half the voters.

One more thing. Consider this scenario: the Orange Party cracks Green voters in a way that gives Orange a landslide win of 75%–25% in every district. So in each district, Orange has 25% wasted votes (the amount over 50% which they didn't need) while Green also has 25% (because they lost, all their votes are wasted). This happens in every district. Totaling across the districts, we find that Orange and Green have the same percentage of waste, namely 25% of all the votes. Because

* Calculus, for example, really only works for things that change continuously without erratically jumping around. Luckily, few processes are naturally discontinuous.

the efficiency gap subtracts one from the other, the final result is zero. No gerrymandering! Well, that's not true because our design used some big-time cracking. In other words, the efficiency gap likes the 75%–25% landslide and will not flag it as potential gerrymandering.

It will, however, flag proportional representation! Check out this funky example:

	District 1	District 2	District 3	District 4	District 5
Orange Party	15,000	15,000	14,000	8,000	8,000
Green Party	5,000	5,000	6,000	12,000	12,000

The Orange Party wins in three of five districts, or 60% of them. They also win

$$15{,}000 + 15{,}000 + 14{,}000 + 8{,}000 + 8{,}000 = 60{,}000 \text{ votes.}$$

The total number of votes is 100,000, so that means the Orange Party won 60% of the votes. We have perfect proportional representation—each party won the same percentage of seats as the percentage of votes.

But let's look at the efficiency gap. Here are the wasted votes:

Wasted Orange votes: $5{,}000 + 5{,}000 + 4{,}000 + 8{,}000 + 8{,}000 = 30{,}000$

Wasted Green votes: $5{,}000 + 5{,}000 + 6{,}000 + 2{,}000 + 2{,}000 = 20{,}000$

The efficiency gap is

$$EG = (30{,}000 - 20{,}000) \div 100{,}000 = 10{,}000 \div 100{,}000 = 0.1.$$

We just measured a 10% efficiency gap *in favor of the Green Party*. In other words, the efficiency gap thinks that the election result shows gerrymandering *against* the Orange Party, even though we have perfect proportional representation.*

* This and the previous issue can be recast as consequences of an alternative description of the efficiency gap: if we assume equal voter turnout in all districts (there are appropriate modifications for an unequal turnout), the efficiency gap (EG) for a party is a function of its vote margin (VM; percentage of votes won over 50%) and the seat margin (SM; percentage of seats won over 50%): EG = SM − 2 × VM. As explained in the paper "A Formula Goes to Court: Partisan Gerrymandering and the Efficiency Gap," by Mira Bernstein and Moon Duchin, one

In their paper "The Measure of a Metric: The Debate over Quantifying Partisan Gerrymandering," Nicholas Stephanopoulos and Eric McGhee address some of the issues various authors have raised in the flurry of research and publishing that followed the inception of the efficiency gap. In particular, they explain the efficiency gap's incompatibility with proportional representation by saying that honest proportional representation is not realistic because of the seats bonus and that the efficiency gap captures this practical experience.

There is one aspect of voting that the efficiency gap—and any other measure described here, for that matter—fails to acknowledge or take into account: those who *didn't* vote. For example, uncontested races can throw the efficiency gap off because had there been another candidate, some voters from the other side would have shown up and cast their ballots. Then when their candidate lost, those votes would have been added to the waste, which would have impacted the final calculation.

Another concern is voter suppression and the way it can mask gerrymandering. In the aftermath of the 2020 presidential elections, scores of "voter integrity" laws, which relied on baseless claims of voter fraud, were proposed or passed in a number of states. In 2021, at least nineteen states passed thirty-four laws imposing stricter conditions for voting. If a party that gerrymanders also creates election conditions that suppress the likely vote for the opponent, the efficiency gap will not paint the full picture. The lower turnout among gerrymandered voters will reduce the measured efficiency gap. It would have taken about 130,000 fewer Democratic voters in Wisconsin in 2012 for the efficiency gap to drop below the red-flag cutoff value of 7%. That seems like a lot of votes to suppress, but according to one estimate, 300,000 Wisconsin residents lack photo ID and a large percentage of them would be deterred from voting if having such an ID were mandatory. In states where the

---

can deduce from this formula, for instance, that the efficiency gap and proportional representation are incompatible—as evidenced by the example above—because they both say something about the seats-votes curve, but what they say is irreconcilable.

efficiency gap is not much higher than 7%, it would take even fewer no-show voters to bring it below this threshold and obscure gerrymandering.

The efficiency gap is, well, efficient. It's a tidy single number that seems to be capturing something. Mathematicians agree that it's . . . okay. Not great, because we think it's oversimple and flawed, but if the stars align and the right circumstances are present in the particular state and the particular districting, it can be useful, as it was in the *Gill v. Whitford* case. When it is used, it should be in conjunction with other methods and never as the sole barometer—and never ever as the *definition*—of gerrymandering.

THE GEOMETRY OF DISTRICTING

So far we've been doing the equivalent of blood testing the body politic, extracting some numbers from the district voting data. Now we're going to start the imaging, looking at the *shape* of the districts. According to the time-honored and obviously highly scientific eyeball test, if the shape of a district looks weird, it might be time to consider filing a lawsuit. But how do we make "weird" rigorous and quantifiable? Legislation points us to the criterion of compactness, but this concept is too vague and nonprescriptive. And it must be vague and nonprescriptive because a lot of funkiness can occur with borders for legitimate reasons. As we saw with the Illinois earmuff district, a peculiar shape might be perfectly justified, not only because of geography but also because of various legal requirements.

So we don't want to be like Iowa and name our three favorite shapes and declare them to be the ultimate list of compact things.* Instead, to catch a gerrymanderer we need to think like one. A better approach is to recognize that deliberate, granular gerrymandering usually produces

---

* To be fair, Iowa law does say more and authorizes two tests for evaluating compactness: *length-width compactness* and *perimeter compactness*. Neither is a good measure of gerrymandering, however.

a meandering border that has limbs or blobs connected by thin areas. The conjecture is that partisan mapmakers carefully surround their desired voters as they move down blocks, streets, neighborhoods, roads, and farms. A snaking border makes for a lot of perimeter, while the thin parts result in low area.

The central idea in compactness-based arguments that try to uncover gerrymandering is to *compare* the two quantities—perimeter and area. We've already seen some districts that are known to have been unscrupulously gerrymandered in figure 12.2. You'll notice they have a great deal of perimeter and not much area by comparison.

You might say borders like these could occur naturally; maybe rivers and mountain ranges are forcing the twists and turns. And you'd be right—sometimes. When that happens, compactness measures give a false positive for gerrymandering. But geography does not fully explain winding borders in many cases, as can be seen by historical districting progressions. For example, figure 12.7 shows the evolution of Maryland's 3rd Congressional District, called a "praying mantis" (or, as one judge described it, a "broken-winged pterodactyl lying prostrate across the center of the state").* I doubt that mountains rose and rivers changed course enough over the past sixty years to explain these transformations.†

Let's add some math to this eyeballing exercise. Comparing the perimeter to the area of a district starts to capture what we mean by a weird shape. For example, a district that looks like a rectangle 1 mile wide and 49 miles long would make us wonder why it is so thin and long. To quantify this insight, we calculate its perimeter as $1 + 49 + 1 + 49 = 100$ miles and area as $1 \times 49 = 49$ square miles. On the other hand, a square district whose sides are 25 miles long would not arouse suspicion. Everyone would agree that the square looks compact. This square also has a perimeter of 100 miles, but its area is $25 \times 25 = 625$ square miles—a much greater area for the same perimeter. A way to capture the variance is to

---

* This impressive creature isn't around anymore; Maryland's 3rd was redrawn in 2021, into a lumpy rectangle.

† If you're interested in how your own district has changed over time, check out the paper "A Two Hundred-Year Statistical History of the Gerrymander," by S. Ansolabehere and M. Palmer.

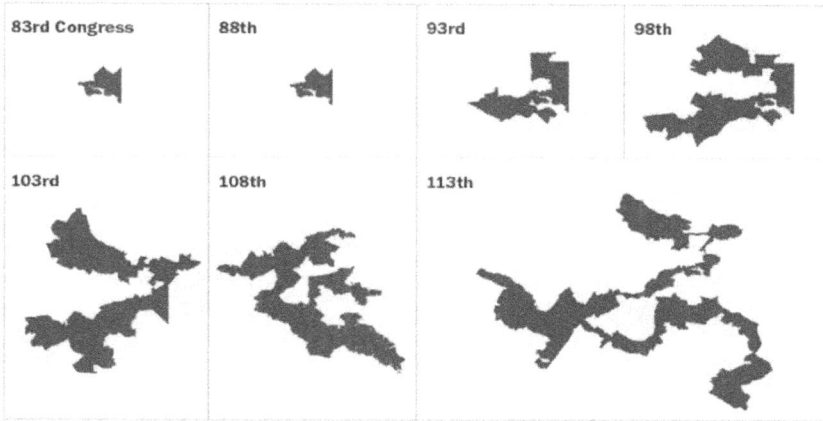

FIGURE 12.7. Evolution of Maryland's 3rd district, 1953–2013.

note that the first district has a smaller area-to-perimeter ratio than the second.

One caveat. Taking the ratio of area to perimeter is not quite what we want. Suppose we're looking at two square districts but one is smaller than the other; say one has sides 10 miles long and one has sides 12 miles long. The first district's perimeter is 40 miles and its area is 100 square miles, so its ratio of area to perimeter is 100 ÷ 40 = 2.5. The second district has a perimeter of 48 miles and an area of 144 square miles, giving a ratio of 144 ÷ 48 = 3.0.

But both of these squares are nice and compact. Who cares that one happens to be bigger than the other? We wouldn't flag either for gerrymandering, so they should be assigned the same numerical attribute. It might be that one sits in a thickly populated urban area and the other in a more rural part of a state. In order for them to include the same number of people, one naturally has to be smaller than the other.

Put mathematically, we want a quantity that is *invariant when scaled*. This means that if we shrink or expand a district—if we put it into a giant copy machine and hit a number like 80% or 120%—our measure of compactness should come out the same.

One way to accomplish this is to take the perimeter and *square* it, then take the ratio.* If we denote the perimeter of a district by p and its area by a, what we're homing in on is the quantity

$$a \div p^2.$$

Calculating this quantity for the two squares we just talked about gives 0.0625 in both cases, so we're getting somewhere.

The expression above is well known in mathematics. It plays the lead role in a famous theorem that places restrictions on how the area and the perimeter of a region in the plane can be related. This is the prettiest part of this story as far as I'm concerned, so bear with me.

Isoperimetric inequality theorem: If R is a region in the plane that has area a and is bounded by the closed curve of length p, then

$$(4 \times \pi \times a) \div p^2 \leq 1.$$

The theorem is attributed to the Swiss mathematician Jakob Steiner, who worked on it in the late 1830s. His proof wasn't quite complete and was finished and extended by many mathematicians. In its contemporary form, the result says something about shapes in spaces of any dimension, not just in the two-dimensional plane.

The inequality can be rewritten as $a \div p^2 \leq 1 \div (4 \times \pi)$ (divide both sides by $4 \times \pi$), so a good way to think about it is that it gives a limit, or a ceiling (an *upper bound*, in mathematical jargon) on how big the ratio $a \div p^2$ can be: $1 \div (4 \times \pi)$. In other words, no matter what kind of shape you draw on a piece of paper, if its boundary has no breaks, then its area divided by its perimeter squared will be less than $1 \div (4 \times \pi)$.

Let's try something easy: suppose R is a square with sides of length s, where s could be any positive number. Then the area a is s^2 and the perimeter p is $4 \times s$, so $p^2 = 16 \times s^2$. Plugging these values into the left side of the inequality gives

$$(4 \times \pi \times a) \div p^2 = (4 \times \pi \times s^2) \div (16 \times s^2) = (4 \times \pi) \div 16 = \pi \div 4.$$

---

* From the unit-canceling point of view, squaring the perimeter makes sense because then the units on the top and the bottom of the ratio will both be miles squared. What's left after cancellation is a number with no units, and that's exactly what we want.

Because π is roughly 3.14 and we're dividing it by 4, a larger number, π ÷ 4 must be less than 1 (it's about 0.785),* which is what the isoperimetric inequality theorem predicts. So we've verified the theorem in one special case. The left side came out to be π ÷ 4, which makes no mention of the side length s. This is exactly what we want—the quantity $a ÷ p^2$ does not depend on s; it is the same for a square of any size. And because any square can be obtained from any other square by scaling, $a ÷ p^2$ is invariant under square-scaling. This will be true for any shape we can dream up.

Let's do one more. Suppose R is the region inside the circle of radius r (again, r can be any positive number). Because the area of the circle is $π × r^2$ and its perimeter is the circumference $2 × π × r$,

$$(4 × π × a) ÷ p^2 = (4 × π × π × r^2) ÷ (2 × π × r)^2$$
$$= (4 × π^2 × r^2) ÷ (4 × π^2 × r^2) = 1$$

(numerator and denominator are the same so they cancel out).

We just discovered that the circle achieves the equality part of the isoperimetric inequality theorem. The theorem says that no matter what the shape is, the quantity $(4 × π × a) ÷ p^2$ is going to be less than *or equal to* 1 and the circle is a case where it equals 1. As it turns out, the circle is *the only* shape for which that is true. (This requires more thinking.) Again observe that it didn't matter how big the circle was because r (the radius) got canceled out in the calculation; any circle will give the same answer.

This calculation shows that the circle is the most "efficient" shape in the sense that it encloses the most area for a given perimeter. In other words, if I give you a loop of string and ask you to lay it on the ground so it encloses the most possible area, you'd want to make a circle. If you make any other shape, the number $(4 × π × a) ÷ p^2$ will always be less than 1. That means that you will have less area for the given perimeter than you will with the circle.

Applying this idea to the situation where the regions R are voting districts gives the *Polsby-Popper (compactness) score*, named after two

---

* This is consistent with what we found for the two squares discussed above because this number is precisely 0.0625 multiplied by $4 × π$.

political scientists who wrote about it in their 1991 paper on compactness.* To obtain a district R's Polsby-Popper score, which I'll call $PP(R)$, we measure its perimeter p and area $a^\dagger$ and compute

$$PP(R) = (4 \times \pi \times a) \div p^2.$$

If the number is close to 1, the perimeter is intuitively "taut," not too big relative to the area it encloses. If the number is closer to 0, the district has lots of perimeter for its area, and that suggests premeditated meandering that is indicative of gerrymandering.

According to Polsby-Popper, Maryland's 3rd was the most gerrymandered congressional district in the country before it was redrawn; its PP score was only about 0.04. The national average across all congressional districts is 0.18; the median is 0.26. The district I live in, Massachusetts' 5th, has a Polsby-Popper score of 0.13. The best in the country is 0.77 for the state of Wyoming, but that doesn't count because Wyoming is a single-district state and doesn't have the opportunity to gerrymander. Among states with more than one district, the best Polsby-Popper score is 0.66. Half of U.S. congressional districts have PP scores of 0.19 to 0.36.

By the way, here's another plug for independent districting commissions: Canada has a median PP score of 0.42 with a low score of 0.13 and half of its districts falling in the range of 0.33 to 0.52. Those scores are better than U.S. scores. This good performance can be attributed to their 1964 federal mandate of independent redistricting commissions. In 2018, Canadian prime minister Justin Trudeau, speaking about U.S. districts, said: "We get actual, reasonable-looking electoral districts, and not some of the zigzags that you guys have." That's snarky, but I'm sure one of the signature Trudeau smiles was immediately deployed to diffuse the jab.

The Polsby-Popper score can be visualized as the ratio of the area of the district R to the area of the circle whose circumference is the same as the perimeter of R. See figure 12.8, where the district is the Ohio 4th

---

* The fact that this number carries the name of two people who didn't really play a role in its development is not uncommon. I imagine that few people who have heard of (or even use) the Polsby-Popper score could name any of the mathematicians behind it.

† These quantities are available through the United States Census Bureau.

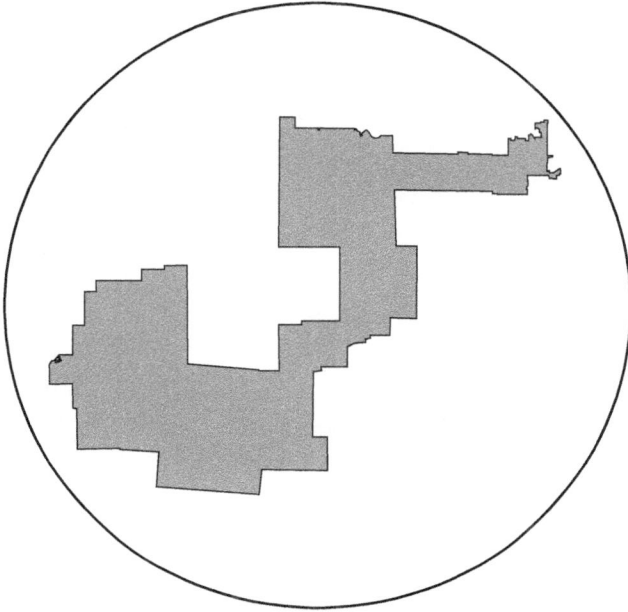

FIGURE 12.8. Polsby-Popper: the circumference of the circle is the same as the perimeter of the district.

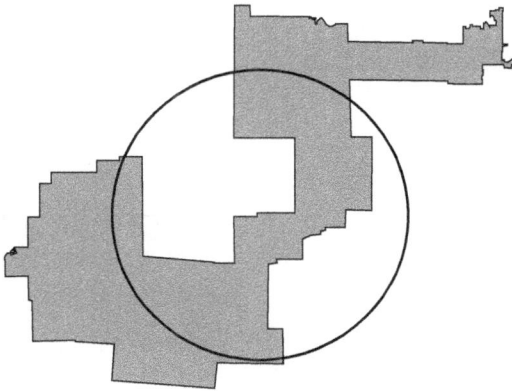

FIGURE 12.9. Schwartzberg: the areas of the circle and the district are the same.

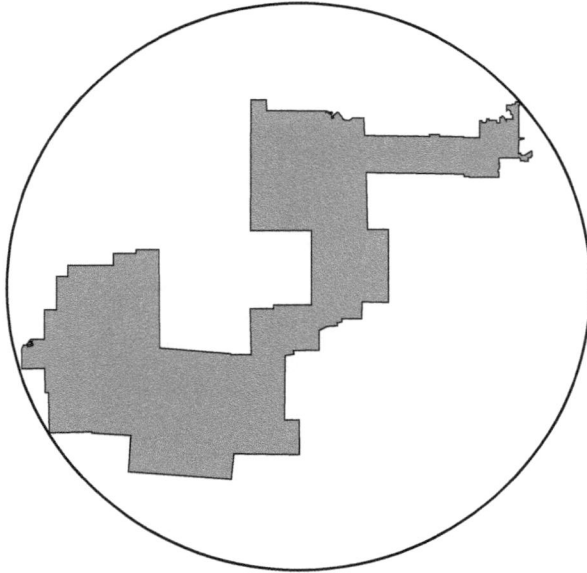

FIGURE 12.10. Reock: the district is inscribed inside a circle.

duck (2013–2023). The big discrepancy between the two areas indicates the district might have been gerrymandered.

Related compactness measures also use a comparison with a circle. For example, the *Schwartzberg score* takes the circle whose area is the same as the area of the district and compares the two perimeters by taking their ratio (figure 12.9). Polsby-Popper and Schwartzberg can be expressed in terms of each other, so they score compactness in the same way and can be used interchangeably.

The *Reock score* fits the district into the smallest possible circle and then takes the ratio of the area of the district to the area of the circle (figure 12.10). The circle will always have greater area than the district it contains, but a more compact district will fill out more of the circle, so this ratio will be greater (but always less than 1). Thus, a Reock score that is closer to 1 indicates better compactness. A low Reock score indicates *dispersion*, or the existence of branches or thin tentacles that spread out far and wide. Michigan mandates the Reock score as a test of compactness.

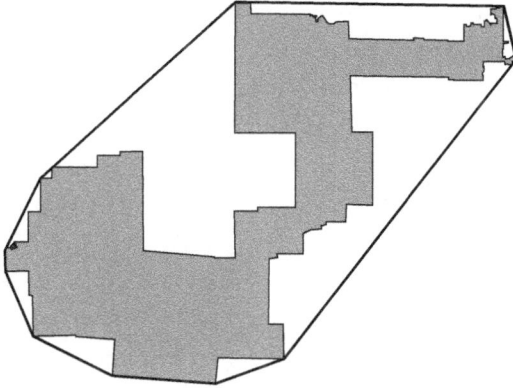

FIGURE 12.11.   Convex hull: the district is inscribed inside the smallest convex polygon.

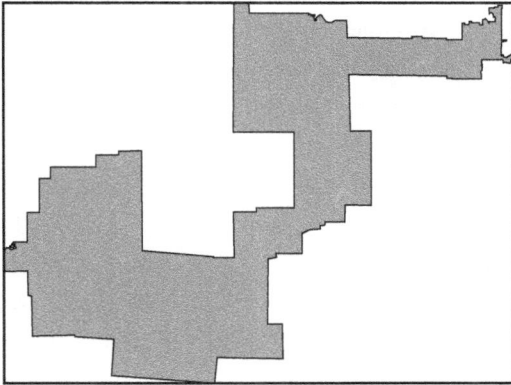

FIGURE 12.12.   Length-width: the district is inscribed inside a rectangle.

The *convex hull score* does something similar, but it inscribes the district in the minimum convex polygon enclosing the district and then divides the areas of the two shapes (figure 12.11). (A shape is *convex* if the line segment between any two points inside the shape also stays inside the shape.) A variant of this measure looks at populations instead of areas. This *population polygon* measure divides the population living in the district to the population living in its convex hull.

The *length-width score* inscribes the district in a rectangle and then compares the ratio of its length to its width (figure 12.12). If the ratio is big, that means the district is long and thin, which might point to gerrymandering. A smaller ratio indicates that a district is closer to fitting into a square, which can be interpreted as more desirable. Iowa and several other states use the length-width test for evaluating compactness.

Now comes the familiar refrain—compactness scores have drawbacks, mostly having to do with geography. For example, Maryland's 6th district is not considered to be gerrymandered, but it has a bad Polsby-Popper score of 0.071 because of the long meandering state border that defines almost half its boundary. States or districts that are partly defined by coastlines will also score poorly on this test. And sometimes it's simply impossible to create fair districts from the standpoint of representation. The Princeton Gerrymandering Project, which calculates various geometric and partisan scores for every district in the United States and gives a letter grade based on the results to every state, assigned an F to Massachusetts.* But it's hard to say who is earning this F: there's nothing that can be done to give representation to the Massachusetts Republicans with districts limited by geography.

A district with an outer boundary that looks mostly circular will do well on the Reock test even if there are thin interior slivers gerrymandered out of it that significantly increase its perimeter. You can draw a tight almost-circle around it, but there is a lot more to its perimeter. A district that looks like a circular saw blade would also do well even though it would have a lot of perimeter. Same for the convex hull, because this score only cares that the district's outer edges fill what is close to a convex polygon and does not care what the district looks like inside or how wiggly on the small scale its outer boundary is. A district that is thin but meanders tightly up and down and left to right, like a snake stuffed into a shallow box, will have a good length-width score.

---

* Interested in the grade your state got? Visit the Gerrymandering Project's "Redistricting Report Card" at https://gerrymander.princeton.edu/redistricting-report-card/.

There are lots of other issues, like the sensitivity to the amount of "zooming in," namely the granularity of measuring the length of the border. Or blindness to large swaths of land that are sparsely or not at all populated yet figure consequentially in the calculations of the scores.[*]

And then there is the basic question: What's a good or a bad compactness score anyway? What does a Polsby-Popper score of 0.30 mean? How much worse is 0.25? And on what criteria has this been decided?

There is no agreement on which of these compactness measures (or several others in use) is the best or even which subset of them is the most useful. Even though thirty-seven states have some kind of compactness requirement, they run the gamut from prescribing particular tests to vaguely suggesting that compactness should be a consideration. Some states even have different standards for congressional and state districts. Some do their own thing, such as California, which weighs the population densities throughout the district, or Maine, which takes into account driving distances across districts.

Judicial disarray doesn't help either. The courts have been using Polsby-Popper more frequently than other compactness tests, but they are not sure how to interpret any of them. They sometimes don't even know which measures they want. For example, court orders from 2018 regarding the Pennsylvania gerrymandering case say that "any redistricting plan the parties choose to submit shall include a report detailing the compactness of the districts according to each of the following measures: Reock; Schwarzberg; Polsby-Popper; Population Polygon; and Minimum Convex Polygon." Throw in a set of steak knives while you're at it.

The perennial problem is that all the quantitative measures we've talked about, whether they are based on geometry or population, are too simple. Too coarse. They take little information as input and produce a single number as output. Contrast this with the situation on the ground, where there are googols of variation in geography, demographics, and

---

[*] A compactness measure that pays attention to the structural units such as precincts or census blocks is the *cut edge score*, which is obtained by counting the number of adjacent units that are assigned to different districts.

politics within and between states. Our spare measures are insufficient to reliably flag gerrymandering when there are so many variables in play. To say they don't paint the whole picture is an understatement.

Still, to a mathematician, the Polsby-Popper score is a many-splendored thing—a result in theoretical math reincarnated as a transformational force in the real world. When that happens, it's a gratifying validation of the prescience of mathematics. Like Leonhard Euler enabling secure online communication and Perron-Frøbenius powering Google's PageRank algorithm, Steiner and his isoperimetric inequality theorem rightfully belong on the list of mathematical landmarks that have taken on new life with reaffirmed purpose.

RANDOM BREWERY HOPPING
IN THE SPACE OF MAPS

One of the key pieces of evidence in the 2016 *Gill v. Whitford* case that convinced the three federal judges to declare the Wisconsin map unconstitutional was 200 other randomly generated districting maps of Wisconsin with one common feature: they had much lower efficiency gaps than the map on trial.

The maps were produced by Jowei Chen, a professor at the University of Michigan, in his 2017 paper "The Impact of Political Geography on Wisconsin Redistricting." In 2013, Chen, along with coauthor Jonathan Rodden, a Stanford political scientist, pioneered the idea of comparing a problematic map with heaps of other maps that could have been. They started from the familiar premise that urban centers packed with Democratic voters tend to produce biased-looking results even in the absence of partisan gerrymandering. Then they asked a question: How much of the districting bias is due to human geography instead of intentional gerrymandering?

Their innovative idea was to randomly produce many hypothetical nonpartisan districting plans that respect equal population size, contiguity, compactness, and preservation of county lines. Then they measured these maps for partisan lopsidedness. When all of that data is in hand, the final step is to see how the actual map drawn by lawmakers stacks up against all the random alternatives in terms of some measure of gerrymandering.

A way to visualize this is to imagine the map of a state colored with blue and red to indicate how the vote in some actual election turned out in each precinct. That gives us fine-scale information on voting tied to geography, the necessary underlying data for the analysis that follows. Then overlay the district borders—these could be state or congressional districts—for each new hypothetical nonpartisan map and measure whatever seems interesting. Chen looked at the efficiency gap. But even something as simple as counting the votes and noting which party would win in each district could be illuminating. Different maps might produce different winners in these hypothetical districts, so the total number of Republican and Democratic delegates to the legislature might also vary from map to map.

Why is this a good idea? Any direct attempt to answer the question "Is this a fair map?" is bound to get pummeled in a mathematical, political, or judicial brawl. But the method of looking at lots of random maps gracefully avoids the fray by circumventing that question. The method is not prescriptive. It simply compares the contentious map to lots of other maps without asserting that one is better than another. The question then becomes this: If a nonpartisan map was drawn at random, is there a good chance it would exhibit the same characteristics as the map on trial? Or, in statistical language: If the null hypothesis is that there was no gerrymandering, what is the probability of observing an election outcome like the one under scrutiny? If the probability is tiny, then the map is an outlier. We can say with confidence that such a map could only have been premeditated and crafted with intent.

For this approach to work, we have to test the map on trial against lots of random maps. Many more than 200. Billions. To illustrate, let's talk beer. Suppose I visit one of eighteen craft breweries in Portland, Maine (I'm getting ready to visit there so I've been doing my research), and the beer is so bad that I hypothesize that it was made using cheap brewing kits of the sort sold on Amazon. The way I plan to prove this is by going to one other brewery, consuming a flight of beer, observing that it is delicious compared to the first brewery, and declaring that the first brewery used kits. Nobody would take me seriously. What if I just happened to walk into a better brewery the second time? Out of seventeen other breweries, I randomly went to one that's better. So what? The element of chance is obviously too great for me to be able to make any kind of generalization about the first brewery. But if I continue my brewery hopping and the first brewery is still terrible compared to other breweries, then my argument starts to carry more weight. There is less chance that I'm making a mistake when I accuse the first brewery. And if I manage to visit all eighteen breweries (Is that a challenge? I accept!) and the first one is still the worst, I can without reservation expose this aberration to beer lovers worldwide.

Now, map hopping to visit *all* possible districting maps is out of reach. The number of maps in a state with p voting precincts and d districts is d^p, which in this context almost always generates a huge number.* Imposing the restrictions of contiguity, compactness, and so on would reduce the number significantly, but still not to a computationally manageable set of choices. In math language, the *space of all maps* is too large.†

There are ways around this difficulty. One is to create an *ensemble* of maps that have been sampled from the space of all maps. This is fast becoming the gold standard in the algorithmic study of gerrymandering. The ensemble should contain a computationally tractable number of maps. Then we do with that sample whatever it was we wish we could have done with all the maps. At the end of the day, we'll have just an approximation of the solution to the original problem, but maybe the ensemble will have all of the salient characteristics of the whole map space. And maybe the picture the ensemble paints is clear enough that we believe that the conclusions it supports would hold up even if we somehow managed to take into account all possible maps. (Or, more important, the picture is clear enough that the conclusions would hold up in court.)

Two issues arise immediately: How do we choose the ensemble? and How do we know the ensemble is representative of the whole space?

---

* With ten districts and eighty-five precincts, the number of possible maps is larger than the number of atoms in the universe.

† Even with an easy definition of "fair," calculating whether a fair districting exists is known as an *NP-hard* problem, basically meaning that it's . . . very hard.

FIGURE 12.13. Actual 2016 Pennsylvania districting (left) and the districting after 2^{40} Markov chain steps (right).

A way to settle the first question is to deploy the *Markov chain Monte Carlo (MCMC)* algorithm. Such algorithms have existed since the 1940s, but they first entered the redistricting scene in 2014 with the paper "Redistricting and the Will of the People" by Jonathan Mattingly, a professor at Duke University, and Christy Vaughn, then an undergraduate at Duke. The idea is to start with some map and then randomly choose another that's "close" to it. Then we compute some measure of gerrymandering for the new map and move on to choose yet another map that's "close" to the previous one. Repeat. Billions of times. This is called a *random walk* through the space of maps.

There are a couple of ways for one map to be "close" to another. We can get from one map to the next by performing a *flip*, which is accomplished by moving a randomly chosen voting precinct located on the border of a district so it now lies in the adjacent district, while making sure that both of the affected districts still satisfy any required conditions such as contiguity and roughly equal populations. Although the process takes close steps, that doesn't mean that the configuration of borders stays near the original map; the enormous number of iterations ensures that we get a large variety of maps over the long run. An example is given in figure 12.13.

A second way to define "closeness" is to get from one map to the next by combining two adjacent districts and then splitting them up again in a way that creates two districts that are different from the original two. This is called *recombination*. It was developed by Daryl DeFord from Washington State University, Moon Duchin from Tufts University, and Justin Solomon from MIT in their 2019 paper "Recombination: A Family of Markov Chains for Redistricting." Duchin is the founder of the Metric Geometry and Gerrymandering Group at Tufts and one of the leading mathematicians who are working on bringing sophisticated mathematics to bear in the study of districting.

How about the second question? The algorithm needs instructions that tell it which random steps are better than others. This is accomplished using a *probability distribution function* that contains information about what's permitted and what's not, including standard criteria such as contiguity and compactness. When the MCMC process chooses the next map on its walk, the distribution ensures that it chooses one that satisfies the criteria for a legitimate nonpartisan map with higher probability. Such maps are therefore sampled much more frequently along the walk and a representative sample is eventually achieved. MCMC also needs a seed, a map from which to start its walk, and that's usually the actual disputed map.

If the map being examined was indeed gerrymandered, this process can demonstrate—well beyond a reasonable doubt—that it had to have been carefully drawn to produce such a high degree of partisan slant. In other words, the chances that a map like the one being examined was drawn with only the standard rules in mind and not with any kind of political purpose will be vanishingly small. The map will be an eccentric outlier in the universe of all possible maps, a lonely special case sitting at home with a nonalcoholic beer in hand while the rest of the party is brewery hopping in Portland.

A telling picture usually emerges from this sort of analysis, by which I mean literally a picture that tells it all. Figure 12.14 is a graphic from the paper "Evaluating Partisan Gerrymandering in Wisconsin," written by Mattingly and two coauthors. They looked at an ensemble of 19,184 state district maps and examined how many State Assembly seats each party would have won in 2012 according to every map. The horizontal axis gives the number of seats Republican candidates won of ninety-nine possible seats. The vertical axis gives the frequency of occurrence of each number of wins—roughly how many of the almost 20,000 maps resulted in that number of Republicans being elected. The tallest bar therefore gives the most "typical" result. It says that for about 0.21, or 21%, of the maps, Republicans would have won fifty-five seats (and Democrats would have won forty-four). The actual Wisconsin map, which gave Democrats

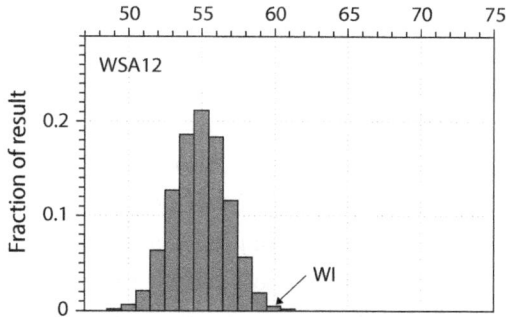

FIGURE 12.14. The breakdown of Republican wins across the ensemble of 19,184 state district maps of Wisconsin.

thirty-nine seats and Republicans sixty, is an outlier and has less than 1% chance of being created without intentional gerrymandering.

Ensemble sampling continues to deliver such startling results across many states. It also provides evidence in favor of independent district-ing commissions. Another paper, also coauthored by Mattingly, uses an ensemble of 66,544 congressional redistrictings of North Carolina to study the results of the 2012 and 2016 elections. It shows that the actual maps are outliers. It also shows that maps built by an independent panel of former judges, as part of a Duke University program called Beyond Gerrymandering, are much more typical of nonpartisan dis-tricting and always fall under one of the tallest bars in the graph. This is another plug for independent districting commissions.

———

Of the methods currently in use, ensemble sampling is the most promis-ing. It uses lots of input data and sophisticated mathematics and it pays attention to geography and demographics.* It can look locally and es-tablish outliers in individual districts, unlike, for example, the efficiency

---

* Another recent and promising method that uses both map and election data is *GEO Metric* due to Marion Campisi, Tommy Ratliff, Stephanie Sommersille, and Ellen Veomett. The algo-rithm finds which additional districts a party "should have won" (but didn't, likely due to ger-rymandering) by considering vote swaps between neighboring districts.

gap, which only says something about the entire state. Again, this is important because many gerrymandering cases rely on exhibiting harm done to an individual in a particular district.

Ensemble sampling is also morally closer to the craft of mathematics. The proposition of objectivity is best honored if mathematicians follow the maxim "show, don't mandate." Metrics such as partisan bias and efficiency gap imply a built-in value judgment about what's "right" and "wrong," but at best, those judgments can be out of touch with the situation on the ground, and at worst, they can be arbitrary. Ensemble sampling is about nonhierarchical comparisons with no benchmarks for "good" or "bad." The contentious subjectivity of the concept of "fairness" is taken out of the picture.

But ensemble sampling, like all the other methods explained here, isn't flawless. Subtle mathematical issues affect what happens under the hood. And like all the other methods, it is not foolproof at detecting gerrymandering or finding the line where mere political gerrymandering becomes nefarious gerrymandering that disenfranchises voters. The input criteria can be incompatible as well. For example, it might be impossible to honor the Voting Rights Act without splitting counties—in which case it is lawful to split counties. That kind of wiggle room for exceptions to the rules is difficult to build into mathematical models.

And here's the hard truth: there is no algorithm that will divide a state into voting districts to reliably produce the fairest possible map while respecting the judicial constraints of things such as geometry and representational opportunity. We have ways of drawing maps more or less at random and checking whether they make sense. But there is no guarantee that any of the maps produced by any of the algorithms are going to be fair according to the existing accepted gerrymandering metrics, especially the Voting Rights Act. Even if we find a map that seems fine, an even better one could exist. The manufactured maps are also not sensitive to local community preferences because that information is not in the input.*

---

* Incorporating the Voting Rights Act into ensemble sampling has been particularly challenging. One strategy is to assume that the original disputed map respected the Voting Rights

MAY THE MATH PLEASE THE COURT

One of the most heavily gerrymandered maps that made it into the REDMAP cohort was Pennsylvania's. In the 2004 case *Vieth v. Jubelirer*, the Supreme Court rejected the challenge to the Republican-drawn map with the explanation that the Constitution gives states the power to draw district maps and that's that.

The key arbiter in the 5–4 decision was Justice Anthony Kennedy, who sided with the four conservative judges. But he seemed to leave the door open a crack: "While there seems to be no workable standard by which to detect gerrymandering at the moment," he said, "this should not be taken to prove that none will emerge in the future."

Scores of academics tried to please Kennedy after those words were written. Many standards were proposed. Partisan symmetry was one. But in *League of United Latin American Citizens (LULAC) v. Perry*, a 2006 case challenging the Texas map that produced the seats-votes curve we saw before, Kennedy, ever the seeker of rigor, said that partisan symmetry is a "helpful tool" but that "asymmetry alone is not a reliable measure of unconstitutional partisanship." An additional feature that bothered him was that the analysis is based on a hypothetical situation in which voters switch their votes, the *counterfactual* rather than the empirical: "We are weary of adopting a constitutional standard that invalidates a map based on unfair results that would occur in a hypothetical state of affairs."

Well, the Court is apparently weary of the counterfactual *and* the factual. The efficiency gap—as factual as it gets; no hypotheticals, real data only—was used in *Gill v. Whitford* as the core of the plaintiff's case that the 2011 Wisconsin map was gerrymandered against the Democrats (with an efficiency gap of 11.46%). The defense argued that the efficiency gap

---

Act and created the requisite majority-minority districts and then tell the random walk to avoid changing the boundary of those districts to guarantee their preservation throughout the MCMC analysis. But this strategy relies on the assumption that the original map did a good job of creating those districts. Progress has been made in avoiding this assumption. A recent paper, "Computational Redistricting and the Voting Rights Act," proposes a way to build more information into the ensemble and extract maps that are better at complying with the Voting Rights Act.

ignores natural packing and that this purely demographic phenomenon, which exhibits itself as Democrats preferring to live in Milwaukee and Madison, explains the lopsidedness of the maps. The plaintiffs clapped back by showing that the 200 maps randomly generated by Jowei Chen satisfied the basic districting requirements but all had a much lower efficiency gap than the contested map. This supplementary evidence convinced the federal court that the map's skewness could not be a product of chance. The three federal judges threw the map out.

The Supreme Court was next. The plaintiffs felt good about their prospects because for the first time, there seemed to be a strictly mathematical yet understandable measure of gerrymandering that could have lasting judicial relevance. Justice Kennedy was gonna love it!

The hour of oral arguments that were made on October 3, 2017, is well worth your time. It is as edge-of-your-seat as it is ridiculous. The Supreme Court responded to the empirical evidence, including the efficiency gap measure, with the defensive annoyance of high school geometry students demanding to know when they will ever use this math. The justices protested that the arguments were too complicated, too newfangled, and "baloney." Chief Justice Roberts declared the efficiency gap to be "sociological gobbledygook." (The American Sociological Association scolded him publicly afterward.) *Gill* was remanded to the lower court and never made it back to the Supreme Court.

In 2019, *Rucho v. Common Cause* sealed its fate. With Kennedy gone, *Rucho* was adjudicated by a majority that was not equipped or willing to do math. This time, ensemble sampling was the name of the game, demonstrating with essentially indisputable certainty (meaning an extremely high probability) that the North Carolina maps were exceptionally gerrymandered. There even was an amicus brief signed by eleven experts, most of them mathematicians, the first ever of its kind.*

But the justices were resistant. They kept wanting to lead the discussion back to the familiar scrutiny of proportional representation—is it true or not that the percentage of the vote a party received statewide is about equal to the percentage of the seats that party won in the legislature? The answer

---

* These "friend of the court" briefs are a way for experts to weigh in on cases.

didn't really matter to the justices; the only thing that mattered was that the arguments in the case be redirected to proportional representation because a legal precedent for that kind of argument already exists. In the 1986 case *Davis v. Bandemer*, the Court concluded that failure of proportional representation isn't good grounds for arguing for gerrymandering (true, it's not), so the Court could quickly dismiss *Rucho* if only they could find a direct legal path back to *Davis*. Justice Gorsuch kept asking about the "deviation from proportional representation" and the court collectively failed to acknowledge that the mathematical method at hand did not care about that rejected yardstick at all but was an entirely novel tool.*

The highest court's unwillingness to engage with the mathematics is troublesome. Maybe it's simply a rejection of the empirical in favor of the ideological, a performance that shields the justices when they rule according to their beliefs. Or maybe it's something as simple as law schools not teaching enough math. But as society continues to advance technologically, interdisciplinary scientific sophistication built on complex and often abstract math is bound to rise to the top of more issues that will come before the courts. Will the judges resist mathematical evidence in cases involving artificial intelligence, cryptocurrency, and quantum computing?

———

There may never be a simple and categorical detection instrument for gerrymandering; the problem might just be *that* difficult, its mathematics so inextricably tangled with politics, law, and history that the only way to consider it is holistically. The line between politics and malice is more elusive than ever, blurred by the computational and algorithmic power at anyone's disposal.

---

* One of the signatories of the amicus brief was Jordan Ellenberg of the University of Wisconsin, a terrific pop math writer and a stellar mathematician. In his *Slate* piece "The Supreme Court's Math Problem" and in the chapter on gerrymandering in his book *Shape: The Hidden Geometry of Information, Biology, Strategy, Democracy, and Everything Else,* he describes what happened in the *Rucho* hearings. The hilarity of his account (which involves a reenactment using sandwiches instead of math) is eclipsed only by the demoralizing recognition that the Supreme Court is kind of innumerate.

But still, we've come a long way. Before mathematics got involved, gerrymandering was essentially on a par with pornography in judicial eyes; Supreme Court Justice Potter Stewart's 1964 "I know it when I see it" criterion for what constitutes obscenity might as well have been uttered with gerrymandering in mind. "We believe that reapportionment is one area in which appearances do matter," said Justice Sandra Day O'Connor in 1993, when the "eyeball test" was still the gold standard for detecting gerrymandering.* But research has shown that there is a discrepancy between how people visually perceive noncompactness (i.e., strangeness of shape) and how the standard geometric measures such as Polsby-Popper, Schwartzberg, and Reock score it. This indicates that there is a disconnect between the way politicians, lawmakers, and judges subjectively evaluate gerrymandering and how mathematics might tell them to do it. Moreover, the speed of computing and the sophistication of mapping software is enabling smarter gerrymandering that is pleasing to the eye but is still extreme, rendering the eyeball test even less useful.

It is imperative that practicing mathematicians be involved in these issues. Judgment must shift toward objective evidence and away from faulty eyeballs. And mathematicians *are* getting involved, even if it annoys the Supreme Court. When Jonathan Mattingly first started applying ensemble sampling to gerrymandering, he had no idea he would end up in federal court as a witness in *Rucho v. Common Cause*. He was asked to write a court declaration and was subsequently called to testify. He did so for hours and was apparently convincing because the panel declared North Carolina's 2011 map unconstitutional, citing Mattingly extensively.

Wesley Pegden, a professor at Carnegie Mellon University, was equally surprised when lawyers for the League of Women Voters of Pennsylvania called him to be an expert witness for the case they had filed against their state's legislature. He had cowritten a paper about Markov chains that used Pennsylvania's 2011 districting as an example. The paper's MCMC analysis

---

* She also said in 1986 that gerrymandering was a "self-limiting enterprise" because in order for a party to give itself an advantage in a district, it would necessarily have to dilute its dominance elsewhere. That is no longer true; the combination of the extreme polarization that makes voting predictable and sophisticated computational methods has turned gerrymandering into an exact procedure that is superior at optimizing and cutting losses.

demonstrated that the degree of partisanship exhibited by the 2011 map shows up on average in only one of every 2.5 billion maps. The case moved on to the Pennsylvania Supreme Court, which ruled in favor of the plaintiffs and ordered the Pennsylvania legislature to redraw the map.

Pegden, along with Jowei Chen and Jonathan Rodden, wrote an amicus brief for *Gill v. Whitford*. His analysis showed that only about 1 in 5 million Wisconsin state district maps were as partisan as the disputed one. Pegden and Rodden, along with Sam Wang of Princeton University, also cowrote an amicus brief for *Rucho v. Common Cause*. The brief I mentioned before, the one by Moon Duchin of the Metric Geometry and Gerrymandering Group Redistricting Lab that was signed by eleven mathematicians, also had ensemble sampling as its centerpiece. Mattingly and Pegden testified in the 2019 case *Common Cause v. Lewis* against North Carolina's state map, aided by a court declaration by Gregory Herschlag, another Duke professor. The result: new maps had to be drawn in 2020.

After the Pennsylvania Supreme Court threw out the Republican state legislature's map and ordered a new one to be made, Governor Tom Wolf enlisted Duchin to examine the redrawn version. The new map looked "prettier" in the sense that it passed the eyeball compactness test, but Duchin found that it was just as gerrymandered. Governor Wolf shut down the new map and the Pennsylvania Supreme Court moved to appoint an independent advisor to draw the maps yet again.

So when the 2020 census kicked off a gerrymandering free-for-all around the country, these mathematicians were ready. Redistricting Lab has been hard at work communicating with the public and advising politicians, commissions, and activists. A number of states are using their open-source districting software, Districtr, to collect map suggestions from the public. (The software also collects data about communities of interest.) Duchin was an expert for the plaintiffs in the landmark *Merrill v. Milligan* case challenging Alabama's 2021 congressional map.* Daryl

---

* In a surprising decision, in June 2023 the Supreme Court ruled in favor of the plaintiffs, upholding a District Court decision that Alabama's congressional map was racially gerrymandered. Duchin was cited extensively in both the majority and the dissenting opinions.

DeFord and Justin Solomon were two of the writers of an amicus brief for that case.

Duchin's Lab at Tufts, Mattingly's Quantifying Gerrymandering group at Duke, Wesley Pegden and collaborators at Carnegie Mellon, the Princeton Gerrymandering Project led by Sam Wang, the Stanford Redistricting Project, the Institute for Computational Redistricting at University of Illinois Urbana-Champaign, the Public Mapping Project by faculty at the University of Florida and MIT—the proliferation of academic involvement by researchers dedicated to eradicating gerrymandering has the makings of a movement. Citizen-scientists are creating open-source computational tools that are available to all. They are finding new ways to put high-powered mathematics to public use, math that in the past was reserved for academic elites. They are taking districting out of the back rooms, away from the select few, and putting it into the hands of the public, where it belongs.

CHAPTER 13

Proportional Representation

IF I believed in such things, I would ascribe the serendipitous events of today to some divine power or a glitch in the Matrix. I had just sat down to start writing this chapter when my wife handed me the mail she had grabbed from the mailbox. In it was an envelope that had traveled all the way from Bosnia, the country I grew up in. The return address was Central Election Commission, Sarajevo. In the envelope were several sheets of paper—absentee ballots—for the elections that would take place there in October 2022. I was to fill them out and mail them back no later than the day of the elections.

Now, as the Matrix would have it, only a few days earlier I had mailed my ballot for the Democratic primary election here in Massachusetts. (Double voting duty, one of the perks of dual citizenship.) That ballot asked me to select a candidate for each position up for grabs, including governor, lieutenant governor, and the representative from my congressional district. When I receive the general ballot before the November election, it will again have me choose among some names, one of them the winner of the Democratic primary I just voted in, one the winner of the Republican primary, and possibly some third-party candidates. The final champion of the congressional race—and there can be only one—will become the delegate to the House of Representatives from the 5th District of Massachusetts, where I live. Millions of people across the United States will do the same come November, sending one and only one person from their district to the House.

One of the Bosnian ballots has the same goal—to elect people to its House of Representatives. But that's about all it has in common with its Massachusetts counterpart. The two ballots look nothing alike. They might as well be from different planets, electing delegates to the Intergalactic Council.

To start, the Bosnian ballot has twenty-four political parties listed on it (figure 13.1). Under the name of each party is an ordered list of the candidates the party is putting forth as potential representatives. The lists are as short as one name and as long as nine, the maximum allowed.

But the really big difference is that I am voting to elect four delegates to the Bosnian House. All districts in Bosnia, including mine, are *multimember* (or *multiseat*, or *multiwinner*) districts. Each is choosing some greater-than-one number of representatives. And here is the serendipitous part I alluded to at the start: the winners are going to be decided using *proportional representation*, the topic of this chapter.

SHARING IS CARING

We had a close encounter with proportional representation in chapter 12, as the central idea in the design of symmetry-based gerrymandering metrics. Proportional representation simply means that the number of seats allocated to a party (in a state, a country, or a multimember district) should be proportional to the number of votes that party gets. In other words, the percentage of seats a party receives (its seat share) should be equal or at least as close as possible to the percentage of votes it wins (its vote share).

If, for example, in this Bosnian election in my four-delegate district, Party A gets 50% of the votes and parties B and C each get 25% of the vote, that would mean that 50% of the available seats, namely two, would go to A while B and C would get 25% of the seats, one each. Of course, the numbers almost never work out this nicely, and that's why we need to add more math to our toolbox.

PROČITAJTE NAPUTAK/UPUTSTVO NA PREDNJOJ STRANICI! **ПРОЧИТАЈТЕ упуТСТВО НА ПРЕДЊОЈ СТРАНИ!**

KOALICIJA DRŽAVA - MIRSAD HADŽIKADIĆ - PLATFORMA ZA PROGRES, NB
1 SENAD ŠEPIĆ
2 DRAGANA GLUMAC
3 SANEL ŠARIĆ
4 RAZIJA MUJANOVIĆ
5 NIJAZ NANIĆ
6 EMINA DRINJAKOVIĆ
7 ELMIR PRŠEŠ
8 NERMA MUJKOVIĆ

SAVEZ ZA BOLJU BUDUĆNOST BIH
1 NERMIN DŽINDIĆ
2 EDITA ĐAPO
3 SUAD KURTĆEHAJIĆ
4 ANIS KRIVIĆ
5 INDIRA PINDŽO
6 SEAD ŽELJO
7 ZIKRETA TUBIĆ
8 DAVOR KLJAJIĆ
9 DŽANA MUJIĆ BEKTEŠEVIĆ

HDZ BIH,HSS,HSP BIH, HKDU,HSPAS,HDU, HSPHB,HRAST
1 NADA SPASOJEVIĆ
2 MARJAN RAJIC
3 DRAGAN PRANJIĆ

BOSANSKOHERCEGOVAČKI BH ZELENI
1 MERSIHA SALIHIĆ

КРуГ
1 МЕЛДИЈАНА СИНАНОВИЋ

HDZ 1990 - HNP
1 IVAN TOMIĆ

SAVEZ ZA NOVU POLITIKU
1 AMAR MUHIBIĆ

УЈЕДИЊЕНА СРПСКА
1 ГОРДАНА ТРАПАРА

SMS
1 VLADIMIR PUŠKAREVIĆ

САВЕЗ НЕЗАВИСНИХ СОЦИЈАЛДЕМОКРАТА - СНСД - МИЛОРАД ДОДИК
1 МИРКО ТОДОРОВИЋ

ЛИЈЕВО КРИЛО
1 АРМИН БИСТРИВОДА

SDA - STRANKA DEMOKRATSKE AKCIJE
1 ALMA ČOLO
2 SAFET KEŠO
3 BISERA TURKOVIĆ
4 AMOR MAŠOVIĆ
5 ADMIR MULAOSMANOVIĆ
6 RABIJA BOSNIĆ
7 HAZIM BAHTANOVIĆ
8 SEFIK DŽELOVIĆ
9 MAJDA OMEROVIĆ

NES - ZA NOVE GENERACIJE
1 ZLATKO MILETIĆ
2 NAIDA HUBANIĆ
3 DUŠAN ČULUM

NAROD I PRAVDA
1 DENIS ZVIZDIĆ
2 MIA KARAMEHIĆ-ABAZOVIĆ
3 AHMET SEJDIĆ
4 AMAR DOVADŽIJA
5 VILDANA TATLIĆ
6 ZULFO AHMETOVIĆ
7 DŽELALUDIN HODŽIĆ
8 ELVEDINA VUGIĆ
9 NERMINA MEMIĆ

BOSS - BOSANSKA STRANKA - MIRNES AJANOVIĆ
1 MELIHA ĆILIMKOVIĆ

SDBIH SOCIJALDEMOKRATE BOSNE I HERCEGOVINE
1 NIHAD ČOLPA
2 LJUBICA TODOROVIĆ
3 SAFET ZEC
4 HILMO KOVAČ
5 MARIJA DUMENČIĆ
6 LJUBICA DOLOVIĆ
7 EDIN ŠATROVIĆ

SDP - SOCIJALDEMOKRATSKA PARTIJA BOSNE I HERCEGOVINE
1 SAŠA MAGAZINOVIĆ
2 LIDIJA KORAĆ
3 DERVO SEJDIĆ
4 HARIS VRANIĆ
5 EDINA PAPO
6 DIJANA BRIGA
7 JASMIN DURAKOVIĆ
8 BENJAMIN SOKO

РЕ-БАЛАНС
1 АЛМИР ХАСАНОВСКИ

NAŠA STRANKA
1 SABINA ĆUDIĆ
2 PREDRAG KOJOVIĆ
3 ŽELIMIR BUKARIĆ
4 AMELA KUSKUNOVIĆ
5 STRAJO KRSMANOVIĆ
6 JASMIN IMAMOVIĆ
7 BELMA ŠKORIĆ AVDIĆ
8 VLADIMIR PODANY
9 JASMINA MRŠO

HRVATSKA REPUBLIKANSKA STRANKA
1 ANTONELA IVKOVIĆ

ПДП - ПАРТИЈА ДЕМОКРАТСКОГ ПРОГРЕСА
1 ДУШАН ВУКАШИНОВИЋ

STRANKA ZA BOSNU I HERCEGOVINU-SBIH
1 ZLATKO HADŽIDEDIĆ
2 AMIRA BAJRAMOVIĆ
3 DINKO REMIĆ
4 IZETA JAHIĆ
5 SANEL PODRUG
6 EMINA ALIBAŠIĆ
7 ADNAN ADEMOVIĆ

ŽELJKO KOMŠIĆ - ZA GRAĐANSKU DRŽAVU - DF/GS
1 DŽENAN ĐONLAGIĆ
2 DALIA ŠABIĆ
3 SENADIN LAVIĆ
4 MILAN DUNOVIĆ
5 BILJANA BEGOVIĆ
6 DAMIR PEKIĆ
7 LANA VELIĆ
8 SIFET PODŽIĆ

BOSANSKOHERCEGOVAČKA INICIJATIVA - KASUMOVIĆ FUAD
1 JASMIN KADRIĆ

KRAJ LISTE.
КРАЈ ЛИСТЕ.

FIGURE 13.1. Bosnian election ballot for the House of Representatives, 2022.

Single-member plurality districts, which dominate U.S. elections, are the nemesis of proportional representation. The winner-take-all system undermines any hope for proportionality. Massachusetts is the perfect example. The discrepancies aggregate to national trends, with the Republican Party typically taking a greater proportion of House seats than its share of the popular vote because of its dexterity with gerrymandering (although the 2022 midterm elections appear to have bucked that trend).

The divergence between vote shares and seat shares is even more apparent in multiparty systems that use single-member districts. The UK is a good case study. Each of its 650 single-member districts (or *constituencies*) elects a single Member of Parliament (MP) from several party candidates. In 2019, the Conservatives won 43.6% of the national vote but ended up with 56.2% of the seats. The Liberal Democrats won 11.5% of the votes but got only 1.7% of the seats. In Scotland, the Scottish National Party won 45% of the votes but got 81% of the seats. Labour was the only party whose seat shares (31%) and vote shares (32%) were approximately equal. The UK has been aware of this predicament for decades. In a 1987 video, the timelessly hilarious John Cleese advocated for proportional representation on the heels of an election in which the percentages of votes won by the Conservatives, Labour, and the SDP-Liberal Alliance were 42%, 28%, and 26%, respectively, but the percentages of seats these parties won were 61%, 32%, and 3.5%. "This is ludicrous," Cleese proclaims, "or, as children would say, not fair. Not fair to us, the voters." The proportional representation system Cleese was promulgating would have replaced the 650 single-member constituencies with 140 constituencies that would elect four or five MPs each. Under this scheme, the seat shares in the 1983 election that inspired the video would have been 42%, 28%, and 25%, respectively, much closer to the vote shares.

Proportional representation is not unknown in the United States. The Democratic Party decides the winner of its presidential primaries according to proportional representation.* Cambridge, Massachusetts,

---

* When primaries are held in the individual states, delegates are awarded to candidates who receive at least 15% of the primary votes and they are awarded in proportion to the popular vote.

has been using proportional representation successfully for over eighty years, as have many other cities and congressional districts.

In fact, apportionment for the House of Representatives is nothing but a proportional representation system in disguise.* Think of the fifty states as fifty political parties, and think of the population of each state as the number of people who vote for that party. The purpose of apportionment of the House is to allocate the 435 seats to the "parties" (the states) according to the proportion of the "votes" they got (the size of their population). This means that the apportionment systems we already know—Hamilton, Jefferson, Webster, Huntington-Hill—are all proportional representation methods. In fact, two of the world's most popular proportional representation systems, D'Hondt and Sainte-Laguë, are nothing but camouflaged Jefferson and Webster methods. We'll see why in a minute.

Proportional representation is used in the vast majority of the world's democracies. Most European countries use it. One of the exceptions is the UK, from which the U.S. blindly copied the plurality system back in the day, but even it used proportional representation to elect its representatives to the European Parliament before Brexit. The apportionment of the 751 seats of the European Parliament also follows a proportional representation system, albeit an ad hoc way with *degressive proportionality*, an allocation in which smaller countries get more seats than their population proportion. That's a shared trait with the U.S. House, which favors small states because of a combination of the Huntington-Hill method and the rule that every state must have at least one representative.

As new nations have formed in Europe, all have adopted proportional representation. None of them wanted anything to do with single-member districts. In 1993, New Zealand switched to proportional representation from single-member plurality districts. The result was a

---

This system of proportional allocation is essentially the Hamilton method. The threshold can produce quirkiness, as can the fact that most delegates are awarded districtwise. The threshold can also cause the *elimination paradox*, whereby a candidate's delegate count might be reduced when another candidate drops out of the race.

* By contrast, the Senate is about as unproportional as a system can be.

transition to a multiparty system that represents the preferences of the electorate more faithfully. There are timid whispers about a switch to proportional representation in Canada.

Proportional representation is a simple and sensible concept. It's great when it's done right. It makes mathematical sense—and it makes common sense. It jibes with most people's notion of fairness: political representation should be commensurate with the public's desire for it, expressed through voting. The portion of power, as the share of total power, that's awarded to some voters through representation should be the same portion as the strength of their vote, seen as the share of the total vote. Proportional representation is what we should be doing in the United States, in combination with multimember districts. Let's see how it works and then we'll return to its many benefits.

PARTY LOYALTY

The Bosnian ballot is an example of a *party list proportional representation* method. Most proportional representation countries, about 80% of them, use some kind of party list structure. In such systems, parties are the dominant political unit. Candidates are important, but parties run the show.

How am I supposed to vote on that ballot? I have three options.

The first is to throw my vote completely behind a party by simply checking the box next to its name. That choice implies that I am okay with the candidates and their ordering as the party decided it. If the party gets one or more seats, they'll be given to its people according to the order on the list.

I can also check a party box and check any number of names on the party's list. This means that I want the party to win but I prefer certain candidates to others within the list. If enough people check a candidate to pass some preordained threshold, the candidate is bumped up the list.*

---

* In one of many Bosnian political machinations, a candidate has to receive at least 20% of the vote to be bumped up, a practically unreachable percentage. This guarantees that voters can't mess with a party's ordering.

The last option allows me to check some candidates on a single party's list but not check the party name. It turns out this is the same as the second option, because voting for one or more candidates on a party's list automatically counts as a vote for that party. This option exists to avoid declaring lots of ballots invalid when people simply forget to check the party box.*

I am not allowed to check two parties or candidates from two separate lists. My vote is therefore necessarily a vote for a party, even if I also select individual candidates as my favorites. Candidates are always identified with their party, even if they do well and are bumped up the list. Party supersedes its candidates.

This is an example of an *open list proportional representation* method. The "open list" part means the voter can still say something about the candidates while voting for a party. Open list proportional representation is used in almost fifty countries, including Iceland, Norway, Austria, Croatia, Brazil, and Japan. There are over a dozen ways to configure and implement it, depending on how the lists are organized, what options voters are given, how thresholds are set that candidates must reach in order to move up the list, or how party-only votes are distributed to the candidates. Variations include ballots where the number of candidates that can be voted for is limited, candidates do not have to come from the same list, a person can lodge a negative vote against a candidate, and so on.

But proportional representation can be even more about the party. It can be *all* about the party. An even more restrictive party-centric proportional representation system is *closed list proportional representation*. In such a system, the voter casts a vote for a party but has no choice about the arrangement of the candidates. The ordering is decided by the party, and that's that. This system is used in over forty countries, including Algeria, Argentina, Portugal, Israel, Turkey, and South Africa.

---

* I find this option misleading because it suggests that a voter can support a candidate without supporting their party. The thinking is "I like someone enough to forgive them their membership in a party I dislike, but I don't want my vote for them to count as a vote for the party." But that is not the effect of such a vote.

And finally, there are *mixed electoral systems* that use some combination of proportional representation and plurality. Over thirty countries use such a mixture. One flavor is the *mixed-member proportional representation* that is used in Germany (where it originated), New Zealand, and five other countries. Here voters cast two genuinely distinct votes in their districts—one for a candidate and one for a party, and the party can be different from their chosen candidate's party. This system offers the best of all worlds: votes for candidates are used to select a single plurality winner in the district, providing geographically mindful representation (although plurality is still bad and should not be used), and votes for parties are used for *compensatory* seats, which are allocated in proportion to parties' total vote counts across the country.

Bosnia uses yet another hybrid method, essentially a two-stage proportional representation system. My ballot actually records two votes—one for the candidates I select and another for the party those candidates come from (so I am providing a *single mixed vote*). I am voting in one of five districts and the five districts will elect twenty-one representatives. But after the representatives are elected, there are an additional seven seats to be allocated in proportion to the overall percentages of votes the parties win. The idea is that these seats will counterbalance any errors in proportionality that might occur in each of the five districts and account for any wasted votes accumulated across the districts. When there is a threshold that must be crossed for a party to win a seat in a district (it's usually 3–5%), it sometimes happens that a party doesn't clear this bar in any single district or clears it in only a few districts. However, the party might do well overall, with votes spread across lots of districts. In that case, the party could win a compensatory seat.

Voters typically have no say about choosing who will fill the compensatory seats. This is sometimes (mis)used as a back door for certain candidates, usually those the party wants to bring into the legislature for whatever strategic reason but who have little chance of getting elected directly. These people get a seat in the government with no input from the voters.

Most party list proportional representation systems have a quota that must be achieved in order for a party to receive any seats. In Germany, for example, the quota is 5%. This threshold prevents a situation where too many parties enter the legislature, which can create gridlock or allow small parties to dictate the formation of coalitions. In Israel, a single-district country, the quota to get a Knesset seat is 3.25%, raised from 1% in 1992. The lower quota brought too many parties into the legislature, creating unmanageable fragmentation. But a quota set too high reduces participation by smaller parties and produces many more wasted votes. A high quota can also be a way for big parties to keep smaller players out of the legislature. In Turkey, for example, the threshold is 10%, and millions of votes are wasted on parties that don't make the cut. There is talk about reducing the quota to 7%.

What is common to party list systems is that some number of seats is allocated to parties in a proportional manner. There are many ways to do this, just as there are many ways to apportion the House of Representatives. Let's discuss the two main methods, which we've already seen in a different guise. We'll ignore the math of everything else that might happen—how candidates move up the list, the threshold for seat eligibility, the mixing of methods—because there are too many variants and studying them would take us too far afield. Ignoring those details will not make the math any less illustrative or valid.

The D'Hondt Method

Belgium was the first country in Europe to use a proportional representation party list system. It was developed in 1878 by the mathematician and lawyer Victor D'Hondt (1841–1902) as an attempt to provide representation for his country's many linguistically and culturally diverse groups. The D'Hondt method remains the most popular proportional representation system to this day. It is used in over forty countries, including most of Europe. The process gives out seats one at a time—in as many rounds as there are seats—with the goal of moving toward equal representation ratios across parties. In each round, one party gets a seat and the total votes it won is divided by the number of seats it has received so far plus one. This is called the party's *quotient*. If a party, for example, receives its third seat in a round, its vote total will be divided by four going into the next round. To determine which party gets the seat in a round, we look for the largest quotient. The process continues until all the seats have been handed out.

As usual, we're in need of an example.

Suppose five parties, A, B, C, D, and E, are running in a multimember election to fill 7 seats. Say the numbers of votes they won were 100,000, 80,000, 70,000, 44,000, and 27,000, respectively. The Round 1 column in the table below contains these numbers. At this point, the parties' quotients are the same as their vote totals. The second number in each cell, to the right of the backslash, is the number of seats each party has at the end of the round. Since Party A has the the biggest quotient (which is the same as having most votes at this initial stage), it gets 1 seat, marked in bold. Whenever a new seat is allocated, the updated number is bolded. Moving to Round 2, all the quotients stay the same, except for Party A, whose votes now get divided by two; this is Party A's new quotient.

In Round 2, Party B has the highest quotient, so it receives a seat. Going into Round 3, Party B's quotient is updated—(80,000 ÷ 2)—and the other quotients stay the same. In Round 3, it's C that has the highest quotient and gets a seat. This process continues for seven rounds because that's how many seats there are.

	Round 1	Round 2	Round 3	Round 4	Round 5	Round 6	Round 7
Party A	100,000 / **1**	50,000 / 1	50,000 / 1	50,000 / **2**	33,333 / 2	33,333 / 2	33,333 / 2
Party B	80,000 / 0	80,000 / **1**	40,000 / 1	40,000 / 1	40,000 / 1	40,000 / **2**	26,667 / 2
Party C	70,000 / 0	70,000 / 0	70,000 / **1**	35,000 / 1	35,000 / 1	35,000 / 1	35,000 / **2**
Party D	44,000 / 0	44,000 / 0	44,000 / 0	44,000 / 0	44,000 / **1**	22,000 / 1	22,000 / 1
Party E	27,000 / 0	27,000 / 0	27,000 / 0	27,000 / 0	27,000 / 0	27,000 / 0	27,000 / 0

The final allocation is in the last column: 2 seats for A, 2 seats for B, 2 seats for C, 1 seat for D, and 0 seats for E. This feels okay from a proportionality point of view. Party A won 100,000 of a total of 321,000 votes, or 32.1% of the votes. It received 2 of 7 seats, which is about 28.6% of the seats. Parties B and C also received 28.6% of the seats, but with 24.9% and 21.8% of the votes: Party C seems to be getting more than its share. Party D had 13.7% of the votes and received 1 seat, or 14.2%, so that feels right. Finally, Party E won 8.4% of the votes and received 0% of the seats.

If we think of parties as states and their votes as the populations of those states, then the problem of allocating seven seats to the parties is exactly the same as the problem of apportioning seven seats to the states, something we know a lot about. Let's solve this reformulation of the problem with the Jefferson method. The only thing we'll change is that we won't automatically give each party one seat; the Constitution demands this when apportioning seats to states, but it's not a requirement here. We first compute the standard divisor:

SD = voting population ÷ number of seats = 317,000 ÷ 7 ≈ 45,286.

Then we determine how many times SD goes into each party's number of votes to find the standard quotas. Rounding down gives us the lower quotas. The table shows the lower quotas adding up to 4, which is less than the number of available seats. So we have to decrease SD. After some trial and error, we find that SD = 35,000 works. The fifth column gives the modified quotas and the last column gives the Jefferson apportionment, obtained by rounding the modified quotas down.

Party	Votes received	Standard quota	Lower quota	Decrease divisor to 35,000 and recalculate quota	Final Jefferson apportionment
A	100,000	2.21	2	2.88	2
B	80,000	1.77	1	2.29	2
C	70,000	1.55	1	2.00	2
D	44,000	0.97	0	1.26	1
E	23,000	0.51	0	0.66	0
Total			4		7

Notice anything? Coincidence? Math thinks not.

The D'Hondt and Jefferson methods are one and the same. One clue (besides the identical allocations) is that the D'Hondt example ended with the last seat allocated to the party with a quotient of 35,000. This is the same as the modified divisor in the Jefferson calculation. The difference between the two methods is that D'Hondt doesn't need the divisor (aka quotient) up front because it adjusts at each stage of the process, allocating seats as it goes along. So, for example, the Round 5 column simulates what would happen if the Jefferson divisor were 44,000. The modified quotas would be 2.27, 1.82, 1.75, 1, and 0.5, respectively, for A, B, C, D, and E, and rounding down would produce the allocation that appears in that column: 2, 1, 1, 1, 0. The iterations end when all the seats have been given out, which produces the final divisor. Jefferson works backward—or, really, outsources the iterations to the user's trial-and-error search for a divisor that works.

Since D'Hondt and Jefferson are equivalent, everything we know about the latter applies to the former. Jefferson's method avoids paradoxes but might fail the quota rule, and this is true for D'Hondt as well. But more important, just as Jefferson's method favors big states, D'Hondt favors big parties. We know from our study of apportionment that if we want to avoid this bias, the Webster method is the way to go. Our second party list proportional representation system agrees.

The Sainte-Laguë Method

In 1910, French mathematician André Sainte-Laguë (1878–1950) offered this modification of the D'Hondt method: instead of dividing by successive integers 1, 2, 3, 4, . . . to determine which party gets a seat at each stage, let's divide by *odd* integers 1, 3, 5, 7, . . . It sounds like a frivolous change that shouldn't require a bona fide mathematician to figure out. But that's like saying "my child could have drawn a bunch of squares" when looking at a Mondrian painting. Sainte-Laguë's insight is surprisingly original and sneaky.

Here is the example from before, now recomputed with the Sainte-Laguë method. The only difference is that if a party gets a seat at some round and now has s seats, we divide its number of votes by $2s + 1$ (the $[s + 1]$th odd number) rather than $s + 1$ (the $[s + 1]$th number).

	Round 1	Round 2	Round 3	Round 4	Round 5	Round 6	Round 7
Party A	100,000 / **1**	33,333 / 1	33,333 / 1	33,333 / 1	33,333 / **2**	20,000 / 2	20,000 / 2
Party B	80,000 / 0	80,000 / **1**	26,667 / 1	26,667 / 1	26,667 / 1	26,667 / 1	26,667 / **2**
Party C	70,000 / 0	70,000 / 0	70,000 / **1**	23,333 / 1	23,333 / 1	23,333 / 1	23,333 / 1
Party D	44,000 / 0	44,000 / 0	44,000 / 0	44,000 / **1**	14,667 / 1	14,667 / 1	14,667 / 1
Party E	27,000 / 0	27,000 / 0	27,000 / 0	27,000 / 0	27,000 / 0	27,000 / **1**	9,000 / 1

The final allocation is different from D'Hondt's: it takes one seat from Party C and gives it to E. Remember that the vote shares for the parties were 31.2%, 24.9%, 21.8%, 13.7%, and 8.4%, respectively. The respective seat shares now are 31.2%, 24.9%, 14.2%, 14.2%, and 14.2%. Party C seems to have gotten the short end of the stick, but the total discrepancy between the seat and vote shares is actually slightly smaller with this allocation (by 1 percentage point), verifying that, at least in this example, Sainte-Laguë does better at achieving proportional representation than D'Hondt.

Sainte-Laguë gave a seat to what appears to be a smaller party. That's not surprising; this method is the same as Webster's, which in the apportionment game pays more attention to small states than Jefferson's method does.*

Sainte-Laguë is not as widely used as D'Hondt. It has been adopted in Bosnia, Germany, Norway, Sweden, New Zealand, Nepal, and a few other countries. It is generally regarded as fairer than D'Hondt, although it does have a freaky flaw: it might assign fewer than 50% of seats to a party (or a coalition of parties) that wins more than 50% of the votes (it's happened in several elections in Denmark, for example). We would never observe this glitch in the Webster apportionment of seats in the House of Representatives because no state has the majority of the U.S. population.†

CANDIDATES BEFORE PARTIES

Party list proportional representation systems, especially closed list ones, put the party ahead of its candidates. The proponents of such systems claim that most people care only about the party anyway and that

---

* Without getting into too much detail, division by odd numbers in Sainte-Laguë corresponds to rounding at the usual 0.5 cutoff in Webster, in contrast to d'Hondt, which corresponds to rounding down like Jefferson.

† Although the phenomenon of a party (or really, a party's candidate) winning the popular vote in the U.S. presidential elections yet losing the Electoral College could be regarded as a manifestation of such an inversion.

party lists keep things simple. But what if the voters want full control of who gets elected? That sounds more like democracy.

Enter *candidate-centered* proportional representation systems. If you've ever voted in a school committee, city council, or board of directors election where several members are to be elected at the same time, you participated in such a system.

Warning: the "proportion" part of proportional representation might be obscured in what's coming. The candidates are supposed to be chosen in proportion to what? The answer is in proportion as it relates to the underlying parties. In other words, if the candidates belong to political parties, then the end result is supposed to produce the same or a similar proportion of candidates from each party as the proportion of the votes won by that party's candidates.

Just as with party-centric proportional representation systems, there are many ways to proceed. I'll describe two systems, *bloc voting* and *single transferable vote*, because they are indicative of the options. These two are also relevant for the United States: bloc voting is what we used to do and single transferable vote is what we should do.

Bloc Voting

Bloc voting is a collection of methods in which voters select some number of candidates but without ordering their choices in any way. The simplest version is *plurality bloc voting*, which is essentially a bunch of simultaneous plurality elections. If someone says "bloc voting," they probably mean plurality bloc voting. Every voter gets as many votes as there are seats to be filled and selects that many candidates. Thus, if a multimember district is to select three representatives, a voter can circle or check up to three names on the ballot. They cannot allocate more than one vote to a candidate. This is like approval voting but with a preset number of approvals, namely three. The winners are the three candidates who get the most votes.

Bloc voting has all the problems of plurality and then some. The most serious is that a minority of the voters can decide the entire slate of winners. To convince you, let's first see if you believe this: if slightly more than half the voters decide to select the same candidates, then those candidates are the winners. This is simply because each of them will get the majority of the votes. So even if just over half the voters in a multimember district are, say, Democrats, they can force the entire slate of delegates to be Democrats. In fact, this is often what people think of when they hear of bloc voting—a subset of voters acting as a voting bloc, aligning their votes to push their preference through the entire election.

Practically, even fewer than a majority of the voters operating in tandem can decide the results. As long as the rest of the electorate votes in some nonaligned way, spreading their votes out across the list, the bloc can win and impose their candidates on everyone. For example, suppose there are thirty-six voters electing three of six candidates. The profile is as follows:

	15	7	7	7
Candidate A	✓	✓		
Candidate B	✓		✓	
Candidate C	✓			✓
Candidate D		✓	✓	
Candidate E		✓		✓
Candidate F			✓	✓

In this scenario, fifteen of thirty-six voters, or 42%, decided to vote together and jointly pushed candidates A, B, and C. Since not enough voters gathered around any of the remaining candidates (each won 14 votes), the election churned out the three candidates this coalition backed. If A, B, and C were from the Orange Party and the other three were from the Green Party, Orange would have shut Green out of representation, even though Green candidates received 42 of 108, or 39% of the vote. Definitely not proportional.

Plurality bloc voting sits at the heart of an unpleasant chapter of U.S. history. In 1787, because the Constitution does not specify the method of elections at any level, the states were left to their own voting devices. They initially held multimember elections to choose representatives at large, meaning the entire state voted. The method was plurality bloc voting. But some states noticed that the results tended to leave large geographic swaths without representation, so they decided to carve themselves into districts and use single-member plurality to guarantee regional representation. In each of a few holdout states, however, one party ran the show and retained the system that gave them a monopoly on elections by voting in blocs. In 1842, Congress banned at-large voting because of this manipulability. The ban was then revoked and reinstated several times until the final fatal blow in the form of the Uniform Congressional District Act of 1967, which mandates single-member districts. At that point, plurality bloc voting had become an issue of racial justice because white populations in southern states were using it to shut the Black population out of the legislature. Justice Ruth Bader Ginsburg even said that she considered bloc voting to be one of the most pressing dangers facing minority representation, along with gerrymandering.

The North didn't smell like roses either. Plurality bloc voting was used in many of its cities to exclude minorities or ethnic groups. In fact, in 2017, in Lowell, Massachusetts, not far from where I sit, a coalition of Asian American, Hispanic, and Latino voters sued the city on the grounds that these groups constituted 41% of the city's population yet were almost completely excluded from the city council and the school committee.* Which, as you can guess, were mostly white. Two years later, the city agreed to change its voting system. For city council elections, Lowell is now divided into eight districts and has three at-large seats. Even though single-member plurality districts are far from the best alternative, the 2021 elections still managed to produce the most diverse, most representative city council in Lowell's history.

Multimember districts with bloc voting are used for state House elections in seven states: Arizona, Maryland, New Hampshire, New Jersey, North Dakota, South Dakota, and Vermont (which also uses them for state senate elections). Most districts in those states elect two representatives to their state legislature.

---

* The Metric Geometry and Gerrymandering Group from Tufts University, which you'll remember from their work on gerrymandering, provided expert analysis in this case.

Plurality bloc voting can be tweaked to give voters more freedom to express preferences and strength of support. A common variant is *cumulative voting*. We saw a single-member version of this in chapter 5. In the context of this chapter, a voter has as many votes as there are candidates to be elected, but instead of having to give those votes to distinct candidates, they can give multiple votes—even all of them—to any candidate. The candidates with the most votes win.

The method can be amended in companies when shareholders are electing members of the board of directors. The number of votes each shareholder gets is not just the number of directors to be chosen; it is the product of that number and the number of shares the shareholder owns. This is a "weighted" system in which the voting power of a shareholder is proportional to the fraction of the company they own.

Yet another version of plurality bloc voting removes the limit on the number of votes a voter can cast, so they can check any number of candidates, but only once each. This is precisely approval voting, except that instead of filling one seat, multiple seats are filled by the candidates with the highest number of votes. This method is appropriately called *multiwinner approval voting* or *bloc approval voting*.

All of these bloc voting systems, however, are vulnerable to a minority of the voters gaining disproportionately large representation by voting in concert.* Let's look at a system built to resist this.

Single Transferable Vote

The 2021 Academy Awards ceremony was a big event in my house. A childhood friend of mine, Jasmila Žbanić, was up for an Oscar in the foreign film category. She was the director of *Quo vadis, Aida?*, a movie depicting the genocide the Serb army committed in July 1995 in the Bosnian town of Srebrenica. True to the weight of its subject, Jasmila had managed to craft a solemn, gut-wrenching film without reaching for

---

* Cumulative voting gets there part of the way because it tends to produce results that are somewhere between plurality and proportional representation. This is why cumulative voting is sometimes characterized as a *semi-proportional* system.

melodrama and cheap sentimentality. A compelling personal story was interwoven with a chronicle of severe cruelty and the world's indifference to it. The movie was important and Oscar worthy. But I knew that Jasmila's chances of winning depended not only on the quality of the film but also on how the Academy of Motion Picture Arts and Sciences votes. So I went digging.

———

Just as plurality bloc voting is a generalization of plurality to the multi-winner setting, *single transferable vote* generalizes instant runoff.* The idea is to do the usual instant runoff with a twist: as soon as a candidate crosses a certain threshold or quota, they are declared one of the winners and are taken off the ballots, while—and here is the crucial part—their excess votes are transferred to the remaining candidates in proportion to the number of second-place votes those candidates received from voters who put the winner first.

Let's say we have a multiseat district with three candidates to be elected of five who are running. All voters fill out a standard ranked ballot, ranking as many of the candidates as they want according to how much they like them. Suppose the profile looks like this:

8	6	8	5	3	3
A	A	C	B	E	E
B	C	E	D	C	A
E	D	D	E	D	B
D	B	A	C	A	D
C	E	B	A	B	C

The first thing we need in single transferable vote is the quota. The most common is the *Droop quota*, named after the English mathematician Henry Richmond Droop (1832–1884), who was the first to use it in

* It is sometimes called the *generalized Hare method* or *proportional ranked choice voting* or *choice voting*.

conjunction with single transferable vote.* (There are also murmurs that he was the first to notice Duverger's law.) The Droop quota (DQ) is calculated like this:

$$DQ = \lfloor \text{total number of voters} \div (\text{seats to be allocated} + 1) \rfloor + 1.$$

The notation $\lfloor x \rfloor$ means that we round x down to the nearest integer. In this example, the number of voters is thirty-three and the number of seats is three, so

$$DQ = \lfloor 33 \div (3+1) \rfloor + 1 = \lfloor 33 \div 4 \rfloor + 1 = \lfloor 8.25 \rfloor + 1 = 8 + 1 = 9.$$

Any candidate who clears 9 votes will thus be declared one of the winners. Here's the cool thing: with this quota, there can only be three winners, which is what we want. If there were four, then they would each need to collect at least 9 votes, but the total number of voters would then have to be $4 \times 9 = 36$, and we don't have that many. Droop constructed the quota formula to give precisely the same number of winners as there are seats to be filled, given the number of voters.

If we look at the first-place row, candidate A has already cleared the quota because they have 14 top votes. So they're in. But now we look at how many extra votes A has beyond the requisite 9, and that's 5. Instead of wasting these votes, we transfer them to the second-place candidates in the columns where A was the top scorer, namely columns one and two. So the 5 excess votes will be passed along to candidate B in the first column and candidate C in the second column.

We're also going to be smart about how we pass these 5 votes down. We'll do it in proportion to the number of voters who had B as second choice and C as second choice. In other words, we'll transfer the 5 votes in the ratio 8:6. Accordingly, B gets

$$(5 \div (8+6)) \times 8 = 2.86 \text{ votes}$$

and C gets

$$(5 \div (8+6)) \times 6 = 2.14 \text{ votes}.$$

* Another option is the *Hare quota*, which simply divides the total number of votes by the number of seats, but this quota is seldom used.

It's worth pausing to get used to the idea that we're now dealing with fractional (or decimal) votes. We can no longer associate these votes to living human beings because they now only exist in the non-integer realm of mathematics, but that's where we want to be. So many of our difficulties have been rooted in working only with integers and the consequent need for rounding.

To keep track of what we've done, we remove A from the profile and shift everything up, as with instant runoff, and record the transferred votes for B and C in the first and second columns.

2.86	2.14	8	5	3	3
B	C	C	B	E	E
E	D	E	D	C	B
D	B	D	E	D	D
C	E	B	C	B	C

Repeat. Looking across the top row, we ask whether anyone has cleared the quota of 9 votes. Yes: C now has $2.14 + 8 = 10.14$, so they are elected as the second winner. C's excess above the 9 necessary votes is 1.14, so we distribute those votes to the second-place candidates in the two columns where C is on top. These extra 1.14 votes have to be distributed in the 2.14:8 ratio. The calculation for the second column is

$$(1.14 \div (2.14 + 8)) \times 2.14 = 0.24 \text{ votes}$$

and for the third, it is

$$(1.14 \div (2.14 + 8)) \times 8 = 0.9 \text{ votes}.$$

We remove C from the profile, shift up, and record the transferred votes to get this result:

2.86	0.24	0.9	5	3	3
B	D	E	B	E	E
E	B	D	D	D	B
D	E	B	E	B	D

Where do we stand now? Nobody has 9 votes at this point. That's okay, because we've built up some experience. We do the familiar instant runoff thing and eliminate the candidate with the lowest first-place score, which would be D, with a score of 0.24. Removing D and shifting everyone up gives this result:

2.86	0.24	0.9	5	3	3
B	B	E	B	E	E
E	E	B	E	B	B

Now B has 2.86 + 0.24 + 5 = 8.1 votes and E has 6.9 votes. Neither candidate has cleared 9 votes, so we again eliminate the lowest scorer, which is E. That leaves B with 15 votes, and B becomes the third candidate elected.*

Single transferable vote is a little complicated, I'll grant you that, but who cares? A machine would do the calculations anyway. What's important is that *for the voter* it's no more difficult than any of the standard ranked methods. It's a good system with all the benefits of instant runoff: it avoids spoilers and vote splitting, it discourages negative campaigning, and it encourages participation by smaller parties and candidates. In fact, the participation effect is even stronger than with instant runoff, and we can quantify it.

Let's examine the Droop quota "in general" without reference to a particular election or specific number of voters, by working with percentages of voters. What I mean is that if there are n seats, then

$$DQ = (100 \div (n + 1))\% + 1.$$

---

* Here's a strange thing: suppose we wanted only one winner from the original profile, treating instant runoff as a special case of single transferable vote. In that election, E wins. However, E is not one of the three winners of the single transferable vote we just ran. This phenomenon is called the *committee size paradox*: if we enlarge the winner pool of a single transferable vote election, the winners of the smaller pool might not be a subset of the winners of the larger pool. In other words, the same profile might elect different winners depending on how many winners are being selected. David McCune and Adam Graham-Squire analyzed 1,079 Scottish single transferable vote elections and found that this paradox occurred nine times as they tweaked the size of the winner pool.

This formula gives the minimum number of first-place votes needed to secure a seat, expressed as a share of the total number of voters (plus 1 vote to get over the bar). We can tabulate the Droop quota for some values of n:

n	DQ
1	50.0% + 1
2	33.3% + 1
3	25.0% + 1
4	20.0% + 1
5	16.7% + 1
6	14.3% + 1
7	12.5% + 1

Look at the case $n = 1$. There is one seat to be filled and we're back to a single-member election. Single transferable vote reduces to instant runoff and the DQ says that to win for sure, you need 50% + 1 voters to capture first place. Or you need to place high enough for enough voters that you eventually have a majority (but there is no clean and easy way to describe what that entails in terms of placement across ballots).

As the number of seats to be filled increases, the threshold for getting elected drops. If $n = 3$ candidates are to be elected, all it takes is to appeal to about 25% of the voters to be guaranteed entry into the legislature. For five seats, the number is about 16.7%. This is a totally different ball game than plurality. Third parties are discouraged from participating in plurality systems because the prospects are daunting when you're up against behemoths like the Democrats or the Republicans. But if you only need to be the favorite of 16.7% of the voters, that's a lot more manageable. You can run on particular issues that appeal to pockets of voters in a niche campaign. Or you can appeal to a population within a district that's aligned according to some geographic trait and is looking for someone to represent their local interests. This allows independence. In a proportional representation election, you do not have to court the big guys or be afraid of them. When you win a seat or two, they'll pay attention. They might even ask you to enter a coalition or court you to earn your legislative support because in a proportional representation system, coalition building happens *after* the election, so each candidate can offer the voters a distinctive platform. These were exactly my arguments, back in the gerrymandering chapters, for why we should ditch contiguity as a restriction on district shape and use multiwinner districts as a superior alternative.

———

I know you've been waiting to hear about the Oscars. In 2009, the Academy of Motion Picture Arts and Sciences changed the way its members vote. It introduced single transferable vote to the process of choosing the nominees for each category. The goal was to ensure that nominated films would have the broadest support possible. So far, so good, because single transferable vote and its baby brother instant runoff are designed to do precisely that. The members of the academy's seventeen branches of film production rank up to five movies in their category (after an initial winnowing to fifteen contenders). The exception is Best Picture, for which any member can submit a ranked list of five to ten films. Because people might not be submitting lists of the same movies, a selection of the final five (or 5–10 in case of Best Picture) has to be made. This is done by single transferable vote. The nominees in each category are then announced to the world. (This announcement has become a pre-Oscars event unto itself.)

Now that we have the nominees, how are the winners selected in the second part of the process? Best Picture is chosen by instant runoff, a good option because it is designed to collect the widest consensus possible. Some complain that Best Picture winners have become bland crowd-pleasers and that the process shuts out the small, daring movies that the Academy Awards were created to recognize nearly a century ago. But the standard argument in favor of instant runoff and against the old method, plurality, is still more compelling: with ten nominees, a film could win with only a little over 10% of the votes with a plurality vote count. That hardly seems like the voice of collective recognition.

What about the other categories? How do they, in particular Foreign Film, pick the winner? Those, for some reason, are still decided by plurality. C'mon Academy, you were doing so well!

Jasmila's movie lost. The winner for Best International Feature Film in 2021 was *Another Round*, a Danish movie with the badass Mads Mikkelsen in the lead role. It was a pleasant feature about midlife crisis and drinking. But it was nothing we hadn't seen before, a low-stakes charmer, exactly the kind of a movie that would make a sufficient number of academy voters feel warm and fuzzy enough to put it down as their favorite. For all we know, the movie could have won with just over 20% of the votes. Had Foreign Film used instant runoff, the outcome could have been different for *Quo vadis, Aida?* Regardless, whoever might have won in an instant runoff, the result would have more fully reflected the academy voters' true preference. As goes the usual refrain with plurality voting, we will never know.

PR FOR PR

A first glance at the history of proportional representation in the United States suggests that it was dropped like a hot potato, essentially given a vote of no confidence all the way back in 1842, when it was first banned for congressional elections. It didn't matter that John Stuart Mill, one of the most important political thinkers of all time, sang the praises of single transferable vote in his 1861 essay *Considerations on Representative Government*. And in a letter to Hare, the method's inventor, he wrote: "You appear to me to have exactly, and for the first time, solved the difficulty of popular representation; and by doing so, to have raised up the cloud of gloom and uncertainty which hung over the futurity of representative government and therefore of civilization." Damn!

Even though proportional representation was banned for congressional elections, U.S. reformers continued to dabble in it. Starting in the late nineteenth century, Progressive Era civic leaders brought it to several cities in a backlash against manipulable single-member districts. In the early 1900s, over twenty cities, including New York, Boulder, Kalamazoo, Sacramento, Cleveland, and Cincinnati, were using single transferable vote. With it came fairer representation and greater voter satisfaction. At one point, New York City elected people from five different parties to its city council instead of the usual suspect, the Democratic Party.

And then every city except Cambridge, Massachusetts, scrapped it. The reasons for this are many, but the most common were attacks on the system by members of the major parties who were forced to share power, insufficient PR (as in public relations), and, as is the case with anything that promotes the rights and participation of minorities, racism. The Cold War–era Red Scare also played a role; accusatory rhetoric of the 1950s painted single transferable vote as a Kremlin import. Single transferable vote now survives in Cambridge and Minneapolis and at some universities, including MIT and the University of California, Berkeley. There are occasional blips of activity, like its recent adoption for municipal elections in Eastpointe, Michigan, and Albany, New York, and in the two Portlands, Oregon and Maine, but such cases are rare.* Internationally, single transferable vote is used in only a handful of places—Ireland, Scotland, Wales, Malta, and Australia—but that's not surprising, because most democracies have multiple active parties and use some kind of party list system.

In the United States, proportional representation seems to be relegated to the history books because it worked too well. It was a thorn in the side of the powerful. And to be sure, it does have objective drawbacks. Multimember districts are often large in terms of geography and population and large size can hinder connection and personalized interaction between representatives and their constituents. In the Netherlands and Israel, for example, the entire country is one district with many seats. On the other hand, if the seats for each district are too few, as in Chile, where all districts elect two members, the entry bar is high. Experts say that the district *magnitudes* that seem to work the best for single transferable vote range from three to five.

Coalition building is a must in proportional representation systems, and that necessity can produce instability, balkanization, and some

---

* The cumulative voting story is even shorter. The most significant example is Illinois, which used it from 1870 to 1980. Other than that, it pops up occasionally and sporadically, mostly on the scale of city council elections across the country.

marriages from hell. Since parties usually decide who their candidates are—there are typically no primaries in proportional representation systems—this can lead to intraparty discord. Oddballs and extremists are sometimes given a voice they do not deserve.

All of these objections derive from examples rooted in very specific circumstances. Those definitely exist—look no farther than, well, the Balkans—but evidence of a consistent failure of proportional representation on any of these counts is lacking.

On the other hand, the benefits of proportional representation are manifold. For one thing, the opportunity for greater voter satisfaction and wider political participation cannot be overstated. Groups of voters who form a respectable but not necessarily huge percentage of the electorate have a chance of being represented. In a five-seat district using single transferable vote, all you need is about 17% of the votes to be elected. A cynic might say that nobody wins this percentage anyway except Democrats and Republicans, but that's because in the current plurality system, smaller parties are discouraged from entering races and spending money on campaigning in elections where only two parties have a chance at clearing the high bar to first place. If the threshold were reachable, there would be a lot more diversity of opinion, ideology, and policy vying to cross it.

The lowered entry bar would spur the creation of small parties dedicated to the interests of racial or ethnic minorities and communities of interest. Winning big would no longer be the one and only object, and political platforms could truly be dedicated to the interests of those the parties purport to serve. Politicians could afford to be part of the community and speak more directly with their constituents. The reaction would be an increase in engagement because people would see themselves reflected in the political arena in terms of appearance and concerns. Ireland has used single transferable vote in its general elections since 1918.* In the five elections of the twenty-first century, the number

* Single transferable vote was introduced by the UK before Irish independence in an effort to contain Sinn Féin's disproportionate success in plurality elections.

of parties who have won seats in the lower house has ranged from seven to ten, continuing a robust history of political diversity.

People would also see themselves reflected in the results. In Cambridge, where single transferable vote was carried into the city from the campuses of Harvard and MIT decades ago, 77% of voters see their first choice elected, 91% their first or second choice, and 95% one of their first three choices. Compare these figures with single-member elections, where often large majorities of voters feel unrepresented because their one and only candidate lost. Higher voter turnout (which, at around 70–80% in proportional representation countries blows the United States out of the water) would necessarily follow the greater chance of being represented by one's actual choice.

Proportional representation would also deal a crippling blow to gerrymandering. To see why, take the Triangle Party and Circle Party example at the top of chapter 11 and instead of dividing the population into five single-member districts, create two districts with thirty and twenty people, respectively, who elect three and two representatives. Try some packing and cracking. You will be less successful than before. For instance, you will not be able to gerrymander the two districts so the Triangle Party wins all the seats.

For multiseat districts that elect five or more representatives, proportional representation essentially eliminates gerrymandering and is successful at curbing it in elections for fewer seats. In countries that use proportional representation, districting is a mundane activity about as exciting as listening to someone recount their dream. Nowhere else is this chore elevated to the blood sport that it is in the United States. A reduction in gerrymandering also implies more competitive races and fewer wasted votes.

In tandem with gerrymandering, the single-member system amplifies the urban-rural divide. In the United States, geography matters more than it should. Voters clustered in cities elect Democrats by large margins while the sprawl of Republican voters gives that party more opportunity for wins in a larger number of less populated districts. In proportional representation systems, districts are bigger and thresholds for getting elected are lower. To cross the threshold, votes from a geographically clumped population and those from voters spread around the district are treated equally.

Another purpose of gerrymandering in single-member districts is to protect incumbents. Under proportional representation, incumbency is not sheltered. The larger the district and the larger the number of representatives it elects, the more sensitive the system is to changing voter sentiments. Small shifts in preferences can flip a seat. According to an analysis by Rank the Vote, a ranked choice advocacy organization, implementing proportional representation with single transferable vote in North Carolina would result in five Democrat and five Republican seats and three seats would be too close to call. Massachusetts Republicans would get three of nine seats, exactly what their numbers deserve. Maryland Republicans would get a seat. Ohio would elect three more Democrats, aligning the seats ratio with the votes ratio. Texas would elect another Democrat and four more seats in that state would become competitive. In California, Republicans would gain two more seats and nine more seats would become competitive.

In multiparty proportional representation systems, negative campaigning is less common. Because parties might need to form coalitions after elections, they have a compelling incentive to play nice. If in addition single transferable vote is used, negative campaigning diminishes further for all the reasons we talked about in chapter 3: a candidate wants to place high on as many people's ranked ballots as possible and trash talk risks backlash and consequent lower placement.

Campaigning is also more honest in proportional representation systems. Because a win in a two-party system means complete control of the government, the winner has no problem backtracking on promises made during the ancient history of the campaign trail. Accountability is greater when more parties participate in the legislature. Leadership in a multiparty system means compromise, moderation, and building coalitions. All of this reduces polarization because accusations and vilification are bad strategies.

Proportional representation is better for women. The evidence comes from countries that use mixed-member proportional representation—meaning that some seats are single member and some are decided by proportional representation. For example, in Germany and Australia, the percentage of women elected by proportional representation is higher than in single-member districts. Half of Mexico's Chamber of Deputies are women because of the combination of proportional representation and quotas that mandate a certain share of female representatives. Overall, women are twice as likely to be elected in proportional representation systems than in plurality systems. In 1994, women in Sweden forced the major parties to nominate more women by threatening to form their own party. As corroborated in research by RepresentWomen, this is not an isolated example—with proportional representation, more women are nominated and are elected as parties strive to strike gender balance (or at least appear to do so).

Finally, what about all the protections for minorities the Voting Rights Act etched into the districting system? This is proportional representation's crowning strength. It turns out that proportional representation would do much better than the current system at protecting the interests of minority voters. A simulation by FairVote shows that with multimember districts using single transferable vote, the number of Black representatives would increase by 25% and the number of Latino representatives would increase by 40%. A new report prepared by FairVote and the Metric Geometry and Gerrymandering Group titled "Modeling the Fair Representation Act" confirms this effect and demonstrates that instituting multimember districts in tandem with single transferable vote would result in better outcomes for underrepresented groups. But here is the truly astonishing finding: this would be true even if the districting process ignored the race-conscious border drawing mandated by the Voting Rights Act, an undertaking that frequently becomes mired in politics and litigation.

I hope the message is clear: proportional representation is to winner-take-all single-member districts what a Lamborghini is to a Yugo (staying with the Balkan theme). A Lamborghini for the people. Even if we replace plurality with something smarter such as instant runoff, the winner might not be the first choice for the majority of voters as long as single-member districts are in place. Upgrading to multimember districts with representatives elected using single transferable vote would create a fully functioning electoral system. And toss in independent commissions to make sure the borders are drawn with no safe seats or incumbent protection.

The Fair Representation Act, which is currently in Congress, would achieve this. This bill was first introduced in 2017 and then again in 2019 and 2021, sponsored by Representative Don Beyer of Virginia. It is a progressive piece of legislation that includes much of the soundest mathematics of democracy. Everything suggested in it is constitutional and would not require an amendment.

Congress, however, appears less than enthusiastic about the electoral utopia the Fair Representation Act envisions. The proposal has only eight cosponsors, all Democrats. It is painfully obvious that most politicians cannot conceive of a system different from the one that elected them. The task ahead is formidable, and impetus will have to come from enlightened citizens demanding action from their representatives.

A good start would be to repeal the Uniform Congressional District Act of 1967 that requires single-member districts. States would then be able to do their jobs as the laboratories of democracy and begin experimenting with multimember districts.

It will be a long road, but hey, as Nirvana bassist Krist Novoselic, one of the most outspoken advocates of proportional representation, said: "When I was in a band in the late 1980s and the early 1990s, we came out of this grunge, alternative world. We never thought we were going to be on MTV or a huge band. But things change."*

Change on this scale has happened before. The advances of the Progressive Era that built America's modern civic infrastructure can serve as an inspiration, as can the Voting Rights Act of 1965. We can also look at other countries: Italy, Japan, and New Zealand have all instituted ambitious reforms in recent decades.

---

* While we're in the music world, David Byrne also supports ranked choice voting. It doesn't get cooler than that if you ask me.

We have the know-how and we face the moment. The Supreme Court seems poised to strip away protections for underrepresented groups, gerrymandering is rampant, and the two-party system is crushing our democracy, unreflective as it is of the myriad ways we are getting more and more diverse. Moving away from plurality single-member districts and toward proportional representation multimember districts would be transformative. Our representation would truly represent—turning a mirror toward America and showing faithfully what its citizens look like and where they want to be headed.

PART III

Civic Infrastructure

DEMOCRACY AS a whole is a vast human undertaking requiring extensive civic infrastructure, of which systems of voting and representation are an important part. To function well, these systems must have strong mathematical foundations, but the math doesn't operate in isolation. As part of the larger infrastructure, math is enmeshed with overlapping systems—political, social, and legal—that constrain choices and magnify and multiply effects, good and bad. Tracing the dysfunctions of the U.S. Electoral College gives a vivid picture of the challenges of legacy systems but also identifies possibilities for improvement informed by a mathematical point of view.

CHAPTER 14

The Electoral College

TO FOCUS on the mathematics of systems of voting and representation—the two pillars of democracy—we've examined methods and practices close up and (arguably) almost out of context. Our efforts to identify benefits and drawbacks and even neutral consequences—such as whether a voting method discourages negative campaigning or how various courts have reacted to measures of gerrymandering—have drawn us further into the surrounding context, but even then in an orderly way that gives only a taste of how messy things can be when the math meets the human, only a hint of how complex our civic infrastructure has become.

Pulling the camera back in the hope of getting the full picture would make the details indecipherable. But there is one part of the civic infrastructure, a unique and uniquely archaic institution that sits at the core of American democracy, in which an amazing amount and variety of mathematical dysfunctions converge: the Electoral College. There we stand a chance of taking it all in. Buckle up for a wild ride.

CHARACTERS PREEMINENT FOR ABILITY AND VIRTUE

The abbreviated story goes something like this.* When it came time to decide how to elect a president, delegates to the Constitutional

* For details, I recommend two books: *Let the People Pick the President* by the New York Times writer and editor Jesse Wegman and *Why Do We Still Have the Electoral College?* by Harvard professor Alexander Keyssar.

Convention of 1787 could not agree on anything. By the time it was settled, the matter had been chewed over (and under and around) for twenty-one days and voted on thirty times, more than any other. The myriad proposals included election of the president by state governors or by (some subset of) the Congress. Election by *popular vote*, meaning that every citizen votes and everyone's vote carries the same weight, was also on the table. All the options were rejected. Finally, the seemingly unresolvable issue was passed to the Committee on Unfinished Parts* (obviously named by a time-traveling Monty Python) consisting of a representative from each of the eleven states.

As Jordan Ellenberg describes it in his book *Shape*, "If you have ever sat in a long meeting as day care pickup got nearer and nearer, knowing you couldn't go home until the meeting produced a policy document everyone there could make themselves grumblingly sign, you have a pretty good idea of how the Electoral College came to be." James Madison, one of the framers of the Constitution, put it in terms that were probably equally funny back in the day: the creation of the Electoral College was "not exempt from a degree of the hurrying influence produced by fatigue and impatience."

The committee finally put forward a solution: "Each state shall appoint in such a manner as its Legislature shall direct, a number of electors equal to the whole number of senators and members of the House of Representatives, to which the State may be entitled in the Legislature." In other words, each state will choose *electors*, and there will be as many of them as is the size of the state's delegation in Congress—the number of state's representatives to the House plus two for the senators. The electors will be the ones casting votes for the president. Whoever they pick will go to the White House.

---

* The creation of this committee was suggested by Representative Roger Sherman of Connecticut. Google him; he looks like someone who liked committees. And isn't *every* committee a Committee of Unfinished Parts? Isn't the purpose of every committee to finish some unfinished parts? Or maybe it is also to *create* unfinished parts so it can exist in perpetuity? Can you tell I'm currently the chair of my department?

The Electoral College solution was a compromise between electing the president by popular vote and electing the president by the Congress. This struck the middle course in several ways. First, it appeased the southern states, which had already scored a major win with the *Three-Fifths Compromise*. Here is another number, 3/5, a singularly appalling fraction that stains U.S. history and seems to have been pulled out of a hat. According to the agreement reached at the Constitutional Convention, each slave would count as 3/5 of a person for the purposes of taxation and apportionment of the House of Representatives. Slaves could not vote, but including them gave southern states more congressional representatives. Various other proposals had been floated around, including counting slaves as 1/2 or 3/4 of a person or even a full person before 3/5 was accepted.

Because of the compromise, southern states suddenly became much more populous and thus secured more representation. Virginia, for example, received six more House seats than Pennsylvania even though Virginia had the same number of eligible voters (who were all white men at the time). The South was on a winning streak and wasn't about to relinquish this power in presidential elections. Madison alone among the southern delegates recognized the perils of the Electoral College system. He advocated for the national popular vote and for putting democracy above local interests. But a popular vote would have significantly reduced the outsized influence of the southern states because slaves could not vote. If the body deciding presidential elections could be modeled after the Congress, on the other hand, then southern states would have proportionally the same inflated sway in selecting the president as they did in the House. This is why the Electoral College is structurally a copy of the Congress—the Congress is mirrored in the number of electors in each state. This is yet another way in which slavery's legacy haunts and impedes our democracy to this day.

Small states sided with the southern states; they appreciated the benefit of having electoral votes corresponding to senators. Because each state gets two senators and at least one representative, the minimum

number of electoral votes is three. All states, including the least populous ones, would in this way hold some power in presidential elections.
That, after all, was the essence of the federalist way. In the end, this
coalition of southern and small states tipped the scales in favor of the
Electoral College.

The Electoral College achieved yet another compromise. Having
Congress choose the president was seen as too centralized and too susceptible to machinations. It also interlaced two branches of government
unacceptably. On the other hand, the popular vote was also manipulable
because ordinary people were deemed malleable and easily swindled by
campaign promises. A popular vote would potentially be dominated by
big states, and that seemed highly impractical because there was no easy
way for people of one state to be informed about candidates from other
states. Moreover, the heavy responsibility of electing the president
should not be left to the uneducated populace; the Electoral College
would consist of learned and informed men who understood the ways
of politics and could rise above it. As Alexander Hamilton put it in Federalist No. 68, the electors would be "men most capable of analyzing the
qualities" and "most likely to have the information and discernment"
when considering the presidential candidates. The subtext is very near
the surface: ordinary people are too ignorant for such a consequential
task. The Electoral College, by contrast, would be so competent that it
would select "characters preeminent for ability and virtue" as presidents,
Hamilton argued.

States decide how they choose their electors. (My home state of Massachusetts has changed its mind about how it does this eleven times.)
Initially, electors were chosen by state legislatures, although states gradually moved to popular elections. Once chosen, electors were supposed
to weigh and deliberate about each of the presidential options in an
objective and rational manner.

That rectitude lasted for all of two elections. The two-party system
arose quickly despite the framers' hope that it wouldn't, and electors fell
under its sway. Parties would choose only electors who were certain to
vote for their candidate, and in turn, candidates for the role of elector

began to run along party lines. Electors quickly became proxies of the political parties, as they are to this day.

In the first four presidential elections, Electoral College delegates cast two votes. The candidate with the most votes became president and the runner-up became vice-president. But clouds had already started to gather in 1796, when John Adams was elected president and his nemesis, Thomas Jefferson, became vice-president. Four years later, all hell broke loose. The two-party system had already taken hold and pairs of candidates ran as a ticket, but the electors still cast two separate votes. So in the 1800 election, Thomas Jefferson and Aaron Burr, his running mate from the Democratic-Republican Party, both received 73 electoral votes (at least 12 of Jefferson's electoral votes came from states with an inflated number of representatives due to the 3/5 clause). The Constitution says that a tie in the Electoral College shall be resolved by a vote in the House, where each state's delegation gets a single vote (more about this in a moment). The House tied thirty-six times before electing Jefferson. The complexity and political intrigue of this era is covered in a number of books whose titles contain phrases like "the most tumultuous election" and "the most controversial election."

The only other time the House had to intervene was in 1824, when it elected John Quincy Adams over Andrew Jackson. The Constitution says that to become the president, a candidate must receive a *majority*, not a plurality, of the electoral votes. Neither Adams nor Jackson had received a majority, and the House elected Adams. The irony is that Jackson had received more popular votes *and* more electoral votes than Adams. He subsequently became an advocate for winner-take-all statewide plurality elections for delegates and was instrumental in getting states to adopt them.

The cataclysm of 1800 led to the passage of the Twelfth Amendment in 1803, which said that electors must specify which one of their two votes is for the president and which for vice-president. That is the only successful amendment ever to address the insufficiencies of the Electoral College.

By 1832, all states except South Carolina were using the popular vote to choose electors, marking a departure from an expression of state sovereignty. When legislatures picked electors, they could do so in keeping with the state's politics and preferences. But when people vote for electors, they vote for the party and its candidate, regardless of the state they live in.

Around the same time, a transition to statewide winner-take-all elections occurred. Initially, electors were chosen districtwise; each district selected one delegate. Under this method, some of the state's electors might vote for one candidate and others for someone else. But gradually, states switched to a *general ticket*, or winner-take-all, plurality system according to which the election took place statewide and *all* the electors were pledged to the state plurality winner. This maximized the bang for the buck; it guaranteed that the state would be relevant to the candidate it supported because it would be giving them all its electoral votes. Madison and many of his fellow founders were dismayed. This system was not what they had imagined. They worried that it would promote sectionalism, strengthen political parties, and disenfranchise voters in districts that went a different way from the state's plurality.

They were right to worry. This is exactly where we are now. All states except Maine and Nebraska, which went back to the districtwide system in 1972 and 1992, respectively, use the statewide winner-take-all system. As does Washington, DC, which the Twenty-Third Amendment gave three electoral votes to in 1961. The electors are chosen by their political parties, sometimes in primaries, sometimes in state party conventions. Being made an elector is a sign of gratitude for party loyalty or activism or an acknowledgment of proximity to the party's head honchos. On Election Day, each party has a slate of electors ready to deploy should their home state elect their candidate. Then on the first Monday after the second Wednesday in December, the electors of the winning party meet in each state and cast their votes. On January 6th, envelopes containing these votes are opened and tallied by the vice-president, who officially announces the winner. And that's how the president of the United States is elected. By the Electoral College. Not by you and me.

No other democracy has anything resembling the Electoral College. Historical examples are also few and far between. From about the thirteenth century, the king of the Holy Roman Empire was chosen by a handful of the greatest princes. Brazil, Argentina, Paraguay, and Chile used some version of an electoral college at some point, but all have replaced it by direct popular vote elections. At the moment, in all democracies except the United States, the head of government or the combined head of state and head of government, as in the United States, is elected either by popular vote or by the national legislature. Many countries hold runoff elections between the top two candidates. One might object and say that indirect elections by members of a legislature—as when the German president is elected by the Bundestag and delegates from state parliaments or when the winning party in the UK appoints the prime minister—are essentially a form of electoral college. And that's true, but that is neither the only nor the primary item in their job descriptions. They already serve in the government and voting for president is just one of the things they do. The sole job of the U.S. Electoral College is to act as an intermediary in the presidential elections. Shouldn't it be a red flag that *nobody else* uses an electoral system like this?

But here's the truly shocking part: nothing in the Constitution says the people can or should elect the president. The Supreme Court affirmed this in its *Bush v. Gore* ruling in 2000. The Constitution does not even say that the people should vote. *The Constitution does not give citizens a voice in presidential elections.* It gives all the power to the electors, who are in turn elected in any way the state legislatures see fit. States technically don't even have to hold elections; they could simply declare whoever they want as electors. In Donald Trump's multiprong attempt to overturn the 2020 election, exploiting this technicality was one of the key ingredients. Trump and his backers, baselessly arguing that the elections were illegitimate, pressed Republican members of several state legislatures to ignore Joe Biden's popular vote wins and appoint a Republican slate of electors.

In thirty-two states plus the District of Columbia, the law mandates that electors must vote according to the popular vote. But seventeen of those states have no enforcement mechanism. An elector who votes against the popular choice is called *faithless*. Of some 24,000 electoral votes cast in fifty-nine elections, there have been only twenty-four such "deviant" votes cast for a member of the opposite major party, and they have never affected the outcome.* There were seven faithless electors in the 2016 elections, five who had originally voted for Clinton and two who had voted for Trump. Three gave their votes to Colin Powell, and four others gave one vote each to Ron Paul, Bernie Sanders, John Kasich, and Faith Spotted Eagle. This was the largest number of faithless electors in history that were not reacting to the death of a candidate. There were no faithless electors in 2020.

The Electoral College does not favor one party over another. There are temporary appearances of bias due to the demographic distribution of voters at any given moment. Democrats seem currently to be disadvantaged, but that's because they tend to carry more urban states by large margins so their popular vote tally is inflated compared to the electoral one (more on the difference between the popular and the electoral vote shortly). Zooming out from the most recent elections and considering a longer period shows that, overall, the Electoral College exhibits no party preference.

Even the most ardent of constitutionalists would admit that some degree of error might be built into the Electoral College as a consequence of the arduous and ad hoc labor that delivered it at the Constitutional Convention. It also doesn't hurt to be reminded of the obvious: status as the oldest written constitution in operation today does not guarantee that it has no flaws, especially considering that the founders could never have anticipated the political, demographic, and technological circumstances we find ourselves in today.

But trying to guess how the founders would feel about the current state of affairs is speculative and open to interpretation, and by extension, to ideology and politics. Mathematically, however, it's a different story: the Electoral College is deeply problematic on at least three separate and indisputable counts.

---

* There were 90 faithless electors, but 63 of those votes were cast for a different candidate due to the death of their own candidate.

SPINNING PLURALITIES INTO MAJORITIES

California, the most populous state, has fifty-two representatives and two senators, so it has the most electoral votes, 54. Then it's Texas, New York, and so on. Because the distribution of representatives potentially shifts after each census (as per the decennial Huntington-Hill reapportionment), the same is true of electors. After the 2020 census, for example, Texas gained two seats, so it will have two more electors in the 2024 presidential elections. California lost one. Six states (Alaska, Delaware, North Dakota, South Dakota, Vermont, and Wyoming) and Washington, DC, have three electoral votes, the minimum number (figure 14.1).

Because the total number of members of the House of Representatives is stuck at 435, the number of electors is also fixed at

$$435 \text{ (House)} + 100 \text{ (Senate)} + 3 \text{ (District of Columbia)} = 538.$$

But the real magic number everybody talks about every four years is 270, half of 538 plus 1, because the Constitution mandates that a candidate must receive the *majority* of Electoral College votes in order to win.

What happens when no candidate receives a majority? That is not an unfathomable scenario. All you need is a tight election between two candidates and a third candidate winning at least a few electoral votes. For example, if Ralph Nader had won any state George W. Bush carried, say Alaska, the Electoral College tally would have been Bush 268, Gore 266, Nader 3 (the total is 537 because one of the electors from the District of Columbia abstained from voting). Or if Gary Johnson had somehow won Texas instead of Trump in 2016, the allocation would have been Trump 266, Clinton 227, Johnson 38. No majority winner. Then what?

The election would then have been decided by the House, as happened in 1800 and 1824. The absurdity here is manifold. First of all, the lame-duck House imposes its preference for the next four years, regardless of what the incoming House might look like. Not only that, but each state's delegation has only one vote! So there are fifty voters, all equal. California

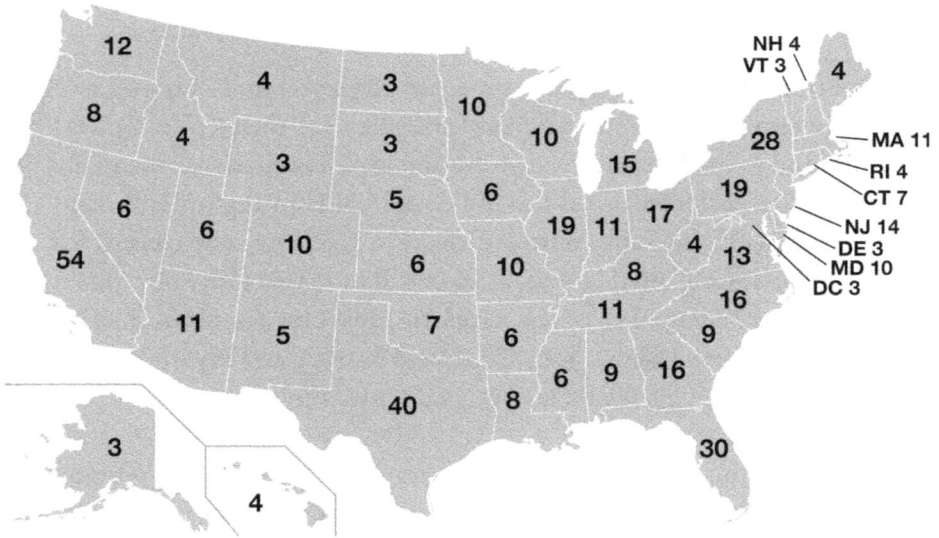

FIGURE 14.1. Electoral College map after the 2020 census.

and Wyoming have the same power at this stage. As do each of the other forty-eight states. Never mind that California has about sixty-seven times more people than Wyoming and fifty-two times the number of representatives in the House as Wyoming. If more states favor a Republican or a Democrat, that's who will become the president, and that choice can be different from what the popular vote—or the next House—demands.

If a state delegation can't agree or is split down the middle, it loses its vote. When the states vote, a *majority* is required for a win. This is not guaranteed to happen if there are more than two viable candidates. Then what? Gridlock that would have to be resolved by wheeling and dealing among the House members. Unlikely, yes, but the possibility should still make us uncomfortable.

By the way, in this situation the Senate elects the vice-president. A majority is required there as well. So a fun scenario (if constitutional crisis is your jam) is that the House elects the president from one party and the Senate elects the vice-president from another.

In contrast to these majority requirements, a candidate only needs a *plurality*, not a majority of the votes, to win electoral votes in a state-level election. Presidential elections are fifty-two separate plurality elections (statewide plurality in forty-eight states plus three districts in Nebraska that choose electors separately plus the District of Columbia) and three instant runoff elections (Maine's two districts that choose electors separately and Alaska's single district that will use instant runoff starting in 2024). Presidential elections are thus fifty-two potential manifestations of all that's bad about plurality. Even the most basic stumbling block, the fact that plurality does not mean majority, shows up regularly because third-party candidates will often win a decent portion of the votes. In 2016, voters in some states saw third-party candidates such as Jill Stein (Green), Gary Johnson (Libertarian), and Evan McMullin (Independent) on their ballots. In 2020, besides Joe Biden and Trump, there were assorted others such as Howie Hawkins (Green), Jo Jorgensen (Libertarian), and my personal favorite it-burns-my-eyes-but-I-can't-look-away-candidate, Kanye West (Independent).* Because of these additional candidates, in 2016, Trump won six states with fewer than 50% of the votes, collecting 101 electoral votes from them. Clinton won seven states by plurality but not majority. That's thirteen of fifty states plus the District of Columbia in which there was no majority winner, or 25% of the fifty-two mini presidential elections.

This shortfall can scale up to national outcomes. In 2016, neither Clinton nor Trump won an overall majority of the countrywide vote; Clinton got 48.2% while Trump received 46.1%. Or take the 1992 election in which Bill Clinton became president with only 43% of the popular vote. The only state he won with a majority of the vote was his home state of Arkansas.

Then there's the clear and present danger of spoilers. The most dramatic example remains the Bush-Gore-Nader calamity in Florida in 2000. Small spoilers can also add up to a big one. Jill Stein might have spoiled the election for Hillary Clinton in three key states (Michigan,

* He got 60,000 votes across twelve states.

Pennsylvania, and Wisconsin). If that is what happened, she took away enough electoral votes to cost Clinton the election.

But the starkest repercussion of the plurality system in presidential elections is the absolute domination of the two-party system, an ironclad duopoly that sets the entrance bar so high that few dare approach it. In recent history, Ross Perot fared best as a third-party candidate; he got 19% of the popular vote in 1992. And he received zero electoral votes. The only time a third-party candidate beat even one of the major candidates was in 1912, when Theodore Roosevelt received more votes than William Howard Taft. In the last hundred years, only four third-party candidates have received any electoral votes at all.

Implementing instant runoff voting, as Maine and Alaska already do, would alleviate these problems. Each of the fifty-five distinct elections would be won by eventual majority, there would be no spoilers, third-party and independent candidates would have a shot, and the voters could feel that the winner in their state better captured their collective preference.

THE BARE MINIMUM

The command to work smarter, not harder is irksomely familiar, and the Electoral College gives fertile opportunities to put it into practice. By looking at the 2016 and the 2020 presidential elections, we can begin to see the possibilities. In 2016, the main candidates were Hillary Clinton, a Democrat, and Donald Trump, a Republican. In 2020, they were Joe Biden, a Democrat, and Donald Trump again. The results are summarized in figure 14.2.

In 2016, Trump won the states in dark gray and earned 304 electoral votes. That's more than 270, so he won the presidency. Biden was able to flip a few states from Republican to Democrat in 2020 and win 306 electoral votes, taking the presidency away from Trump.

But let's look at these results alongside the popular vote tally, meaning the count of votes each candidate won nationwide.

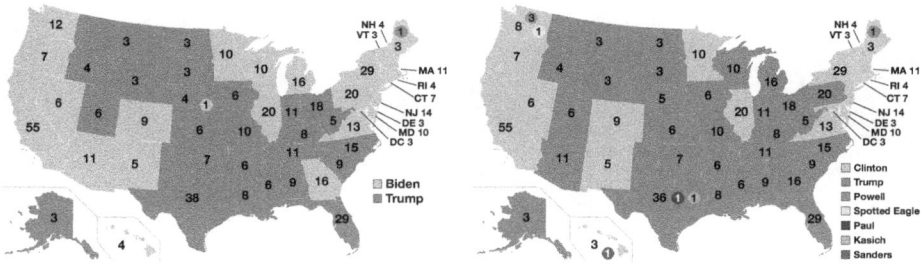

FIGURE 14.2. 2016 and 2020 U.S. presidential election results.

2020	Electoral votes	Popular votes
Joe Biden	306	81,268,867
Donald Trump	232	74,216,747

2016	Electoral votes	Popular votes
Hillary Clinton	227	65,853,514
Donald Trump	304	62,984,828

In 2020, the two counts are consistent in the sense that Joe Biden won both the electoral and the popular vote. But check out 2016—Clinton won almost 3 million more votes than Trump but lost the electoral vote and hence the presidency.

This disagreement between the popular and the electoral vote has happened four other times in U.S. history (figure 14.3).*

That is about 9% of U.S. presidential elections. It has *almost* happened several more times. For example, in 2020, if Trump had won Arizona, Georgia, and Wisconsin, he would have won the presidency, taking 37 electoral votes from Biden. He lost those three states by a combined total of about 43,000 votes. So if Trump had been able to flip about 21,500 voters in those states, or about 0.014% of the total votes cast na-

---

* An argument could be made that the 1960 Kennedy-Nixon election also had differing popular and electoral outcomes because of how the electoral votes were allocated in Alabama and Mississippi.

| Andrew Jackson (left) won 10.5% more of the popular vote than elected President **John Quincy Adams** (right) in 1824. | **Samuel J. Tilden** (left) won 50.9% of the popular vote and still had to concede to elected President **Rutherford B. Hayes** (right) in 1876. | **Grover Cleveland** (left) won 0.8% more of the popular vote than elected President **Benjamin Harrison** (right) in 1888. | **Al Gore** (left) won 0.5% more of the popular vote than elected President **George W. Bush** (right) in 2000. |

FIGURE 14.3. Winners of U.S. presidential elections, besides Donald Trump, who did not win the popular vote.

tionally, he would have won. Of course, these votes would need to have been distributed the right way across the three states, which is about 5,230 in Arizona (0.16% of that state's votes), 5,890 in Georgia (0.12% of votes), and 10,340 in Wisconsin (0.69% of votes). Biden would have won over 7 million more votes than Trump, yet he would have lost the election. In 2004, if about 60,000 more people had voted for John Kerry in Ohio, he would have won the Electoral College with 271 electors, even though George W. Bush had 3 million more votes overall.

This means that four of six, or two-thirds, of the presidential elections since 2020, had or nearly had conflicting outcomes between the popular and the electoral votes.* Mathematical modeling suggests that this is not an event that happens only once every 1,000 years. The chance of a mismatch between the popular and the electoral vote is about one in three when the candidates' shares of the popular vote are within two percentage points of each other. The more competitive the election, the more likely it is that the Electoral College outcome will misrepresent the popular outcome.

Why does this happen?

The problem is the winner-take-all system. (We'll ignore Maine and Nebraska for the moment.) All that matters is that a candidate receives a plurality of the votes in a state. The size of the margin by which they win is irrelevant.

---

* The 2016 election was close to going the other way, though. If Clinton had flipped about 11,300 of Trump's votes in Wisconsin, 5,300 in Michigan, and 22,100 in Pennsylvania, she would have won both the electoral vote and the popular vote.

To illustrate, pretend the United States consists of just four states: Massachusetts, Colorado, Idaho, and Wyoming. The table below gives their approximate populations after the 2020 census and their actual electoral vote numbers. Let's also pretend that everyone in these states votes, which of course is not true, but it doesn't affect the point of the example.

State	Population	Number of electoral votes
Massachusetts	6,900,000	11
Colorado	5,700,000	9
Idaho	1,750,000	4
Wyoming	585,000	3

Because there are a total of 27 electoral votes, a candidate needs at least 14 votes to win the presidency. One way to achieve this is to win Massachusetts and Wyoming. So let's say candidate A wins those states, but barely, with about half the votes. To make the counting easier, let's just say A wins a few more than 3,450,000 votes in Massachusetts and 292,500 in Wyoming. So A receives a little over 3,752,500 votes. Let's go all out and pretend that in addition to winning almost half the votes in both Massachusetts and Wyoming, B wins *all* the votes in Colorado and Idaho. So even though B loses the election, they receive nearly

$$3,450,000 + 5,700,000 + 1,750,000 + 292,500 = 11,925,000 \text{ votes.}$$

That's 75% of the total number of votes! In other words, A won the election by winning only about 25% of the popular vote.

This example can be extended to all fifty states plus the District of Columbia and can even be brought closer to reality by incorporating actual voter turnout from a recent election. Do this: list the states according to their number of electoral votes, from highest to lowest— California, then Texas, Florida, New York, and so forth. Starting at the top, add the electoral votes until you get to 270. It turns out that the first eleven states on this list will get you there. Then take the voter turnout in each state and figure out what half of that is. For example, in California, it's about 17,116,000. By winning barely 50% of the vote in those eleven states, you will guarantee yourself the presidency. As a percentage of all the votes cast in 2020, half the votes in these eleven states account for about 27%.

But you can win the presidency with even fewer popular votes. Order the list the other way, from lowest to highest electoral vote count. In that case, you'd need to carry forty-one states to get to 270. But that would only require 23% of the total votes cast.

These schemes seem outlandish and cooked, and they are. But they illustrate a serious deficiency built into the system. A candidate could win anywhere from 23% to 100% of the popular vote and become president. Any mathematical election procedure so insensitive to the input of the popular voting data should be discarded.

ONE PERSON, FOUR VOTES

Imagine this: shortly before the 2020 elections, about 5,000 Democrats living in Arizona decide to move to the neighboring state, California. At the same time, 5,000 Republicans from California move to Arizona. A similar swap happens between 6,000 Democrats from Georgia and as many Republicans from, say, Alabama. Finally, 10,000 Wisconsin Democrats switch places with as many Republicans from any state near it, let's say Minnesota.

What has happened is that about 41,000 people have moved. At the same time, no state's population count has changed. Neither has anyone's political opinion. But those moves change everything about the presidential election. The swaps affect the outcome of the elections in Arizona, Georgia, and Wisconsin. Now Trump carries those states by the tightest of margins and that earns him enough electoral votes to become president instead of Biden. (The winner in California, Alabama, and Minnesota is unchanged because those states were won by larger margins than the change in the number of people voting differently.) The same analysis unfolds in 2000, 2004, and 2016; those elections would have turned out very differently if there had been a handful of voters of a different party than the party that won in an even smaller handful of states.

The Democratic voters who brought Biden the victory in those three swing states were in some sense more "valuable" there. If they had lived somewhere else, their vote would not have mattered as much. This is one way the Electoral College amplifies voices in some states and diminishes them in others.

Here is another way: the Electoral College violates the one person, one vote rule, enshrined into law by the 1964 case *Reynolds v. Sims*. Well, it violates it in spirit, because the law applies only at the state level; it demands that each state respect it within its borders. No law says how this tenet should be regarded, much less enforced, when states are compared to each other. And because the number of electors in each state is determined by the size of that state's delegation in Congress, any flaws of the apportionment process will propagate into the Electoral College.

For example, Montana has about 1,084,000 inhabitants and gets two representatives in the House. Delaware's population is about 1,003,000 and it has one House seat. We saw in chapter 10 that this is a consequence of the rounding procedure of the Huntington-Hill method: states that are close in population can have different numbers of representatives. After adding in the 2 votes of the senators of each state, Montana has 4 electoral votes and Delaware has 3. Let's look at the number of people per electoral vote in each state. This is like the representation ratio we calculated in earlier chapters, but now we're dividing by the number of representatives plus 2 because of the 2 extra electoral votes that correspond to senators.

Montana people-per-electoral-vote ratio = 1,084,000 ÷ 4 = 271,000

Delaware people-per-electoral-vote ratio = 1,003,000 ÷ 3 = 334,333

This result doesn't seem fair. It looks like Montanans get more mileage from their votes than Delawareans. In terms of the Electoral College, every vote from Montana is worth

$$334,333 \div 271,000 = 1.23$$

times a vote from Delaware.

This effect is magnified by the requirement that every state have at least one representative. So take Wyoming, the least populous state, with about 585,000 residents and 3 electoral votes. Its ratio is

Wyoming people-per-electoral-vote ratio = 585,000 ÷ 3 = 195,000.

It takes 1.71 Delaware votes to match 1 Wyoming vote. Put differently, Wyoming has 1.71 times more voting power than Delaware in the Electoral College.

It's even worse when you look at California, population about 39,700,000, which has 54 electoral votes:

California people-per-electoral-vote ratio = 39,700,000 ÷ 54
= 735,000.

Dividing this by Wyoming's ratio gives

735,000 ÷ 195,000 = 3.78.

It takes almost *four times* as many Californians as Wyomingites to match Wyoming's 3 electoral votes.

This is called the *+2 effect*. The two extra electoral votes that correspond to the number of a state's senators amplify imbalances in the representation ratio. Combining this effect with the rounding error from Huntington-Hill gives us a gamut of ratios across the states.

The differing heights of the bars in figure 14.4 mean, in short, that the Electoral College violates the equal representation principle.* Nothing in the Constitution addresses this. Nothing in *Reynolds v. Sims* addresses it. Nothing anywhere in the legal or constitutional framework of the United States addresses it. As far as the courts are concerned, there is nothing to be done.

Math begs to differ.

---

* The discrepancies persist and are even more pronounced when the numerator is not the state's entire population but the size of its voter turnout.

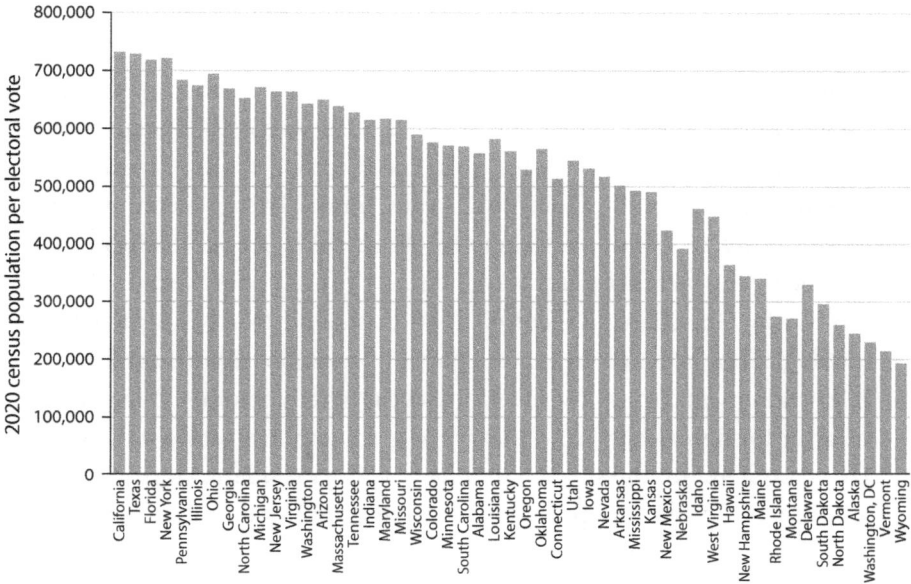

FIGURE 14.4. Population per electoral vote across the states.

ABOLISH, AMEND, AVOID

From the beginning, the Electoral College has been a peculiar piece of infrastructure, and its performance over the past 200 plus years provides plenty of evidence that it's outdated and flawed. But it executes the critical service of electing the U.S. president. So what are the better alternatives? Are they feasible?

Abolish It and Use the National Popular Vote

This strategy isn't feasible. Because the Electoral College is baked into the Constitution, an amendment is required to abolish it. That means that 2/3 from both the House and the Senate would have to vote for the amendment and 3/4 of the states (thirty-eight) would have to ratify it. Mustering this kind of bipartisan political will in Congress has proven to be impossible. Sentiment among voters also shows partisan effects.

According to a 2022 Pew Research poll, 63% of American adults support terminating the Electoral College and switching to the popular vote. Among those who identify as Democrats or who identify as leaning Democrat, support is at 80%; among Republicans, 42%. Because Republicans have benefited twice from the Electoral College system in recent history—in the Gore-Bush and Clinton-Trump elections—it makes sense that fewer are eager to eliminate it. Republicans aren't swayed by the fact that influential members of their party such as Richard Nixon, Gerald Ford, George H. W. Bush, and Newt Gingrich, the ideological architect of the current party, were on board with abolishing it. Even Donald Trump tweeted in 2012 that the Electoral College is "a disaster for democracy" and that it's a "great and disgusting injustice." He was quieter on the subject after he won the presidency in 2016 with a deficit of 3 million popular votes but declared again in 2018 that he would rather have the national popular vote.

There have been over 700 proposed amendments to reform or abolish the Electoral College. There's one sitting in the Congress as I write (proposed by Representative Steve Cohen of Tennessee). The farthest a proposal ever advanced was in 1969, in the wake of a tight election that gave Nixon the presidency by a huge Electoral College margin despite a popular vote difference of less than 1%. The House voted for scrapping the Electoral College overwhelmingly (338 to 70), but the proposed amendment was killed in the Senate by a filibuster spearheaded by three segregationist senators from the South (including Strom Thurmond, our acquaintance from the discussions of supermajorities and filibusters).

That's the Congress, but what about the states? Because many of the requisite thirty-eight would have to be small states that benefit from the attention and power they get through the Electoral College, it would be an uphill battle to convince them to give up this perk. (Or what they think is a perk—presidential candidates already fly over those states because they are not competitive. Nobody campaigns in Vermont or North Dakota.)

Nevertheless, the Electoral College's armor plating shouldn't prevent us from talking about the national popular vote, because we will soon see a workaround.

If we switched to the popular vote,* then all the math problems that filled the preceding pages would go away. The system would embody one person, one vote on the national level. If we are so committed to this principle at the state level, why are we so cavalier about it on the national level? Yes, the United States is a union of states, but the presidency is a national job, so anyone anywhere in the country should have an equal voice in choosing the president.

A popular vote would encourage voter turnout because every vote would matter. In 2016, Trump won 4,484,000 votes in California—a state-level result behind only those of Texas and Florida—but received no electoral votes there. To borrow terminology from gerrymandering, these votes were "wasted" because the Electoral College ignores them. In forty-eight states and the District of Columbia, neither the votes cast for a losing candidate nor the votes cast for the winning candidate over 50% matter, but they should. And they would if we had the popular vote. All those voters in "safe" states who prefer the other candidate would feel empowered. So too would moderate voters who are disillusioned with the vitriolic polarization.

A popular vote would motivate wider campaigning. Candidates would have to talk to more people because every vote counts. There would be no swing states where all the action is concentrated. Attention might pivot to large urban centers that promise the most bang for the campaign buck, but that would still spread campaigning around the country instead of its current concentration in a handful of states. And the urban-rural dichotomy is overblown to begin with: only about 15% of the U.S. population lives in the fifty biggest cities. About the same percentage lives in rural areas, and the two mirror each other in the

---

* There is also a hybrid idea whereby each state would give its overall plurality winner 2 electoral votes and the remaining 438 would be allocated according to the proportion of votes won nationally. So, for example, if candidate A won in 24 states, they would get 48 electoral votes ($2 \times 24 = 48$), and if B won in 26 states, they would get 52 electoral votes ($2 \times 26 = 52$). We then look at the national popular vote. If A won 60% of the popular vote, they would get 263 more votes (60% of 438 is 263). B, with 40% of the popular vote, would get 175 votes (40% of 438 is 175), bringing the totals to 311 for A and 227 for B.

way they vote. The true tension is in the small towns and suburbs that ultimately decide elections.

The most common argument against the popular vote is the view expressed in the Federalist Papers that the government should not allow big states to have outsized power over small ones. The president should consequently be someone who represents all of the states with all their diversities. But contrary to common misbelief, a handful of big states wouldn't have a monopoly on electing the president in a popular vote. California, for example, has about one-tenth of the U.S. population and one-third of it votes Republican. There is no monolith looming over the rest of the country, waiting to swallow up small states. That impression is produced by the Electoral College, which conceals the reality behind crude blocs of state-by-state numbers.

The argument that the founders wanted it this way and that one person, one vote must take a back seat to the necessary interweaving and balancing of federal, state, and local powers and interests is also weak. The rebuttal consists of one word: Senate. That's the mechanism for balancing the influence of small and big states. The Senate has a lot of power. There is no requirement that the Senate place a figurative finger on the scales of a presidential election, and there seems to be no rationale to do so beyond hunger for more influence, especially when mathematics speaks so strongly against it. Moreover, none of the founders' ideas about how the Electoral College would function are realized today. "Majoritarianism" or "populism," as national popular vote is sometimes called by proponents of the Electoral College, is the only natural and permanent fix for what has become a deeply idiosyncratic and flawed system.

The one fundamental drawback of this plan is the possibility that no candidate would win a majority. Instead, a candidate would win only a plurality of the votes. That would undermine the legitimacy of the winner and obscure the will of the people. Unfortunately, people have become used to this because it happens in elections at all levels, but that familiarity doesn't make it desirable. A remedy would be a national instant runoff. All of the benefits of this method would be injected into the biggest, most consequential election of them all. Instant runoff would eliminate national spoilers and vote splitting and would encourage third-party and minority candidates to run for *the* office. And because instant runoff punishes negative campaigning, it would reward the civility and decorum desperately needed in presidential races.

Amend It with Districtwise Allocation

In other words, be like Maine and Nebraska and use the district method. That is how things worked way back when. The switch to statewide plurality was made for strategic and political reasons, not mathematical ones. Maine returned to the district method in 1972 because its citizens did not like that a low-scoring plurality win could carry all its electoral votes, and Nebraska has used it since 1992 in an attempt to matter more politically.

In the district method, electoral votes are allocated according to the winner in each congressional district. The two extra electoral votes corresponding to senators go to the statewide plurality winner. For example, in 2016, Clinton and Trump each won one district in Maine, but Clinton received 3 electoral votes because she won the statewide popular vote. In 2008, Obama won one district in Nebraska and McCain won two and McCain received 4 electoral votes because he won the popular vote.

There are several advantages to this method. For one thing, it is more refined than the statewide winner-takes-all system. It mimics the congressional House elections except that it's an elector from one party or the other that is being determined. And the finer resolution can make a difference.

If this method had been used in 2016, Trump would have received 290 electoral votes to Clinton's 248. If it had been used in 2020, Biden would have won 274 to Trump's 264. In both of those elections, the winner would have been the same as it was with the current statewide winner-take-all method. But in 2012, if the district method had been used, the Romney-Obama electoral breakdown would have been 274–264, handing Romney the win instead, despite his loss in the popular vote. The divergence is not unexpected because plurality still underlies the district method. Obama won many districts by a large margin, so he came out ahead in the popular tally, but he won fewer districts than Romney overall. The district method would also have given George W. Bush a bigger Electoral College win over Al Gore.

Mismatches between the Electoral College and the popular vote are generally not as pronounced with the district method as they are with the current system (although they can occur) because allocation of electors by district is more granular. Instead of fifty-three winner-take-all plurality elections, 436 elections would be conducted. The results would supply a more nuanced answer to the question of voters' preference because a finer distribution of electoral votes is possible at this level of voting. However, it is important to note that the +2 effect would persist because those votes would be given to the *state's* plurality winner, as they are now.

The district method would encourage participation by candidates from third parties and allow them to concentrate their campaign in districts they thought they could win, thereby saving precious resources. Even if minor candidates have no chance of getting to 270, winning some number of electoral votes would send a message to supporters and fellow politicians alike and help shape the landscape for subsequent elections. The corresponding risk is that if more than two candidates secure electoral votes, there might not be a majority winner. The House would potentially be called upon to resolve presidential elections more often according to its own mind-boggling method.

And the districtwise method interacts with political forces in a way that reinforces unhealthy trends. Gerrymandering has reduced the number of competitive districts significantly. If districtwise allocation were used to elect a president, swing districts would play the role of swing states. About 17% of the U.S. population lives in those districts, compared to the 31% that live in traditionally swing states. This means that campaigning would become even more hyperfocused than it is now. (In the 2020 presidential race, 94% of campaigning happened in twelve states.) The incentive to gerrymander would increase, which would create even fewer competitive districts.

States might also be apprehensive about implementing this system. The switch to winner take all was meant to increase the influence of the states in presidential elections, but the district method would dilute that influence, especially if some states made the switch and some didn't. Finally, the party that controls a state legislature has no incentive to change the system that brought it to power and has likely awarded all its electoral votes to one of its own candidates running for president. Transitioning to districtwise allocation could mean relinquishing electoral votes in some districts, and that is not something politicians in power or their voters typically embrace.

Amend It with Proportional Allocation

This will sound familiar. Allocate electoral votes in each state according to the proportion of the statewide popular vote each candidate receives.

For example, in the 2016 presidential elections, California had 55 electoral votes. The popular vote split this way: Hillary Clinton got 61.5% of the votes, Donald Trump got 31.5%, Gary Johnson got 3.4%, and Jill Stein got 2.0%. (Other candidates received votes, but much fewer than the 1/55th, or 1.8%, share needed to earn an electoral vote.) So the proportional allocation would be

Clinton: 61% of 55 ≈ 34 electoral votes
Trump: 31.5% of 55 ≈ 18 electoral votes
Johnson: 3.4% of 55 ≈ 2 electoral votes
Stein: 2% of 55 ≈ 1 electoral vote

The numbers are rounded. I used the Hamilton method, but the official calculation would be more careful. It would use more decimal places and a more sophisticated rounding method such as Webster or Huntington-Hill. Or better yet, use decimal electoral votes! This would mean detaching them from people—namely electors would be no more but would be replaced by numbers. Which I don't see a problem with.

You can already see one advantage of this method over districtwise allocation. It gives votes to candidates who get traction in the state overall but do not necessarily win any districts. This political landscape is most pronounced in states where voters of a certain preference are scattered so that no district contains enough of them. Massachusetts is again the perfect example. In 2020, Biden received 65.6% of the vote and Trump received 32.1% (all others received 0.5% or less). Dividing the 11 Massachusetts electoral votes in these proportions would give Biden 7 and Trump 4. Massachusetts Republicans would have an actual voice in presidential elections. This would not happen with the district method and certainly not with winner-take-all plurality.

Statewide proportional representation does not imply that the total electoral vote count will in the end be proportional to the national popular vote results. Rounding error in each state can aggregate to a different picture (another reason to use decimal electoral votes). In fact, a candidate could still win the Electoral College with a proportional method but lose the popular vote. This would have been the case in the 2016 election; Trump would still have won even with proportional allocation of electoral votes.*

Proportional representation would bring us closer to the popular vote results without eliminating the Electoral College. As a bonus, it would encourage participation by third parties and independent candidates—something that over 60% of Americans say is necessary. It would significantly reduce the danger of spoilers and make every vote matter in each state, much as the national popular vote would make each vote matter nationally.

Because proportional allocation would likely give electoral votes to third parties on occasion (and that should be regarded as a benefit), it risks sending presidential elections to the House for resolution more often, as districtwise allocation would. This would have happened in 2000; with proportional allocation, Bush would have won 263 electoral votes, Gore 262, and Nader 13.

The remaining counterarguments are not mathematical. All of the political challenges that the district method faces are evident here too and are all tied to the surrender of power by the party in charge.

---

* But if the House were bigger—470 seats or more—Clinton would have won the Electoral College under proportional allocation.

With the same drawbacks but greater benefits, proportional allocation is better than the districtwise solution. Not as good as the national
popular vote, but it gets close.

Avoid It with the National Popular Vote Interstate Compact

This solution is my favorite. It's an elegant outmaneuvering of the Constitution by consensus among enough states. The setup is that states in
the compact pledge to give their electoral votes to the winner of the
national vote. So if my state, Massachusetts, is in the compact and if
candidate A were to win the countrywide popular vote, it wouldn't
matter that candidate B might have won here. The 11 Electoral College
votes from Massachusetts would go to candidate A.

The National Popular Vote Interstate Compact (NPVIC) was
dreamed up in 2006 by John Koza, a computer scientist who is the
father of genetic programming, a technique for optimizing computer
programs used in neural networks and machine learning. When he was
a graduate student at the University of Michigan, he was fascinated by
the math of the Electoral College and invented the board game called
Consensus whose object was for players to get as many electoral votes as
possible in a pretend election by appealing to various demographics.
Unsurprisingly, the game didn't take off commercially. But something
else did. A few years later, Koza invented and designed the scratch-off
lottery ticket and, as you can imagine, became very wealthy. He first
learned about interstate compacts when states started working together
to combine their jackpots in order to attract more gamblers.

The trick is that for NPVIC to work, enough states have to join to
create a combined electoral vote count of at least 270. Then the winner
of the popular vote in those states is guaranteed to win the Electoral
College as well. Ingenious, right?

The sad news is that it may never happen. So far, sixteen states plus
the District of Columbia have adopted the compact, and their electoral
vote total is . . . drumroll . . . 205. Not enough. The sixteen states are by
and large the Democratic strongholds of the two coasts, so for the compact

to work, some more of the purple or red states must be convinced to join. That could be hard. For example, it would suffice for Arizona, Florida, Michigan, and Ohio to join. Or Michigan, Minnesota, Ohio, and Pennsylvania. Or it might be easier to convince a bunch of smaller states to join to make up the difference. But there is no easy path. Several Republican-led states were on the verge of adopting NPVIC before the 2016 election, but the current system proved so beneficial to their party that they backed off.

The legal case for NPVIC is that nothing in the Constitution or federal law mandates that electoral votes be allocated in any particular way. States do have their own rules, but those rules can be changed. Electors do not have to cast their vote according to their state's popular vote outcome; that is what faithless electors are all about. Faithlessness is precisely what would make NPVIC work.

Despite these arguments, legal issues might paralyze NPVIC even if the compact hits 270. The winner of a compact-determined election runs the risk of seeming illegitimate and having to defend against a flood of court challenges. There is a lively debate over whether such an election result would even be constitutional because it would shift the balance of power between states or whether it would require congressional approval. Reactions to NPVIC among legislators can be heated and have included phrases like "civil insurrection" and "legal train wreck." The verdict: we won't know whether NPVIC will work until it is implemented.

Political, moral, and ideological opposition to NPVIC aligns with opposition to the national popular vote because the two are equivalent in outcome. For instance, there is the issue of the winner being chosen by plurality and not by majority. (The solution, as before, is national instant runoff.) In addition, all of the objections of the fervent federalists apply here too. And the rebuttal is the same: the equal rights of states are already embodied in the Senate, and that suffices.

———

The Electoral College no longer serves its original purpose. Even if it did, that purpose does not reflect the realities of today. It is an instrument of disenfranchisement that shuns the one person, one vote doctrine and tips the scales in favor of voters living in certain states. The job of the U.S. president should be decided in a national contest in which each citizen has equal say. The federalist distinction between state and federal governance is irrelevant in this arena.

Over 60% of Americans agree that the Electoral College belongs in history books. Replacing it by the popular vote would be ideal. The next best option is NPVIC because it would yield the same result. After that, proportional allocation of electoral votes in each state makes the most sense. Whatever the case, we need a more refined system that strives to put equal power of electing the president of the United States in each voter's hands.

CHAPTER 15

The Citizen-Mathematician

AN ELECTION TOOK place in Bosnia in November 1990 that changed the course of my life. I had no idea that it would; at sixteen, I was too busy growing out my hair, playing in a band (which is still around—go Konvoj!), hanging out with my friends (still my favorite thing to do, and with those same people), and being hopeful about my all-consuming (and generally nonmaterializing) romantic pursuits. I'm not sure I even noticed that an election was happening around me.

Following similar elections in Slovenia and Croatia the previous spring, this was Bosnia's first general election while it was still a constituent republic of Yugoslavia, a country whose socialist-era days were clearly numbered. Three ethnic parties, representing Bosnian Muslims, Croats, and Serbs, swept the election and formed the new seven-member presidency. The dream of Bosnia as a multicultural, multiethnic model of tolerance and coexistence for the rest of Yugoslavia—and Europe—had started to unravel.

On June 25, 1991, Slovenia and Croatia declared independence from Yugoslavia. A brief military intervention in Slovenia and a full-scale war in Croatia quickly unfolded. The military campaign was initiated by Serbia, the dominant Yugoslav republic, which was unwilling to relinquish its centralized and increasingly nationalistic power.

The day after those declarations of independence took place, I was on a plane to Boston on a two-month visit to an uncle who had already lived there for several years. The timing was a coincidence; my family had been planning this trip for a while.* But midway through my visit, seeing that a fate similar to Croatia's was not unfathomable in Bosnia, my mom suggested that I try to stay in Boston—maybe turn this visit into the fun and valuable experience of spending my senior year in an American high school, wait things out a little. I hated the idea. All I wanted to do was go back to my band and my friends. But deep down, I understood that staying was the sensible thing to do. So I stayed.

In April 1992, as I was getting ready to graduate from high school, the war reached Bosnia after a referendum on its independence. Serbia's barbaric aggression, of a magnitude and brutality not seen in Europe since World War II, lasted for three and a half years.

And here I am, thirty-two years later. And there they are, those same parties. They continue to profit from dialing up fear and nationalism during each election cycle, stoking grievances from the past, and skill-fully turning artificial ethnic segregation into streams of money that fill their coffers. Bosnia is a country with no prospects, its only claim to fame the incessant flood of its fleeing youth, who barely glance in the rearview mirror as they search for a better life elsewhere.

While a different voting method would probably not have changed the outcome of that milestone 1990 election, the leading parties have benefited greatly from presidential plurality elections since the war. Many of their candidates were elected with minority support, poten-tially misrepresenting the people's will and perpetuating Bosnia's stag-nation. The hacksaw Frankendistricting, one of many concessions to the three factions enacted by the 1995 peace agreement brokered in Dayton, Ohio, entrenched ethnic divisions that have been calcifying ever since.

Traces of what led to Bosnia's demise are unnervingly apparent in the United States. Our political system is a knotted web of malfunctioning processes that propagate and amplify the perils to democracy from every direction. Crowded primary elections elevate fringe candidates

* I was supposed to visit Boston the year before but had been denied a visa. For some reason, I got it the following year. If I had gotten the visa the first year, I would have gone back home at the end of that summer and, well, everything would have been different.

who compete in winner-take-all races using plurality in districts that have been gerrymandered into uncompetitive insignificance. The process elects candidates who do not represent the true will of the people and the distorted agenda that results is carried out to the dissatisfaction of most Americans. Fanatical bases gather under banners of intolerance and exclusion while vast swaths of the population feel disenfranchised and removed from the political process. The connection between an ordinary citizen and the government that purportedly acts on their mandate has never been stretched this thin.

I have always felt that as a Bosnian and an American and a citizen of the world, I had to do something about this, and math is the only way I know how. I am convinced that mathematics can be a clear-eyed guide to help us repair many of our democracy's weaknesses; hence this book and everything else I do in the space where math and politics overlap. Thankfully, the math community is right there with me. Choose most any topic covered in this book and there will be recent mathematical research on it, scholarly papers fresh off the press. More classes that blend math and politics are sprouting up on college campuses. The organization I started, the Institute for Mathematics and Democracy, is gathering a growing network of people who are working in the field and more and more students and young researchers are joining the ranks. We share the latest research, toss around ideas for bringing political quantitative literacy into classrooms, and organize workshops where we teach each other how to become expert witnesses in court cases. These feel like exciting times. We are learning to be catalysts for empowerment, to supply expertise and authority to the mission of rehabilitating our civic infrastructure. Better yet, we are becoming part of the civic infrastructure.

But it doesn't have to stop with us. During a recent event at the college where I teach, a woman took me aside to tell me about her sixth-grade math teacher who, decades earlier, had informed her that she hadn't scored high enough on some test and could therefore not go on to take more math classes. Ever. "Imagine what I could have done if I had gotten to take more math," she said. I suggested that she still could. I told her what I want to tell you and everyone else: you can become part of the civic infrastructure too. You can build your political numeracy and your confidence with the mathematics of voting and representation. You can engage with and advocate for efforts to introduce better democratic systems. You can educate legislators and encourage them to rock the boat for the public good. You can help us spread the message through the media. There is so much you can do. Your math career is just beginning.

An Infrastructure Plan

HERE IS this book's takeaway in terms of action—my mathematically informed recommendations for what we should change:

- Use instant runoff voting in single-winner elections
- Eliminate the Electoral College
- Use popular vote with instant runoff for presidential elections
- Use multimember districts with single transferable voting
- Increase the size of the House of Representatives
- Use the Webster apportionment method
- Use independent districting commissions
- Introduce more political quantitative literacy into school curricula

ACKNOWLEDGMENTS

MANY PEOPLE are responsible for this book seeing the light of day. They've made it more readable, more accessible, and more relevant. Had a different set of people been involved in its creation, something else, likely worse, would have transpired. But only one group of people is so important that without them there would be no book at all—my family. My wife Catherine has been my most unwavering supporter— even as I perpetually think up things to do that I don't really have to, like write this book—and an invaluably level-headed critic. My children, Andrei and Iris, are my principal motivation and inspiration. Most kids would have rolled their eyes at their dad's impassioned dinnertime monologues about math and politics, but not the two of them. I can tell they hold me and my work in high regard, and there's no greater satisfaction anything could ever bring me.

Diana Gillooly, my editor at Princeton University Press, has been amazing. I can't imagine anyone else who would have guided a novice writer with more patience. She steered me with an expert hand and was spot on with every suggestion and correction. But most important, she taught me how to stop apologizing for being a mathematician and embrace the idea that I might have something to say after all. I am also grateful to my colleague Oscar Fernandez, who suggested that PUP might be the right home for this book and introduced me to Diana.

Diana's editorial assistant, Kiran Pandey, has made all the logistics pleasantly smooth, as has the entire production team at PUP and Westchester Publishing. Copyeditor Kate Babbitt did a stellar job, as did Fred Kameny, who indexed the book.

I am indebted to several of my students. Cece Henderson and Delaney Morgan were there from day one, reading drafts, researching, and fact checking. They were awesome at helping me snap back from "this is stupid, why am I doing this?" to "this could work!" Hanqi Zhu was incredible in dealing with the surprisingly lengthy and taxing process of obtaining figure copyrights.

Thank you to Emily Pallan and Raj Savla, brilliant high schoolers (and now college students) who did some of the research cited in the book. With gifted young people like them, political quantitative literacy has a chance yet.

This book grew out of my Math and Politics class, and all the students I ever taught in its several iterations are responsible for the book's content. They showed me how to think and talk about this stuff clearly and pointed the way toward the interesting and the relevant and to the simplicity underlying the complexity. I thank them for paying it forward to everyone who reads this book and to all my future students.

Many experts helped me understand voting, gerrymandering, and other topics covered here. I am indebted to the anonymous reviewers for their careful reading of an early draft. Their comments and suggestions, especially on the gerrymandering chapters, improved the book immensely. David McCune and Adam Graham-Squire provided indispensable insight into voting theory, and Eric Maskin helped me put ranked choice voting into perspective, both academic and historic. Daryl DeFord's generosity with his time and his knowledge of gerrymandering following my plea for help made those chapters incomparably better. Jordan Ellenberg's writing as well as Moon Duchin's research and the way she has brought it to the public and to the judicial eye have served as inspiration. I am also grateful to Nathan Lockwood for bringing me closer to the way math interacts with politics in the real world.

I also thank Jay Turner for sharing his expertise and experience with the writing and the publishing process. Math is not a "book field," so with this project I entered a new, unfamiliar realm that would have been much more dauntingly unnavigable without his help.

Wellesley College, where I teach, has been nothing but supportive of all my endeavors. A departure from a more traditional research track and a plunge into something this interdisciplinary and public facing would have been shunned at many institutions of higher learning, but that is not the case at Wellesley. I am grateful to all my colleagues in the math department for letting me experiment with math and politics and for adding a class on that topic to our curriculum.

Special thanks go to Stanley Chang for starting the Institute for Mathematics and Democracy with me and to Andy Schultz for running it with us. It's been a full-time job (one of several I seem to be doing), but one of incomparable gratification and meaning. I hope they feel the same way.

Then there are all those who don't even know how much they've influenced this book. I am deeply indebted to my brilliant graduate advisor, Tom Goodwillie, who taught me that working in one corner of one room of the mansion that is mathematics isn't all that interesting. True innovation is trapped in the walls between the rooms and those walls must be torn down to release it.

A shoutout goes to my longest buddies in math (and many other things), Ben Brubaker and Brian Munson, who propped me up in grad school as we jointly discovered that there is a whole lot of suffering to go along with the joy of math. It's questionable whether I would have made it this far without them.

Thank you to all my amazing and accomplished friends at the Bosnian-Herzegovinian American Academy of Arts and Sciences (yes, I'm thanking the academy) for providing the soapbox from which I could try to make a difference in Bosnia, my old home and my current home away from home. If I manage to move the needle on political quantitative literacy by the tiniest bit in that ill-fated country, this will all have made sense.

Thank you to my best friends—from childhood and until forever. The gang of Mi—Boban, Đaja, Edo, Kaja, Samir, Sandrino, and Željko—have taught me how to tell right from wrong, how to power through defeat, and how to truly enjoy life. Sometimes it seems like all I'm really doing is trying to arrange my life so we can all hang out more.

I am grateful to Đozo and Hamo for rooting for me and for still doing what I wish I could have been doing with them all these years. After it all stops, your music will still be playing.

And finally, thank you to my mom Jasna and my sister Ljiljana, simply because any expression of gratitude I might have for anything in my life would not be complete without a mention of their names.

GLOSSARY

Alabama paradox: An apportionment paradox that occurs when the number of legislative seats increases but a state loses a seat.

anonymity criterion: If any two voters switch ballots, the election result does not change.

apportionment problem: A problem arising when some amount of a resource must be distributed but the resource comes in indivisible integer values.

approval voting: Each voter indicates all of the candidates they regard favorably, often by checking a box next to each of their names. The winner is the candidate who receives the most approvals.

arithmetic mean: An average of a set of n numbers calculated by adding them and dividing the sum by n.

Arrow impossibility theorem: The only ranked voting method for three or more candidates that satisfies monotonicity and independence of irrelevant alternatives is dictatorship.

axioms: The basic propositions that are taken to be true that a theory is built on.

Balinski-Young impossibility theorem: There is no apportionment method that satisfies the quota rule and population monotonicity.

bloc voting: Each voter selects some predetermined number of candidates but without ordering their choices in any way.

Borda count: A ranked voting method whereby for an election with n candidates, points ranging from 0 to $n-1$ are assigned to each placement, with highest receiving $n-1$ and lowest receiving 0. The candidate with the most points wins.

bullet voting: A form of strategic voting in which a voter selects a single candidate even though the method allows them to express their opinion on more than one candidate.

burying: A form of strategic voting in ranked elections in which a voter deliberately ranks a candidate low if they perceive them to be a threat to their top choice.

candidate-centered proportional representation: Proportional representation systems whereby voters directly indicate preferences for candidates nominated by political parties.

candidate cloning: A strategy by which several candidates from the same party run in an election, hoping to increase the chances that one of them will be elected.

cardinal voting: Each voter evaluates each candidate individually without explicitly making comparisons between them.

center squeeze effect: Moderate candidates are squeezed out of the race by polarizing candidates on either side of the political spectrum.

closed list proportional representation: A type of party list proportional representation in which each voter casts a vote for a party but has no choice about the candidates the party nominates.

community of interest: A group of people who have shared traits or interests in the context of legislative districting.

compactness: A desirable property of legislative districts that is typically taken to mean that a district shouldn't have a strange shape.

compensatory seats: In a proportional representation system, additional seats that are allocated in proportion to total vote counts across the country.

competitive district: A district in which parties have approximately equal support and elections are won by small margins.

Condorcet cycle: A situation in which a ranked profile yields pairwise voting preferences that form a circle among three or more candidates.

Condorcet method: A ranked voting method whereby candidates are compared pairwise. If there is a candidate who wins all the head-to-head matchups, that candidate is the winner.

Condorcet paradox: See Condorcet cycle.

constituency: A group of voters in a specified geographic area that elect a representative to a legislative body.

contiguous: A desirable property of legislative districts that states that a voter can travel from any place in the district to any other place in the district without leaving the district.

convex: A shape is convex if the line segment between any two points in the shape stays inside the shape.

convex hull score: A district compactness measure that inscribes a district into the minimum convex polygon enclosing the district and then divides the areas of the two shapes. A low convex hull score indicates potential gerrymandering.

cracking: A gerrymandering tactic whereby voters of one party are split across some number of districts in numbers that deny that party wins in those districts.

cube root law: In many democracies, the size of the legislature is approximately the cube root of its population.

cumulative voting: Each voter is given a certain number of points or votes that they can distribute among the candidates in any way they see fit. The winner is the candidate with the most points.

Dean's apportionment method: An apportionment method whereby the divisor is selected so that the corresponding quotas, rounded up or down according to comparison with the harmonic mean, total to the right number of seats. The roundings are the seat allocations. Alternatively, this is the method that allocates seats in a way that minimizes the average deviation of representation ratios.

degressive proportionality: Allocating legislative seats in a way that awards smaller states more seats than their population proportion.

D'Hondt method: A proportional representation method equivalent to the Jefferson apportionment method.

dictatorship method: A voting method whereby one voter, the dictator, controls the outcome of the election no matter how the other voters cast their ballots. Whoever the dictator chooses is the winner.

dispersion: In a legislative district, the existence of branches or thin tentacles that make the district appear less convex or compact.

district: A subdivision of a larger administrative region that is created to provide its population with representation in the larger region's legislature.

divisor apportionment method: An apportionment method whereby the standard divisor is modified to achieve a certain allocation.

double bunking: A gerrymandering tactic whereby district lines are intentionally redrawn so that two similar candidates now live in the same district, guaranteeing that one of them will lose the election. Also known as pairing incumbents.

educational gerrymandering: A variant of gerrymandering that usually takes the form of school secession, where affluent, typically white communities separate from integrated school districts.

efficiency gap: A numerical measure of gerrymandering that calculates the difference in total wasted votes between the two parties expressed as a fraction of total votes cast.

electors: In an American presidential election, the individuals each state chooses to cast votes for the president.

equal proportions method: See Huntington-Hill apportionment method.

faithless elector: An elector who votes for a candidate who did not win the popular vote in their state.

favorite betrayal: A form of strategic voting in which a voter does not rank their favorite candidate first.

Federalist: Shorthand for the Federalist Papers, a series of letters to the public that offered commentary on the U.S. Constitution in 1787–1788. The authors, who used the pseudonym Publius, were Alexander Hamilton, James Madison, and John Jay.

flip (Markov chain): A way to get from one district map to another "close map" in the MCMC algorithm by switching a randomly chosen voting precinct on the border of a district so it lies in the adjacent district.

geometric mean: An average of a set of n numbers calculated by multiplying them and taking the nth root of the product.

gerrymandering: The practice of drawing district lines to achieve a certain political goal.

Gibbard-Satterthwaite impossibility theorem: In elections of three or more candidates, the only ranked voting system that is strategy proof is dictatorship.

greatest divisors method: See Jefferson apportionment method.

Hamilton apportionment method: An apportionment method whereby each state receives its lower quota of seats, then any remaining seats are allocated to the states whose standard quotas have the largest fractional remainders. Also known as the largest remainders method and the Vinton method.

Hare's method: See instant runoff voting.

House monotonicity: An apportionment property that says if the number of seats to be allocated increases, no state should lose a seat.

Huntington-Hill apportionment method: An apportionment method whereby the standard divisor is selected so that rounding all the resulting quotas according to comparison with the geometric mean totals to the required number of seats. The roundings are the seat allocations. Alternatively, this is the method that allocates seats in such a way that quotients of representation ratios between any two states cannot be made smaller by a transfer of a seat from one to the other. Also known as the equal proportions method.

incumbent pairing: See double bunking.

independence of irrelevant alternatives (IIA): If candidate A is the winner, then modifying the ballots in a way that does not change A's position relative to another candidate B should not make B the winner.

instant runoff voting: A type of ranked voting method whereby the tallying is completed by performing a series of runoff elections that eliminates the candidate with the least number of first-place votes at each stage until a candidate has a majority of the votes. Also known as Hare's method.

invariance under scaling: When a property or an attribute of a shape does not change when the shape is scaled, i.e., shrunk or expanded.

isoperimetric inequality theorem: If R is a region in the plane that has area a and is bounded by the closed curve of length p, then $(4 \times \pi \times a) \div p^2 \leq 1$.

Jefferson apportionment method: An apportionment method whereby the standard divisor is changed so that all the resulting lower quotas for each state add up to the required number of seats. Also known as the greatest divisors method.

jungle primaries: See top-two primaries.

largest remainders method: See Hamilton apportionment method.

length-width score: A district compactness measure that inscribes a district in a rectangle and then compares the ratio of its length to its width. A large ratio indicates potential gerrymandering.

lower quota: A state's standard quota rounded down to the nearest integer (in case of U.S. congressional districts, if the rounding gives 0, the allocation is 1); used in apportionment calculations.

major fractions method: See Webster apportionment method.

majority criterion: If a candidate receives a majority of the first-place votes, that candidate is the winner.

majority judgment: A variant of range voting whereby each voter rates each candidate on a non-numeric "satisfaction" scale, such as Poor to Excellent. The winner is the candidate with the highest median score. Also known as majority grading.

malapportionment: Electoral districts with unequal ratios of voters to representatives, creating a disproportionate distribution of power.

May's theorem: Simple majority is the only voting system that satisfies anonymity, neutrality, and monotonicity and that can result in a tie only when candidates have the same number of votes.

MCMC (Markov chain Monte Carlo): An algorithm used for choosing a collection of samples, or an ensemble, out of some distribution. In the context of districting, the distribution is the set of all possible ways to divide a map into districts. Because that set is too large, a sample has to be produced that satisfies some criteria, and MCMC does this job.

minimum quota: See lower quota.

mixed electoral system: A voting method that uses a combination of proportional representation and plurality.

mixed-member proportional representation: A mixed electoral system in which each voter casts two votes in their district—one for a candidate and one for a party (that is possibly different from the chosen candidate's party).

modified divisor: The modified standard divisor that is used to apportion seats using a divisor method.

monarchy: A voting method whereby one candidate, the monarch, wins no matter how the electorate votes.

monotonicity: It is impossible for a winning candidate to become a losing candidate by gaining additional votes or by being moved up in one or more rankings.

multimember district: A contiguous legislative district that elects more than one candidate.

multiwinner approval voting: A version of approval voting for electing more than one candidate.

neutrality: If every voter flips their preference between a winner and some other candidate, the other candidate becomes the winner.

new states paradox: An apportionment paradox that occurs when new seats are added to a legislature to accommodate a new state and this affects the number of seats already allocated to an existing state.

no-show paradox: A voting paradox whereby a candidate would perform better in an election if some of their supporters did not vote.

open list proportional representation: A type of party list proportional representation whereby voters primarily vote for a political party but they can still indicate their preference among the candidates that party has nominated.

ordinal voting: See ranked (choice) voting.

packing: A gerrymandering tactic whereby a party creates a "throwaway district" filled with voters of the other party.

parity: A voting method whereby the winner is the candidate with an even number of votes. If both candidates receive an odd number of votes or both receive an even number of votes, then it is a tie.

partisan bias: A symmetry score that calculates the discrepancy between a party's hypothetical seat share and 50% in the scenario when it wins 50% of the votes.

partisan symmetry: The concept that if a voting system is fair, then flipping the vote shares of two parties should result in their seat shares getting flipped as well.

party list proportional representation: A proportional representation method that allocates to each party a number of seats proportional to the number of votes it gets.

plurality: Each voter selects one candidate and the candidate with the most votes wins.

plurality bloc voting: Each voter selects more than one candidate; essentially multiple simultaneous plurality elections.

+2 effect: States with small populations have more electoral votes per capita because of the two votes each state has that correspond to its two senators.

Polsby-Popper (compactness) score: A district compactness measure based on the isoperimetric inequality theorem; equivalent to Schwartzberg score.

popular vote: A population's vote.

population monotonicity: An apportionment property that says if the population growth of one state is faster than that of another, then the first should not give up a seat to the second.

population paradox: An apportionment paradox that occurs when a seat is awarded to a slower-growing state at the expense of a faster-growing state.

population polygon: A district compactness measure obtained by dividing the population living in a district by the population living in the minimum convex polygon that encloses the district.

positional method: A generalization of Borda count whereby the points do not have to range from 0 to $n - 1$ but can be assigned in any way.

precinct: A subdivision of a district. Precinct-level election results are typically the most granular data available for an election.

prison gerrymandering: A variant of gerrymandering in which districts containing prisons count inmates as residents, giving the districts greater electoral power and increased resources, even if the inmates cannot vote or benefit from the resources.

probability distribution function: In the context of districting, a function that contains information about redistricting requirements that ensures that an MCMC algorithm will choose maps that meet these criteria with higher probability.

profile: A summary of the preferences of all voters in an election; also called a voting profile or election profile.

proportional representation: An electoral system in which the percentage of seats a party receives is (ideally) equal to the percentage of votes it wins.

quota: A threshold number of votes a candidate must receive to win in the supermajority method.

quota rule: An apportionment property that states that each state's apportionment should be one of the integers nearest to the state's standard quota.

quotient: The result obtained by dividing one quantity by another; ratio.

range voting: Each voter gives each candidate a score in a set range. The winner is the candidate with the largest average score. Also known as score voting.

ranked (choice) voting: Each voter ranks the candidates in order of preference.

recombination: A way to get from one district map to another "close" map in the MCMC algorithm by combining two adjacent districts and then splitting them in a certain way, creating two districts different from the original two.

Reock score: A district compactness measure that fits the district into the smallest possible circle and then takes the ratio of the area of the district to the area of the circle.

representation ratio: The number of people per representative.

responsiveness (of the seats-votes curve): A measure of how gains in votes translate to gains in legislative seats.

runoff election: A follow-up election in which the top two candidates from the first election face off against one another.

safe district: A legislative district where one party or a candidate is virtually guaranteed to win an election.

Sainte-Laguë method: A proportional representation method equivalent to the Webster apportionment method.

Schwartzberg score: A district compactness measure that takes the circle whose area is the same as a district's and compares the two perimeters by taking their ratio; equivalent to the Polsby-Popper score.

score voting: See range voting.

seats bonus: The winning party gets a larger percentage of seats than the percentage of votes it won.

seats-votes curve: A graph in which each point plots the share of seats a party would receive if it won a certain share of votes.

simple majority: Each voter selects their favorite candidate and the candidate with the majority (at least 50%) of the votes wins.

single mixed vote: A ballot that records two votes, one for the candidates and another for the party those candidates come from.

single transferable vote: A generalization of instant runoff for elections with more than one winner.

social choice theory: The study of collective decision-making.

spoiler (effect): Informally, a candidate with no chance of winning takes enough votes away from one of the major candidates to cause them to lose the election. More precisely, a non-winning candidate whose removal from all the ballots causes the election result to change.

standard divisor: A country's total population divided by the number of legislative seats to be distributed; used in apportionment calculations.

standard quota: A state's population divided by the standard divisor; used in apportionment calculations.

strategic voting: Voting in a way that does not reflect an honest preference; also known as tactical or insincere voting.

supermajority: A voting method whereby candidates are required to reach a certain number (quota) of votes beyond simple majority to win.

tactical voting: See strategic voting.

Three-Fifths Compromise: The compromise reached at the 1787 Constitutional Convention that stated that 3/5 of a state's enslaved population would be counted toward the state's total population for the purpose of apportioning congressional seats.

top-two primaries: Primary elections in which all candidates are listed on the ballot regardless of party affiliation. The top two scorers move on to the general election.

unanimity: A voting method in which a candidate must receive all the votes to win.

unanimity criterion: If everyone votes for one candidate, that candidate should win.

uniform partisan swing: A simplifying modeling assumption that a statewide change in voting trends is replicated in the same way in all of the state's districts.

variance: A measure of how "spread out," or dispersed, a dataset is.

Vinton method: See Hamilton apportionment method.

vote splitting: When some number of like-minded voters distribute their votes among similar candidates, allowing another candidate to rise to the top.

voting: Expression of political will via a ballot.

voting method: A process or algorithm for selecting a winner (or a group of winners) from a collection of votes cast by individuals.

wasted vote: A vote for a losing candidate or a vote for a winning candidate beyond the threshold number they needed to win.

Webster apportionment method: An apportionment method whereby the divisor is selected so that the corresponding quotas, rounded up or down according to the usual 0.5 cutoff, totals to the right number of seats. The roundings are the seat allocations. Alternatively, this is the method that allocates seats in such a way that differences of representation ratios between any two states cannot be made smaller by a transfer of a seat from one to the other. Also known as the major fractions method.

winner's bonus: See seats bonus.

Wyoming rule: A method of deciding the size of the House of Representatives in which the representation ratio is set to the population of the least populous state.

FIGURE CREDITS AND SOURCES

Fig. 2.1. *Source:* https://ballotpedia.org/Presidential_election_in_Minnesota,_2020.

Fig. 2.2. *Data source:* City of Boston Department of Elections. https://www.boston.gov/departments/elections/state-and-city-boston-election-results#_021.

Fig. 2.3. *Data source:* French Ministry of the Interior.

Fig. 2.4. *Source:* 2019–20 Croatian presidential election. https://en.wikipedia.org/w/index.php?title=2019%E2%80%9320_Croatian_presidential_election&oldid=1165143059 (last visited September 15, 2023). Data *source:* State Election Commission of the Republic of Croatia. https://www.izbori.hr/arhiva-izbora/index.html#/app/predsjednik-2019.

Fig. 3.2. Courtesy of Maine Department of State. *Source:* https://www.maine.gov/sos/cec/elec/upcoming/pdf/REPRCV1Contest20180420.pdf.

Fig. 3.3. *Source:* Ian Fowell, "Eurovision 2021: Controversy and Voting Problems," Esccovers, May 24, 2021. http://www.esccovers.com/eurovision-2021-controversy-and-voting-problems/.

Fig. 3.7. *Source:* Simon Borg, "World Cup Bracket 2022: Final Updated FIFA Knockout Stages," *Sporting News*, December 19, 2022. https://www.sportingnews.com/us/soccer/news/world-cup-bracket-live-updated-knockout-stage-fifa-2022/dsjork6jpvpvosvmjnt5usgw.

Fig. 5.1. *Source:* Laoeuaoeu, Creative Commons Attribution-Share Alike 3.0 Unported. https://commons.wikimedia.org/wiki/File:Completed_Score_Voting_Ballot_version2.png.

Figs. 8.2 and 8.3. *Data source:* U.S. Bureau of the Census, Population Base for Apportionment and the Number of Representatives Apportioned: 1790 to 1990. https://www2.census.gov/programs-surveys/decennial/1990/data/apportionment/tabb.pdf.

Fig. 8.4. *Data source:* U.S. Bureau of the Census, Historical Apportionment Data (1910–2020). https://www.census.gov/data/tables/time-series/dec/apportionment-data-text.html.

Fig. 8.7. *Data source:* ACE Electoral Knowledge Network and published within Lee Drutman, Jonathan D. Cohen, Yuval Levin, and Norman J. Ornstein, *The Case for Enlarging the House of Representatives* (Cambridge, MA: American Academy of Arts and Sciences, 2021). https://www.amacad.org/ourcommonpurpose/enlarging-the-house/section/7.

Fig. 8.8. *Source:* 2016 Population and Legislative Body Size, Select Countries. From "U.S. House Districts Are Colossal. What's the Right Size?" by K. Crespin. November 15, 2017. *StatChat*. University of Virginia Weldon Cooper Center, Demographics Research Group. https://statchatva.org/2017/11/15/u-s-house-districts-are-colossal-whats-the-right-size/. Copyright by University of Virginia.

Fig. 10.1. *Source:* W. M. Steuart (Director, U.S. Bureau of the Census), Memorandum to R. P. Lamont (U.S. Secretary of Commerce), "Statement giving the whole number of persons in each State exclusive of Indians not taxed, as ascertained under the Fifteenth Decennial Census of Population, and the number of Representatives to which each State would be entitled," November 17, 1930. https://www.census.gov/history/pdf/ApportionmentInformation-1930Census.pdf.

Fig. 10.2. *Source:* Michel L. Balinski and H. Peyton Young, *Fair Representation: Meeting the Ideal of One Man, One Vote* (Washington, DC: Brookings Institution Press, 2001).

Fig. 11.2. *Source:* Elkanah Tisdale (1771–1835). Originally published in the *Boston Centinel*, 1812.

Fig. 11.3. *Source: Gomillion v. Lightfoot*, 364 U.S. 339 (Supreme Court of the United States, 1960).

Figs. 12.1 and 12.2. *Source:* National Atlas of the United States.

Figs. 12.3 and Fig. 12.4. *Source:* William T. Adler and Ella Koeze, "One Way to Spot a Partisan Gerrymander," *FiveThirtyEight*, July 9, 2019. https://projects.fivethirtyeight.com/partisan -gerrymandering-north-carolina/.

Figs. 12.5 and 12.6. Courtesy of Jeffrey Shen. Source: Jeffrey Shen, *Exploring the Seats-Votes Curve: A Historical Primer on Congressional Seat-Votes Curves and Partisan Bias, Symmetry, and Responsiveness, from 2000–2016*. https://jeffreyshen19.github.io/Seats-Votes-Curves/.

Fig. 12.7. *Source:* Christopher Ingraham, "What 60 Years of Political Gerrymandering Looks Like," *Washington Post*, May 21, 2014; based on shapefiles maintained by Jeffrey B. Lewis, Brandon DeVine, Lincoln Pritcher, and Kenneth C. Martis, UCLA. https://www.washingtonpost.com /news/wonk/wp/2014/05/21/what-60-years-of-political-gerrymandering-looks-like/.

Fig. 12.13. Courtesy of Maria Chikina, Alan Frieze, and Wesley Pegden, and the National Academy of Sciences of the United States of America. *Source:* Maria Chikina, Alan Frieze, and Wesley Pegden, "Assessing Significance in a Markov Chain without Mixing," *Proceedings of the National Academy of Sciences* 114, no. 11 (February 28, 2017): 2860–2864. https://doi.org/10.1073/pnas.1617540114.

Fig. 12.14. Courtesy Gregory Herschlag, Robert Raviera, and Jonathan Mattingly. *Source:* Gregory Herschlag, Robert Raviera, and Jonathan Mattingly, "Evaluating Partisan Gerrymandering in Wisconsin." arXiv:1709.01596 [stat.AP].

Table 12.1. *Data source:* Maryland State Board of Elections. https://elections.maryland.gov /elections/2016/results/general/gen_results_2016_4_008X.html.

Fig. 13.1. *Source:* Central Election Commission Bosnia and Herzegovina.

Fig. 14.1. *Source:* U.S. Census Bureau.

Fig. 14.2. *Sources:* For 2020, https://commons.wikimedia.org/wiki/File:ElectoralCollege2020.svg; for 2016, Gage, Creative Commons Attribution-Share Alike 4.0 International. https://commons .wikimedia.org/wiki/File:ElectoralCollege2016.svg.

Fig. 14.3. *Source:* Creative Commons Attribution-ShareAlike License 4.0. https://en.wikipedia .org/wiki/List_of_United_States_presidential_elections_in_which_the_winner_lost_the _popular_vote.

Fig. 14.4. *Source:* Chart by Perl coder, Creative Commons Attribution-Share Alike 4.0. https://en.wikipedia.org/wiki/National_Popular_Vote_Interstate_Compact#/media /File:US_2020_Census_State_Population_Per_Electoral_Vote.png.

NOTES

An "n" by the page number refers to a footnote.

Chapter One

Page

15   **math is both.** Kelsey Houston-Edwards, "Is the Mathematical World Real?" *Scientific American*, September 1, 2019, https://www.scientificamerican.com/article/is-the-mathematical-world-real/.

20   **Simple Majority Decision."** Kenneth O. May, "A Set of Independent Necessary and Sufficient Conditions for Simple Majority Decision," *Econometrica* 20, no. 4 (1952): 680, https://doi.org/10.2307/1907651.

24   **of the pope.** BBC News, "Conclave: How Cardinals Elect a Pope," February 21, 2013, https://www.bbc.com/news/world-21412589.

24*n*   **The Berlin Philharmonic uses a conclave election:** Michael Cooper, "Berlin Philharmonic Prepares to Vote on New Chief Conductor," *New York Times*, May 6, 2015, https://www.nytimes.com/2015/05/07/arts/music/berlin-philharmonic-prepares-to-vote-on-new-chief-conductor.html.

26   **the magic number.** Chris Gorski, "The Mathematics of Jury Size," Inside Science, May 15, 2012, https://www.insidescience.org/news/mathematics-jury-size.

Chapter Two

Page

28   **other than plurality.** Bryan Eastman, "How Many Politicians Are There in the USA?" PoliEngine, August 1, 2019, https://poliengine.com/blog/how-many-politicians-are-there-in-the-us.

31   **America's biggest cities.** "Vote Splitting," The Center for Election Science, December 2, 2021, https://electionscience.org/vote_splitting_poc.

33   **in one-on-one contests.** Rob Richie, "New Polls Show that GOP Split Vote Problem Continues," FairVote, March 8, 2016, https://fairvote.org/new_polls_show_that_gop_split_vote_problem_continues/.

37   **by such candidates.** Philip Bump, "How Often Do Third-Party Candidates Actually Spoil Elections? Almost Never," *Washington Post*, October 8, 2014, https://www.washington

post.com/news/the-fix/wp/2014/10/08/how-often-do-third-party-candidates -actually-spoil-elections-not-very/.

39   **the Democratic nominee.** Michael Walsh, "Bloomberg Didn't Want to Be Spoiler Leading to Trump Presidency: Adviser," Yahoo! Entertainment, March 8, 2016, https://www.yahoo .com/entertainment/bloomberg-didnt-want-to-be-spoiler-leading-to-180855743.html.

40   **and contested academically.** Patrick Dunleavy, "Duverger's Law Is a Dead Parrot. Outside the USA, First-Past-The-Post Voting Has No Tendency at All to Produce Two Party Politics," British Politics and Policy at LSE, June 18, 2012, https://blogs.lse.ac.uk /politicsandpolicy/duvergers-law-dead-parrot-dunleavy/.

40   **fifty-four state parties.** "Parliament of India Lok Sabha House of the People," Members: Lok Sabha, 2019, https://sansad.in/ls/members/.

41   **the national level.** William A. Galston, "UK Elections: Where Did Support for the Liberal Democrats Go?" Brookings, May 8, 2015, https://www.brookings.edu/blog /fixgov/2015/05/08/uk-elections-where-did-support-for-the-liberal-democrats-go/.

41   **eke out representation.** Matthew M. Singer, "Was Duverger Correct? Single-Member District Election Outcomes in Fifty-Three Countries," British Journal of Political Science 43, no. 1 (2012): 201–20, https://doi.org/10.1017/s0007123412000233.

41   **at stimulating diversity.** Dylan Difford, "Duvager's Law—More Guidelines than Actual Rules?" Electoral Reform Society, March 9, 2022, https://www.electoral-reform.org.uk /duvagers-law-more-guidelines-than-actual-rules/.

42   **Ireland and Malta).** Neal Jesse, "A Sophisticated Voter Model of Preferential Electoral System," in Elections in Australia, Ireland, and Malta under the Single Transferable Vote: Reflections on an Embedded Institution, ed. Shaun Bowler and Bernard Grofman (Ann Arbor: University of Michigan Press, 2000).

42   **climate change denial.** Bill Tieleman, "Proportional Representation Empowers the Extreme and Bizarre," The Tyee, August 16, 2016, https://thetyee.ca/Opinion/2016/08 /16/Proportional-Representation-in-Canada/.

43   **mudslinging still works.** Wesleyan Media Project, "Political Ads in 2020: Fast and Furious," Wesleyan Media Project, March 23, 2021, https://mediaproject.wesleyan.edu/2020 -summary-032321/.

44   **world's leading democracy.** Drew DeSilver, "Turnout in U.S. Has Soared in Recent Elections but by Some Measures Still Trails that of Many Other Countries," Pew Research Center, November 1, 2022, https://www.pewresearch.org/fact-tank/2022/11/01 /turnout-in-u-s-has-soared-in-recent-elections-but-by-some-measures-still-trails-that -of-many-other-countries/.

44   **of voter apathy.** Denise-Marie Ordway and John Wihbey, "Negative Political Ads and Their Effect on Voters: Updated Collection of Research," The Journalist's Resource, September 25, 2016, https://journalistsresource.org/politics-and-government/negative -political-ads-effects-voters-research-roundup/.

44   **campaigned like this.** Mark Eves and Betsy Sweet, "In Maine, Ranked-Choice Voting Changed Everything," Cape Cod Times, October 21, 2020, https://www.capecodtimes .com/story/special/special-sections/2020/10/22/opinionmy-view-in-maine-ranked -choice-voting-changed-everything/42868213/.

44   **New York mayoral elections.** Michael Gold, Jeffery C. Mays, and Emma G. Fitzsimmons, "Yang and Garcia Announce Plans to Campaign Together," *New York Times*, June 18, 2021, https://www.nytimes.com/2021/06/18/nyregion/eric-adams-mayor-primary.html.

44   **list goes on.** "Candidates Campaigning Together Is Legitimate RCV Strategy," readMedia, June 20, 2021, http://readme.readmedia.com/Candidates-Campaigning-Together-is-Legitimate-RCV-Strategy/17968259.

45   **he won again).** Matthew Brown, "Georgia's Runoff System Was Created to Dilute Black Voting Power," *Washington Post*, December 5, 2022, https://www.washingtonpost.com/politics/2022/12/05/georgia-runoff-history/.

45   **for mayoral elections.** "Runoff Election," Ballotpedia, n.d., https://ballotpedia.org/Runoff_election.

46   **Portugal, and Turkey.** Gianluca Passarelli and Matthew Bergman, "Runoff Comebacks in Comparative Perspective: Two-Round Presidential Election Systems," *Political Studies Review* 23, no. 3 (2022), https://doi.org/10.1177/14789299221132441.

48   **their second choice.** Warren Smith, "Louisiana's Famous 1991 'Lizard versus Wizard' Governor Race," RangeVoting.org, accessed April 4, 2023, https://rangevoting.org/LizVwiz.html.

50   **a lowball estimate.** Zach Mohr, Martha Kropf, Mary Mcgowan Shepherd, Joellen Pope, and Madison Esterle, "How Much Are We Spending on Election Administration?" MIT Election Data Science Lab, n.d., https://electionlab.mit.edu/sites/default/files/2019-01/mohr_et_al_2017summary.pdf.

50   **the first round.** "What's the Deal with Runoff Elections?" RepresentWomen, July 7, 2018, https://www.representwomen.org/what_s_the_deal_with_runoff_elections.

51   **received 11,608 votes.** Sarah McCammon, "Virginia Republican David Yancey Wins Tie-Breaking Drawing," NPR, January 4, 2018, https://www.npr.org/2018/01/04/573504079/virginia-republican-david-yancey-wins-tie-breaking-drawing.

Chapter Three

Page

64   **Begich: 53,810 (28.53%)** "State of Alaska 2022 Special General Election," elections.alaska.gov, accessed March 30, 2023, https://www.elections.alaska.gov/results/22SSPG/RcvDetailedReport.pdf; Adrian Blanco and Kevin Uhrmacher, "How Second-Choice Votes Pushed a Democrat to Victory in Alaska," *Washington Post*, August 31, 2022, https://www.washingtonpost.com/elections/2022/08/31/ranked-choice-totals-alaska-peltola/.

66   **as she could.** Zusha Elinson and Gerry Shih, "The Winning Strategy in Oakland: Concentrate on Being 2nd or 3rd Choice," *New York Times*, November 12, 2010, https://www.nytimes.com/2010/11/12/us/politics/12bcvoting.html.

66   **and third-place votes.** David Sharp, "Ranked Choice as Easy as 1, 2, 3? Not So Fast, Critics Say," *Seattle Times*, October 9, 2016, https://www.seattletimes.com/nation-world/ranked-choice-as-easy-as-1-2-3-not-so-fast-critics-say/.

68   **declared the winner.** "Mayoral Election in New York, New York (2021)," Ballotpedia, n.d., https://ballotpedia.org/Mayoral_election_in_New_York,_New_York_(2021).

68   **most first-place votes.** Erin Durkin, "New York's First Full Ranked-Choice Election
       Changed Campaigns—If Not the Results," *POLITICO*, August 24, 2021, https://www
       .politico.com/states/new-york/albany/story/2021/08/24/new-yorks-first-full-ranked
       -choice-election-changed-campaigns-if-not-the-results-1390428.

68*n*   **Another example of a nail-biter upset:** D. Saari, *Chaotic Elections! A Mathematician
       Looks at Voting* (Providence, RI: American Mathematical Society, 2001).

73   **Fishburn in 1983.** Peter C. Fishburn and Steven J. Brams, "Paradoxes of Preferential
       Voting," *Mathematics Magazine* 56, no. 4 (1983): 207–14.

73*n*   **This is spelled out in detail:** Adam Graham-Squire and David McCune, 2022. "A Math-
       ematical Analysis of the 2022 Alaska Special Election for US House," Papers 2209.04764,
       arXiv.org, revised November 2022.

75   **and socioeconomic background.** Zachary Bleemer, "Increasing Bias in SAT Test
       Scores? A Variation on Simpson's Paradox," UC-CHP Report, February 2020, http://
       uccliometric.org/wp-content/uploads/2020/02/UC-CHP-2020.2-SAT-Paradox.pdf.

75   **to racial bias.** Jeff Suzuki, *Constitutional Calculus: The Math of Justice and the Myth of
       Common Sense* (Baltimore, MD: Johns Hopkins University Press, 2015).

76   **studied military engineering.** "Jean-Charles de Borda—Biography," MacTutor, ac-
       cessed April 4, 2023, https://mathshistory.st-andrews.ac.uk/Biographies/Borda.

76   **what was needed.** " Jean-Charles de Borda," Wonders of the World, n.d., https://www
       .wonders-of-the-world.net/Eiffel-Tower/Pantheon/Jean-Charles-de-Borda.php.

76*n*   **Napoleon was a stalwart:** Michael Molinsky. n.d. "Quotations in Context: Napoleon |
       Mathematical Association of America." Www.maa.org. Accessed August 17, 2023. https://
       www.maa.org/press/periodicals/convergence/quotations-in-context-napoleon.

78   **and NASCAR racing,** Paolo Serafini, *Mathematics to the Rescue of Democracy* (Cham:
       Springer, 2020).

81*n*   **The result appears in:** Donald G. Saari, 1984. "The Ultimate of Chaos Resulting from
       Weighted Voting Systems." *Advances in Applied Mathematics* 5 (3): 286–308. https://doi
       .org/10.1016/0196-8858(84)90011-3.

85   **the Bobek-Bonaly-style flip.** "What Happened When the Figure Skaters Decided Not
       to Use Range Voting," RangeVoting.org, accessed April 4, 2023, https://rangevoting.org
       /OlympBonaly.html.

85   **subject of chapter 5.** Maureen T. Carroll, Elyn K. Rykken, and Jody M. Sorensen, "The
       Canadians Should Have Won!?" *Math Horizons* 10, no. 3 (2003): 5–22.

91   **election is small.** W. V. Gehrlein, "Condorcet's Paradox with Three Candidates," in *The
       Mathematics of Preference, Choice and Order*, ed. S. J. Brams, W. V. Gehrlein, and F. S.
       Roberts (Berlin: Springer, 2009); Scott L. Feld and Bernard Grofman, "Who's Afraid of
       the Big Bad Cycle? Evidence from 36 Elections," *Journal of Theoretical Politics* 4, no. 2
       (April 1992): 231–37; Yves Balasko and Hervé Crès, "The Probability of Condorcet Cy-
       cles and Super Majority Rules," *Journal of Economic Theory* 75, no. 2 (August 1997):
       237–70; Peter Kurrild-Klitgaard, "An Empirical Example of the Condorcet Paradox of
       Voting in a Large Electorate," *Public Choice* 107, nos. 1–2 (2001): 135–45; Adrian Van
       Deemen, "On the Empirical Relevance of Condorcet's Paradox," *Public Choice* 158
       (2014): 311–30.

Chapter Four

Page

100   **Independence after Euclid.** Seelye Martin, "Euclid's Fifth, July Fourth," LabLit.com, June 30, 2010, https://www.lablit.com/article/606.

107   **a genuine science.** William H. Riker, "The Two-Party System and Duverger's Law: An Essay on the History of Political Science," *American Political Science Review* 76, no. 4 (1982): 753–66; William H. Riker, *Liberalism against Populism: A Confrontation between the Theory of Democracy and the Theory of Social Choice* (San Francisco, CA: W. H. Freeman, 1982).

109   **Amartya Sen.** George Szpiro, *Numbers Rule: The Vexing Mathematics of Democracy, from Plato to the Present* (Princeton, NJ: Princeton University Press, 2010).

109   **stronger than necessary,** Partha Dasgupta and Eric Maskin, "On the Robustness of Majority Rule," *Journal of the European Economic Association* 6, no. 5 (September 1, 2008): 949–73; Eric Maskin, "Lecture with Nobel Laureate Eric Maskin: How Should We Elect Presidents?" February 17, 2022, YouTube video, https://www.youtube.com/watch?v=kG-vsBn4p28; Wesley H. Holliday and Eric Pacuit, "Axioms for Defeat in Democratic Elections," *Journal of Theoretical Politics* 33, no. 4 (October 2021): 475–524.

109   **modification of IIA.** Eric Maskin, "Arrow's Theorem, May's Axioms, and Borda's Rule," unpublished article, December 2022, https://scholar.harvard.edu/files/maskin/files/arrows_theorem_mays_axioms_and_bordas_rule_1.19.2023.pdf.

109   **reduce this incompatibility.** Donald Saari, "Geometry of Voting," in *Handbook of Social Choice and Welfare*, vol. 2, ed. Kenneth Arrow, Amartya Sen, and Kotaro Suzumura, 897–945 (Elsevier, 2011).

109   **badly at times."** Phil McKenna. 2008. "Vote of No Confidence." *New Scientist* 198 (2651): 30–33. https://doi.org/10.1016/s0262-4079(08)60914-8.

110   **every single time.** Theodore Landsman, "All RCV Elections in the Bay Area so Far Have Produced Condorcet Winners," FairVote, August 9, 2022, https://fairvote.org/every_rcv_election_in_the_bay_area_so_far_has_produced_condorcet_winners/.

110   **the spoiler effect.** David McCune and Jennifer Wilson, "Ranked-Choice Voting and the Spoiler Effect," *Public Choice* 196 (2023), https://doi.org/10.1007/s11127-023-01050-3.

110   **however, are rare.** Adam Graham-Squire and N. Zayatz, "Lack of Monotonicity Anomalies in Empirical Data of Instant-Runoff Elections," *Representation* 57, no. 4 (2020): 565–73.

110   **exhibited Simpson's paradox.** David McCune and Lori McCune, "Does the Choice of Preferential Voting Method Matter? An Empirical Study Using Ranked Choice Elections in the United States," *Representation*, October 18, 2022, 1–16.

111   **kinds of anomalies.** David McCune and Adam Graham-Squire, "Monotonicity Anomalies in Scottish Local Government Elections," ArXiv.org, May 28, 2023, https://doi.org/10.48550/arXiv.2305.17741.

111   **show monotonicity failure.** Nicholas R. Miller, "Closeness Matters: Monotonicity Failure in IRV Elections with Three Candidates," *Public Choice* 173, nos. 1–2 (2017): 91–108; Joseph T. Ornstein and Robert Z. Norman, "Frequency of Monotonicity Failure under Instant Runoff Voting: Estimates Based on a Spatial Model of Elections," *Public Choice* 161, nos. 1–2 (2013): 1–9.

Chapter Five

Page

113   **book *Approval Voting*.** Steven J. Brams and Peter C. Fishburn, *Approval Voting* (Boston: Birkhäuser, 1983).

115   **of Joe Biden.** Aaron Hamlin, "The Early 2020 Democratic Primary: Comparing Voting Methods," The Center for Election Science, February 28, 2020, https://electionscience .org/commentary-analysis/the-early-2020-democratic-primary-comparing-voting -methods/; Caitlyn Alley Peña, "Gaming the Vote: A Conversation with William Pound-stone," The Center for Election Science, May 28, 2020, https://electionscience.org /events/gaming-the-vote-a-conversation-with-william-poundstone/.

115   **in approval voting.** Donald G. Saari and Jill van Newenhizen. "The Problem of Inde-terminacy in Approval, Multiple, and Truncated Voting Systems," *Public Choice* 59, no. 2 (1988): 101–20. http://www.jstor.org/stable/30024954.

115   **of its advantages.** Steven J. Brams, Peter C. Fishburn, and Samuel Merrill, "The Respon-siveness of Approval Voting: Comments on Saari and Van Newenhizen," *Public Choice* 59, no. 2 (1988): 121–31.

115   **all four criteria.** E. Arthur Robinson and Daniel H. Ullman, *The Mathematics of Politics* (Boca Raton, FL: CRC Press, 2016).

116   **works just fine.** Steven J. Brams and Peter C. Fishburn, "Going from Theory to Practice: The Mixed Success of Approval Voting," in *Handbook on Approval Voting*, ed. J. F. Laslier and M. Sanver, 19–37 (Cham: Springer, 2010).

116   **Texas, and Colorado.** Aaron Hamlin, "Who Uses Approval Voting?" The Center for Election Science, May 21, 2015, https://electionscience.org/voting-methods/approval -voting-progress.

116   **dissolution in 1991.)** Theodore Shabad, "Soviet to Begin Multi-Candidate Election Experiment in June," *New York Times*, April 15, 1987, https://www.nytimes.com/1987 /04/15/world/soviet-to-begin-multi-candidate-election-experiment-in-june.html.

119   **a good thing.** "Why Range Voting Is Better than IRV (Instant Runoff Voting)," Ran-gevoting.org, accessed March 30, 2023. https://rangevoting.org/rangeVirv.html.

119   **using this scale.** I know citing Wikipedia is undesirable, but this is a Wikipedia refence about Wikipedia, so it's meta-Ok: "Score Voting," March 28, 2023, https://en.wikipedia .org/wiki/Score_voting.

120   **a 5-point scale.** Jennifer Viegas, "Bonobos Rate Food on Scale from Bark to Grunt," *NBC News*, April 17, 2009, https://www.nbcnews.com/id/wbna30266532.

120   **sites and nests.** Warren Smith, "Ants, Bees, and Computers Agree Range Voting Is Best Single-Winner System," unpublished article, April 2007, https://rangevoting.org /WarrenSmithPages/homepage/naturebees.pdf.

120   **approval voting does.** "Score Voting," The Center for Election Science, n.d., https:// electionscience.org/library/score-voting/.

120   **of range voting.** "About CRV's Co-Founders," RangeVoting.org, accessed April 4, 2023, https://www.rangevoting.org/AboutCofounders.html.

120   **avoidable human unhappiness."** "Bayesian Regret for Dummies," RangeVoting.org, accessed April 4, 2023, https://rangevoting.org/BayRegDum.html.

120   **rather than strategically.** "Bayesian Regrets Shown Graphically," RangeVoting.org, accessed April 4, 2023, https://www.rangevoting.org/BayRegsFig.html.

123   **two months later.** Peter Applebome, "The Guinier Battle: Where Ideas That Hurt Guinier Thrive," *New York Times*, June 5, 1993, https://www.nytimes.com/1993/06/05 /us/the-guinier-battle-where-ideas-that-hurt-guinier-thrive.html.

123   **in Oregon use.** "How Does STAR Voting Work?" STAR Voting, accessed April 4, 2023, https://www.starvoting.org.

123*n*   **Another simulation similar in spirit to Bayesian regret:** Jameson Quinn, "Voter Satisfaction Efficiency Simulator," n.d., accessed April 8, 2023, http://electionscience.github .io/vse-sim/VSEbasic.

123   **and approval voting.** "How Does STAR Voting Work?"

Chapter Six

Page

127   **at the game."** C. L. Dodgson and F. F. Abeles. 2001. *The Pamphlets of Lewis Carroll, Vol. 3, the Political Pamphlets and Letters of Charles Lutwidge Dodgson and Related Pieces: A Mathematical Approach.* Lewis Carroll Society of North America, New York.

127   **1992 presidential election.** Alexander Tabarrok, "President Perot or Fundamentals of Voting Theory Illustrated with the 1992 Election," *Public Choice* 106, nos. 3–4 (2001): 275–97.

129   **to hurt Gore.** Laura Meckler, "GOP Group to Air Pro-Nader TV Ads," AP News, October 27, 2000.

129   **Clinton the election.** Jane Mayer, "How Russia Helped Swing the Election for Trump," *New Yorker*, September 24, 2018, https://www.newyorker.com/magazine/2018/10/01 /how-russia-helped-to-swing-the-election-for-trump.

130   **voting for Green.** "Democrats and Republicans Deceptively Propped up Third-Party Candidates," OpenSecrets News, November 9, 2018, https://www.opensecrets.org/news /2018/11/democrats-and-republicans-deceptively-propped-up-3rd-party-candidates/.

131   **elections in Slovenia.** Pa Ba, "Raziskovalci o anketah: Zmagalo taktično glasovanje," DELO, April 25, 2012, https://web.archive.org/web/20120425082526/http://www .times.si/slovenija/raziskovalci-o-anketah-zmagalo-takticno-glasovanje--eba9dab4dd -6810500f55.html.

131   **of wasted votes.** Laura B. Stephenson, John H. Aldrich, and André Blais, eds., *The Many Faces of Strategic Voting: Tactical Behavior in Electoral Systems around the World* (Ann Arbor: University of Michigan Press, 2018).

131   **by Brexit considerations.** Joel Selway, "People Thought Tactical Voting Was a Big Deal in Britain's General Election. It Wasn't," *Washington Post*, December 30, 2019, https:// www.washingtonpost.com/politics/2019/12/30/people-thought-tactical-voting-was -big-deal-britains-general-election-it-wasnt/.

132 **2022 midterm elections.** David Martin Davies, "Some Democrats Are Voting in the Texas GOP Primaries. Will It Make a Difference?" Texas Public Radio, March 1, 2022, https://www.tpr.org/government-politics/2022-02-28/some-democrats-are-voting-in -the-texas-gop-primaries-will-it-make-a-difference; Jenny Whidden, "Push to Modify N.H. Primary to Prevent Spoiler Votes," *Concord Monitor*, December 28, 2021, https:// www.concordmonitor.com/Proposed-legislation-would-close-New-Hampshire-s -primary-elections-to-undeclared-voters-44265601.

134 **voting are not.** "Favorite Betrayal Criterion," Electowiki, December 3, 2021, https:// electowiki.org/wiki/Favorite_betrayal_criterion.

134 **their first choice.** Change Research, "Poll of 1,023 Likely Voters in Alaska (5/22–25)," https://twitter.com/ChangePolls/status/1402371014324113411/photo/1; David Leonhardt, "A Guide to Ranked-Choice Voting," *New York Times*, June 16, 2021, https:// www.nytimes.com/2021/06/16/briefing/a-guide-to-ranked-choice-voting.html ?searchResultPosition=3.

136 **camp is secondary.** T. N. Tideman, "Independence of Clones as a Criterion for Voting Rules," *Social Choice and Welfare* 4, no. 3 (1987): 185–206.

137 **meaning of 'meaning.'** "Thinking How to Live," interview of Allan Gibbard by Richard Marshall, *3:AM Magazine*, December 5, 2015, https://www.3ammagazine.com/3am /thinking-how-to-live/.

139 **between 1.1 and 1.4.** "Ranked Choice Voting vs. Approval Voting," FairVote, accessed April 4, 2023, https://fairvote.org/resources/electoral-systems/ranked_choice_voting _vs_approval_voting/#advantages-of-rcv-compared-to-approval-voting.

Chapter Seven

Page

143 **voting, or burying.** Rob Richie, "New Lessons from Problems with Approval Voting in Practice," FairVote, December 14, 2016, https://fairvote.org/new_lessons_from_problems _with_approval_voting_in_practice/; Greg Dennis, "How Is RCV Better than Approval, Score or Condorcet Voting Methods?" FairVote, October 3, 2022, https://fairvote.org/how _is_rcv_better_than_approval_score_or_condorcet_voting_methods/.

143 **authoritarianism and extremism.** Protect Democracy, *Advantaging Authoritarianism: The U.S. Electoral System and Antidemocratic Extremism* (Protect Democracy, January 2022), https://s3.documentcloud.org/documents/21458655/2022-03-21-advantaging -authoritarianism.pdf.

144 **people of color.** "Ranked Choice Voting and Minority Representation," Metric Geometry and Gerrymandering Group, October 7, 2021, https://mggg.org/rcv; "Ranked Choice Voting Elections Benefit Candidates and Voters of Color," FairVote, May 2021, https://fairvote.org/report/report_rcv_benefits_candidates_and_voters_of _color.

144 **in plurality races.** Adam Ginsburg, "New Report Illustrates How Ranked Choice Voting Helps Elect More Women," FairVote, September 30, 2022, https://fairvote.org/new _report_illustrates_how_ranked_choice_voting_helps_elect_more_women/.

144   **before instant runoff.** Drew Penrose, "Ranked Choice Voting and Racial Minority Voting Rights in the Bay Area," FairVote, November 2019, https://fairvote.org/report /rcv_and_racial_minority_voting_rights_in_the_bay_area.

144   **there in 2010.** Darwin BondGraham, "How Well Did Ranked-Choice Voting Work in the 2022 Oakland Mayor's Race?" The Oaklandside, December 1, 2022, https:// oaklandside.org/2022/12/01/how-well-did-ranked-choice-voting-work-in-the-2022 -oakland-mayors-race/.

144   **promotes political civility.** Mark Eves and Betsy Sweet, "In Maine, Ranked-Choice Voting Changed Everything," *Cape Cod Times*, October 21, 2020, https://www.capecodtimes.com /story/special/special-sections/2020/10/22/opinionmy-view-in-maine-ranked-choice -voting-changed-everything/42868213/; "Candidates Campaigning Together Is Legitimate RCV Strategy," readMedia, June 20, 2021, http://readme.readmedia.com/Candidates -Campaigning-Together-is-Legitimate-RCV-Strategy/17968259.

144   **in future elections.** "Exit Poll: New Yorkers Love Ranked Choice Voting," FairVote, June 30, 2021, https://fairvote.org/exit_poll_new_yorkers_love_ranked_choice_voting/.

144   **high satisfaction rates.** BondGraham, "How Well Did Ranked-Choice Voting Work in the 2022 Oakland Mayor's Race?"

145   **the most represented.** Drew DeSilver, Carrie Blazina, Janakee Chavda, and Rebecca Leppert, "More U.S. Locations Experimenting with Alternative Voting Systems," Pew Research Center, June 29, 2021, https://www.pewresearch.org/fact-tank/2021/06/29 /more-u-s-locations-experimenting-with-alternative-voting-systems.

145   **the United States.** "Ranked Choice Voting Information," FairVote, accessed April 4, 2023, https://fairvote.org/our-reforms/ranked-choice-voting-information/#ballot -measures.

145   **universities implement it.** Matthew Oberstaedt, "19 Editorial Boards Have Supported Ranked Choice Voting in 2022," FairVote, September 14, 2022, https://fairvote.org/19 -editorial-boards-have-supported-ranked-choice-voting-in-2022; "A Look Inside the Ranked Choice Voting Movement on College Campuses," FairVote, April 14, 2021, https://fairvote.org/a_look_inside_the_ranked_choice_voting_movement_on_college _campuses/.

145   **most approved it.** "Ranked Choice Voting Just Had Its Biggest Election Day Ever," Fair-Vote Action, November 25, 2022, https://fairvoteaction.org/results-for-ranked-choice -voting-ballot-measures-in-2022/.

145   **educate the electorate.** Oberstaedt, "19 Editorial Boards Have Supported Ranked Choice Voting in 2022."

Chapter Eight

Page

152   **representatives, Texas 38.3.** U.S. Department of Commerce, U.S. Census Bureau, "Table 1. Apportionment Population and Number of Representatives by State: 2020 Census," n.d., https://www2.census.gov/programs-surveys/decennial/2020/data /apportionment/apportionment-2020-table01.pdf.

153 **held by immigrants.** Charles Hirschman and Elizabeth Mogford, "Immigration and the American Industrial Revolution from 1880 to 1920," *Social Science Research* 38, no. 4 (December 2009): 897–920.

154 **a representative legislature."** "Founders Online: The Federalist No. 55, [13 February 1788]." n.d. Founders.archives.gov. https://founders.archives.gov/documents/Hamilton/01-04-02-0204.

155 **the United States."** "Founders Online: The Federalist No. 57, [19 February 1788]." n.d. Founders.archives.gov. https://founders.archives.gov/documents/Hamilton/01-04-02-0206.

156 **number of representatives."** "Founders Online: The Federalist No. 55, [13 February 1788]."

156 **for the House.)** "The Apportionment Act of 1842: 'In All Cases, by District,'" history.house.gov, History, Art & Archives, United States House of Representatives, April 16, 2019, https://history.house.gov/Blog/2019/April/4-16-Apportionment-1/.

157 **lost a seat.** Dan Bouk, "No One Loses at 454 Seats," Census Stories, USA, April 26, 2021, https://censusstories.us/2021/04/26/house-size.html.

157 **to urban areas.** Lee Drutman, Jonathan Cohen, Yuval Levin, and Norman Ornstein, "The Case for Enlarging the House of Representatives," Our Common Purpose, December 9, 2021, https://www.amacad.org/ourcommonpurpose/enlarging-the-house.

157 **and for all.** Geoffrey Skelley, "How the House Got Stuck at 435 Seats," FiveThirtyEight, August 12, 2021, https://fivethirtyeight.com/features/how-the-house-got-stuck-at-435-seats/.

157 **Act of 1929.** Ch. 28, An Act To provide for the fifteenth and subsequent decennial censuses and to provide for apportionment of Representatives in Congress, 71st Cong., 1st sess., June 18, 1929, https://static1.squarespace.com/static/5d0a7c994e51a30001253d0a/t/5e600834fea7b32f9c728f31/1583351861608/The+Permanent+Apportionment+Act+of+1929.pdf.

158 **435 "shackled democracy."** Dan Bouk, *House Arrest: How an Automated Algorithm Has Constrained Congress for a Century* (Data & Society, April 2021), https://datasociety.net/wp-content/uploads/2021/04/House-Arrest-Dan-Bouk.pdf.

160 **of their constituents.** "Opinion: America Needs a Bigger House," *New York Times*, November 10, 2018, https://www.nytimes.com/interactive/2018/11/09/opinion/expanded-house-representatives-size.html; Frances E. Lee and Bruce Ian Oppenheimer, *Sizing up the Senate: The Unequal Consequences of Equal Representation* (Chicago: University of Chicago Press, 1999).

160 **must deal with.** "Congressional Apportionment after the 2020 Census," Ballotpedia, accessed April 4, 2023, https://ballotpedia.org/Congressional_apportionment_after_the_2020_census.

161 **than 655 seats.** Michael G. Neubauer and Joel Zeitlin, "Apportionment and the 2000 Election," *College Mathematics Journal* 34, no. 1 (2003): 2–10.

162 **1929 Reapportionment Act.** Lots of historical background can be found in Jeffrey W. Ladewig and Mathew P. Jasinski, "On the Causes and Consequences of and Remedies for Interstate Malapportionment of the U.S. House of Representatives," *Perspectives on Politics* 6, no. 01 (February 26, 2008).

162   **Arts and Sciences.** Drutman et al., "The Case for Enlarging the House of Representatives."

163   **the current Congress.** Caroline Kane, Gianni Mascioli, Michael McGarry, and Meira Nagel, "Why the House of Representatives Must Be Expanded and How Today's Congress Can Make It Happen," Fordham Law Archive of Scholarship and History, 2020, https://ir.lawnet.fordham.edu/faculty_scholarship/1100.

163   **to about 25%.** "America Needs a Bigger House," *New York Times*, November 10, 2018, https://www.nytimes.com/interactive/2018/11/09/opinion/expanded-house -representatives-size.html.

164   **past forty years.** Kevin R. Kosar, "Congress Can't Keep up with Its Workload," *Boston Globe*, January 13, 2021, https://www.bostonglobe.com/2021/01/13/opinion/congress -cant-keep-up-with-its-workload/.

165   **numbers of representatives.** Bouk, "No One Loses at 454 Seats."

166   **to the House.** Drutman et al., "The Case for Enlarging the House of Representatives."

167   **next seventy years.** "Why 435? How We Can Change the Size of the House of Representatives," FairVote, October 12, 2017, https://www.fairvote.org/how_we_can_change _the_size_of_the_house_of_representatives.

170   **of California, Irvine.** Rein Taagepera, "The Size of National Assemblies," *Social Science Research* 1, no. 4 (1972): 385–401.

170   **with each other.** Skelley, "How the House Got Stuck at 435 Seats."

171   **democracies have not.** Luke Angelillo, Yoo Min Lee, McConnell Bristol, and Jodi Robinson, "The Politics of Numbers: Why 435 Does Not Add Up," *Brown Political Review*, December 22, 2018, https://brownpoliticalreview.org/2018/12/politics-numbers-435 -not-add/.

171   **reflect demographic changes.** "Representation in the House of Commons of Canada," Redistribution: Federal Election Districts, n.d., accessed April 4, 2023, https://redecoupage -federal-redistribution.ca/content.asp?section=info&dir=his%2Frep&document =p1&lang=e.

171   **of its legislature.** Drutman et al., "The Case for Enlarging the House of Representatives."

172*n*   **The cube root law:** Giorgio Margaritondo, "Size of National Assemblies: The Classic Derivation of the Cube-Root Law Is Conceptually Flawed," *Frontiers in Physics* 8 (January 15, 2021), https://doi.org/10.3389/fphy.2020.614596; Emmanuelle Auriol and Robert J. Gary-Bobo, "On the Optimal Number of Representatives," *Public Choice* 153, nos. 3–4 (2011): 419–45.

Chapter Nine

Page

176   **about 30,000 people.** Royce Crocker, *The House of Representatives Apportionment Formula: An Analysis of Proposals for Change and Their Impact on States*, Congressional Research Service Report for Congress, August 26, 2010, https://sgp.fas.org/crs/misc /R41382.pdf.

177   **population was 3,615,920.** The population numbers can be found, for example, in "Apportioning the U.S. House of Representatives 1790," https://www.austincc.edu /hannigan/Math1513/1790HouseOfRep.pdf; and United States Census Bureau, "Schedule of the Whole Number of Persons within the Several Districts of the United States, Taken According to 'An Act Providing for the Enumeration of the Inhabitants of the United States,'" https://www2.census.gov/library/publications/decennial /1790/number_of_persons/1790a-02.pdf. It is important to use the representative population (see footnote in the text for explanation of what this means) and not all the population captured by the census because the two were different for the purpose of apportionment.

178   **in some states.** Margaret Wood, "Apportionment and the First Presidential Veto," Library of Congress Blogs, November 3, 2014, https://blogs.loc.gov/law/2014/11 /apportionment-and-the-first-presidential-veto/. This website links to Jefferson's letter and a transcription.

179   **every thirty thousand.** "Founders Online: Opinion on Apportionment Bill, 4 April 1792." n.d., Founders.archives.gov, accessed August 19, 2023, https://founders.archives.gov /documents/Jefferson/01-23-02-0324.

181   **9,900, and 22,400.** Mathispower4u, "Apportionment: The Population Paradox," YouTube video, n.d., https://www.youtube.com/watch?v=S4LH4t9P1Hg.

181   **of A increased.** This and the next example are borrowed from Daniel H. Ullman and E. Arthur Robinson, *The Mathematics of Politics* (Boca Raton, FL: CRC Press, 2016).

Chapter Ten

Page

183   **by the bill.** "Apportionment Bill," in *Annals of Congress*, House of Representatives, 2nd Congress, 1st Session, 539–40, on "A Century of Lawmaking for a New Nation: U.S. Congressional Documents and Debates, 1774–1875," Library of Congress, accessed April 4, 2023, https://memory.loc.gov/cgi-bin/ampage?collId=llac&fileName=003/llac003 .db&recNum=267.

183   **could possibly differ.** "Washington Exercises First Presidential Veto," History, November 16, 2009, https://www.history.com/this-day-in-history/washington-exercises-first -presidential-veto.

186   **quota was 1.95.** Michel L. Balinski and H. Peyton Young, *Fair Representation* (Washington, DC: Brookings Institution Press, 2010).

186   **divisor method does.** Section 14.3 of Consortium for Mathematics and Its Applications (COMAP), *For All Practical Purposes: Mathematical Literacy in Today's World* (New York: W. H. Freeman, 11 ed., 2021).

186   **apportionment to date.** Section 14.3 of COMAP, *For All Practical Purposes*. That section also contains specific examples for the 2010 census.

187   **good at math.** Much of the history recounted in this section comes from Balinski and Young, *Fair Representation*. See also E. Arthur Robinson and Daniel H. Ullman, *The Mathematics of Politics* (Boca Raton, FL: CRC Press, 2016); Shannon Guerrero and

Charles Biles, "The History of the Congressional Apportionment Problem through a Mathematical Lens," Digital Commons @ Humboldt State University, January 1, 2017, https://digitalcommons.humboldt.edu/cgi/viewcontent.cgi?article=1006&context =apportionment; Dan Bouk, "The Harvard Mimeograph," Census Stories, USA, September 1, 2020, https://censusstories.us/2020/09/01/harvard-mimeograph.html.

187   **what Dean meant.** For a more formal treatment, see Robinson and Ullman, *The Mathematics of Politics*. See also Joseph Malkevitch, "Apportionment—Feature Column from the AMS," American Mathematical Society, accessed April 4, 2023, http://www.ams.org /publicoutreach/feature-column/fcarc-apportionii2.

191   **would have won.** Balinski and Young, *Fair Representation*.

192   **removed from politics.** Bouk, *House Arrest*.

197   **the *Huntington-Hill method*.** E. Huntington, "The Mathematical Theory of the Apportionment of Representatives," *Proceedings of the National Academy of Sciences* 7 (1921): 123–27.

197   **of the quotients."** Bouk, *House Arrest*.

198   **Webster, and Huntington-Hill.** Balinski and Young, *Fair Representation*.

198   **does precisely this.** George Szpiro, *Numbers Rule: The Vexing Mathematics of Democracy, from Plato to the Present* (Princeton, NJ: Princeton University Press, 2010).

200   **voted for it.** Michael Caulfield, "Apportioning Representatives in the United States Congress—Hill's Method of Apportionment," Mathematical Association of America, November 2010, https://www.maa.org/press/periodicals/convergence/apportioning -representatives-in-the-united-states-congress-hills-method-of-apportionment.

200   **the best method.** Szpiro, *Numbers Rule*.

202   **better than Huntington-Hill.** Jeffrey O. Bennett and William L. Briggs, *Using & Understanding Mathematics: A Quantitative Reasoning Approach* (Boston, MA: Pearson Learning Solutions, 2019).

203   **to large states.** COMAP, *For All Practical Purposes*.

205   **method been used.** Jeffrey W. Ladewig and Mathew P. Jasinski, "On the Causes and Consequences of and Remedies for Interstate Malapportionment of the U.S. House of Representatives," *Perspectives on Politics* 6, no. 1 (February 26, 2008).

205   **formalizing political arguments."** Alma Steingart, "Democracy by Numbers," *Los Angeles Review of Books*, August 10, 2018, https://lareviewofbooks.org/article/democracy -by-numbers.

Chapter Eleven

Page

207   **of the district.** Doug Israel, "Why Gerrymandering Must Go," Gotham Gazette, February 14, 2005, https://www.gothamgazette.com/index.php/open-government/2744-why -gerrymandering-must-go.

210   **of the reception."** *Last Week Tonight*, "Gerrymandering: Last Week Tonight with John Oliver (HBO)." YouTube, 2017, https://www.youtube.com/watch?v=A-4dIImaodQ.

211   **there were 72.** Mo Rocca, "Gerrymandering: A Threat to Democracy?" *CBS News*, January 14, 2018, https://www.cbsnews.com/video/gerrymandering-a-threat-to-democracy.

368 NOTES TO CHAPTER 11

211 **are only 35.** Reid J. Epstein and Nick Corasaniti, "'Taking the Voters out of the Equation': How the Parties Are Killing Competition," *New York Times*, February 6, 2022, https://www.nytimes.com/2022/02/06/us/politics/redistricting-competition-midterms.html; Madison Fernandez, "Competitive Congressional Districts Decline," *POLITICO*, February 27, 2023, https://www.politico.com/newsletters/weekly-score/2023/02/27/competitive-congressional-districts-decline-00084506.

211 **race was 28%.** Fernandez, "Competitive Congressional Districts Decline."

211 **win by less."** "Nothing Inflated in Arnold Schwarzenegger's Claim on Gerrymandering," PolitiFact, February 23, 2017, https://www.politifact.com/factchecks/2017/mar/17/arnold-schwarzenegger/true-nothing-inflated-schwarzneggers-claim-average.

212 **1708 to 1832.** Greg Jenner, "9 Unexpected Facts about the History of General Elections," Radio 4 You're Dead to Me, accessed April 5, 2023, https://www.bbc.co.uk/programmes/articles/3qhx6XP4lhcZLGmqCBn9sly/9-unexpected-facts-about-the-history-of-general-elections.

212 **or *incumbent pairing*.** Moon Duchin and Olivia Walch, eds., *Political Geometry* (Cham: Birkhäuser, 2022).

213 **by Elbridge Gerry,** Erick Trickey, "Where Did the Term 'Gerrymander' Come From?" *Smithsonian Magazine,* July 20, 2017, https://www.smithsonianmag.com/history/where-did-term-gerrymander-come-180964118.

214 **tests, and intimidation.** Erik J. Engstrom, "Gerrymandering and the Evolution of American Politics," in Engstrom, *Partisan Gerrymandering and the Construction of American Democracy* (Ann Arbor: University of Michigan Press, 2013), 1–18.

215 **more than 600,000.** Jesse Wegman, *Let the People Pick the President: The Case for Abolishing the Electoral College* (New York: St. Martin's Griffin, 2021).

215 **of the population.** Wegman, *Let the People Pick the President.*

216 **populations about equal.** For a great summary of the racial gerrymandering judicial history, see Ellen D. Katz, "Race and Redistricting: The Legal Framework," in *Political Geometry*, ed. Moon Duchin and Olivia Walch (Cham: Birkhäuser, 2022), 137–62.

216n **The 1929 Reapportionment Act:** Margo Anderson, "An Enumeration of the Population: A History of the Census," American Bar Association, May 22, 2020, https://www.americanbar.org/groups/public_education/publications/insights-on-law-and-society/volume-20/issue-2/an-enumeration-of-the-population--a-history-of-the-census/.

216 **representation of minorities.** Ellis Champion, "Nine Redistricting Cases That Shaped History," Democracy Docket, August 17, 2021, https://www.democracydocket.com/news/nine-redistricting-cases-that-shaped-history/; Arusha Gordon and Douglas M. Spencer, "Explainer: A Brief Introduction to the Voting Rights Act," in *Political Geometry*, ed. Moon Duchin and Olivia Walch (Cham: Birkhäuser, 2022), 131–36.

216 **lines is partisanship.** Erik J. Engstrom, *Partisan Gerrymandering and the Construction of American Democracy* (Ann Arbor: University of Michigan Press, 2016).

217 **district wherever possible."** Colorado Constitution, Article V, Section 47, accessed September 13, 2023, https://ballotpedia.org/Article_V,_Colorado_Constitution.

217   **across the states.** "Communities of Interest," Brennan Center for Justice, n.d., https://www.brennancenter.org/sites/default/files/analysis/6%20Communities%20of%20Interest.pdf.

217   **districting at all.** Heather Rosenfeld and Moon Duchin, "Explainer: Communities of Interest," in *Political Geometry*, ed. Moon Duchin and Olivia Walch (Cham: Birkhäuser, 2022), 235–45.

218   ***Vote Doesn't Count.*** Elizabeth Kolbert, "Drawing the Line: How Redistricting Turned America from Blue to Red," *New Yorker*, June 20, 2016, https://www.newyorker.com/magazine/2016/06/27/ratfcked-the-influence-of-redistricting; "'Gerrymandering on Steroids': How Republicans Stacked the Nation's Statehouses," WBUR Here & Now, July 19, 2016, https://www.wbur.org/hereandnow/2016/07/19/gerrymandering-republicans-redmap.

218   **of 435 districts.** "State Legislative Elections, 2010," Ballotpedia, n.d., https://ballotpedia.org/State_legislative_elections,_2010.

219   **and 9 seats.** "United States Congress Elections, 2012," Ballotpedia, n.d., https://ballotpedia.org/United_States_Congress_elections,_2012.

219   **exactly flipped, 46%–54%.** "State Legislative Elections, 2012," Ballotpedia, n.d., https://ballotpedia.org/State_legislative_elections,_2012.

219   **kept bearing fruit.** Chris Cillizza, "Republicans Have Gained More than 900 State Legislative Seats since 2010," *Washington Post*, January 14, 2015, https://www.washingtonpost.com/news/the-fix/wp/2015/01/14/republicans-have-gained-more-than-900-state-legislative-seats-since-2010/.

219   **into the maps.** Jeffrey Shen, "Exploring the Seats-Votes Curve," accessed April 6, 2023, https://jeffreyshen19.github.io/Seats-Votes-Curves/.

220   **can control Congress."** Karl Rove, "The GOP Targets State Legislatures," *Wall Street Journal*, March 4, 2010, https://www.wsj.com/articles/SB10001424052748703862704575099670689398044.

221   **to modern art."** Alex Isenstadt, "California Incumbents Safe No More?" *POLITICO*, April 12, 2011, https://www.politico.com/story/2011/04/california-incumbents-safe-no-more-052970.

222   **and 2017 maps.** Michael Wines, "The Battle over the Files of a Gerrymandering Mastermind," *New York Times*, September 5, 2019, https://www.nytimes.com/2019/09/04/us/gerrymander-north-carolina-hofeller.html.

222   **of the devil."** Hansi Lo Wang, "Deceased GOP Strategist's Daughter Makes Files Public That Republicans Wanted Sealed," *NPR*, January 5, 2020, https://www.npr.org/2020/01/05/785672201/deceased-gop-strategists-daughter-makes-files-public-that-republicans-wanted-sea.

222   **as racially gerrymandered.** "North Carolina v. Covington," Brennan Center for Justice, June 28, 2018, https://www.brennancenter.org/our-work/court-cases/north-carolina-v-covington.

222   **to bring suit.** For more on Wisconsin's story, see Bridget Bowden and Shawn Johnson, "How the 2011 Political District Map Changed the Game for Wisconsin," Wisconsin Public Radio, October 13, 2021, https://www.wpr.org/mappedout/how-2011-political-district-map-changed-game-wisconsin.

222   **the federal level.** Nina Totenberg, Domenico Montanaro, and Miles Parks, "Supreme Court Rules Partisan Gerrymandering Is beyond the Reach of Federal Courts," NPR, June 27, 2019, https://www.npr.org/2019/06/27/731847977/supreme-court-rules-partisan-gerrymandering-is-beyond-the-reach-of-federal-court.

222   **few months later.** "Gill v. Whitford," Brennan Center for Justice, July 3, 2019, https://www.brennancenter.org/our-work/court-cases/gill-v-whitford.

223   **of state courts),** Patrick Berry, Alicia Bannon, and Douglas Keith, "Legislative Assaults on State Courts—May 2021 Update," Brennan Center for Justice, May 19, 2021, https://www.brennancenter.org/our-work/research-reports/legislative-assaults-state-courts-may-2021-update.

223*n*   **North Carolina's courts:** Michael Wines, "North Carolina Gerrymander Ruling Reflects Politicization of Judiciary Nationally," *New York Times,* April 28, 2023, https://www.nytimes.com/2023/04/28/us/north-carolina-supreme-court-gerrymander.html.

223   **gerrymandering is acceptable.** Totenberg, Montanaro, and Parks, "Supreme Court Rules Partisan Gerrymandering Is beyond the Reach of Federal Courts."

223   **by partisan actors.** Chris Leaverton, "Who Controlled Redistricting in Every State," Brennan Center for Justice, October 5, 2022, https://www.brennancenter.org/our-work/research-reports/who-controlled-redistricting-every-state.

224   **use independent commissions.** Jeffrey O. Bennett and William L. Briggs, *Using & Understanding Mathematics: A Quantitative Reasoning Approach* (Boston, MA: Pearson Learning Solutions, 2019); James Ruley, "One Person, One Vote: Gerrymandering and the Independent Commission, a Global Perspective," *Indiana Law Journal* 92, no. 2 (2017): article 9.

224   **in the courts.)** Nicholas Fandos, "Judge Tosses N.Y. District Lines, Citing Democrats' 'Bias,'" *New York Times,* March 31, 2022, https://www.nytimes.com/2022/03/31/nyregion/judge-new-york-redistricting-gerrymandering.html; Grace Ashford and Nicholas Fandos, "N.Y. Democrats Could Gain 3 House Seats under Proposed District Lines," *New York Times,* January 31, 2022, https://www.nytimes.com/2022/01/30/nyregion/new-york-redistricting-congressional-map.html.

224   **drew its own.** Nick Corasaniti and Reid J. Epstein, "How a Cure for Gerrymandering Left U.S. Politics Ailing in New Ways," *New York Times,* November 17, 2021, https://www.nytimes.com/2021/11/17/us/politics/gerrymandering-redistricting.html.

225   **distributed more efficiently.** Anthony J. McGann, Charles Anthony Smith, Michael Latner, and Alex Keena, *Gerrymandering in America: The House of Representatives, the Supreme Court, and the Future of Popular Sovereignty* (Cambridge: Cambridge University Press, 2016).

225   **the general elections."** Pete Bailey, "A Brief History of How Gerrymandering Distorts U.S. Politics," Population Education, November 13, 2020, https://populationeducation.org/a-brief-history-of-how-gerrymandering-distorts-u-s-politics.

225   **their legislature-drawn counterparts.** Nathaniel Rakich, "Did Redistricting Commissions Live up to Their Promise?" FiveThirtyEight, January 24, 2022, https://fivethirtyeight.com/features/did-redistricting-commissions-live-up-to-their-promise.

225   **14.6% since then.** Matthew Nelson. "Independent Redistricting Commissions Are Associated with More Competitive Elections." *PS: Political Science & Politics* 56, no. 2 (2023): 207–12. doi:10.1017/S104909652200124X.

225   **in the nation.** Michael Li, "Anti-Gerrymandering Reforms Had Mixed Results," Brennan Center for Justice, September 19, 2022, https://www.brennancenter.org/our-work /analysis-opinion/anti-gerrymandering-reforms-had-mixed-results.

226   **didn't contain prisons.** Peter Wagner, "Locked Up, but Still Counted: How Prison Populations Distort Democracy," Prison Gerrymandering Project, September 5, 2008, https://www.prisonersofthecensus.org/news/2008/09/05/stillcounted.

226   **from his neighbor.** Annie Han, "Prison Gerrymandering Disenfranchises Incarcerated People in Political Process," Wilson Center for Science and Justice, August 10, 2021, https://wcsj.law.duke.edu/2021/08/prison-gerrymandering-disenfranchises -incarcerated-people-in-political-process.

226   **than 2 million.** Abdallah Fayyad, "How Prisons Distort American Democracy," *Boston Globe*, March 1, 2021, https://www.bostonglobe.com/2021/03/01/opinion/how -prisons-distort-american-democracy/.

226   **the 2030 census.** Andrea Fenster, "How Many States Have Ended Prison Gerrymandering? About a Dozen!" Prison Gerrymandering Project, October 6, 2021, https://www .prisonersofthecensus.org/news/2021/10/26/state_count/.

227   **lay off teachers.** Benjamin Barber, "School Secession Movement Drives Re-Segregation," Facing South, October 25, 2019, https://www.facingsouth.org/2019/10/school -secession-movement-drives-re-segregation.

227   **the period 2000–2017.** EdBuild, *Fractured: The Accelerating Breakdown of America's School Districts* (N.p.: [EdBuild], 2019), https://edbuild.org/content/fractured/fractured -full-report.pdf.

227   **have been excluded."** Alvin Chang, "School Segregation Didn't Go Away. It Just Evolved," *Vox*, July 27, 2017, https://www.vox.com/policy-and-politics/2017/7/27 /16004084/school-segregation-evolution; Erika Wilson, "The New School Segregation," *Cornell International Law Journal* 49, no. 3 (2016): 139, https://scholarship.law .cornell.edu/cgi/viewcontent.cgi?article=1887&context=cilj.

Chapter Twelve

Page

229   **for state districts.** Doug Spencer, "Where Are the Lines Drawn?" All About Redistricting, n.d., https://redistricting.lls.edu/redistricting-101/where-are-the-lines-drawn.

230   **districts be compact.** Spencer, "Where Are the Lines Drawn?"

232   **for congressional districts.** Spencer, "Where Are the Lines Drawn?"

232   **part to another.** Spencer, "Where Are the Lines Drawn?"

232   **sense these days.** Royce Crocker, *Congressional Redistricting: An Overview* (Washington, DC: Congressional Research Service, November 21, 2012), https://sgp.fas.org/crs/misc /R42831.pdf.

233   **to represent them.** Wikipedia, "Reserved Political Positions," https://en.wikipedia.org
      /wiki/Reserved_political_positions.

233   **over the country.** "Fair Representation Act," n.d., FairVote, https://fairvote.org/our
      -reforms/fair-representation-act/. See also Steven Hill, "'Representation for All'—for
      Real?—Here's What It Looks Like," DemocracySOS, July 15, 2022, https://democracysos
      .substack.com/p/representation-for-all-for-real-heres.

233   **or three candidates.** Nikhil Garg, Wes Gurnee, David Rothschild, and David Shmoys,
      "Combatting Gerrymandering with Social Choice," in *EC '22: Proceedings of the 23rd ACM
      Conference on Economics and Computation* (New York: Association for Computing
      Machinery, 2022), 560–561.

234   **House *cannot exist*.** Moon Duchin, Taissa Gladkova, Eugene Henninger-Voss, Ben Klin-
      gensmith, Heather Newman, and Hannah Wheelen, "Locating the Representational
      Baseline: Republicans in Massachusetts," *Election Law Journal: Rules, Politics, and Policy*
      18, no. 4 (2019): 388–401, https://doi.org/10.1089/elj.2018.0537.

235   **to some extent.** Molly E. Reynolds, "Republicans in Congress Got a 'Seats Bonus' This
      Election (Again)," Brookings, January 27, 2017, https://www.brookings.edu/blog/fixgov
      /2016/11/22/gop-seats-bonus-in-congress/.

235   **called *partisan symmetry,*** For a review of the history and the issues with partisan sym-
      metry, see Daryl DeFord, Natasha Dhamankar, Moon Duchin, Varun Gupta, Mackenzie
      McPike, Gabe Schoenbach, and Ki Wan Sim, "Implementing Partisan Symmetry: Prob-
      lems and Paradoxes," *Political Analysis* 31, no. 3 (2021): 1–20, doi:10.1017/pan.2021.49.

242   **to the victory.** Nicholas Stephanopoulos and Eric McGhee, "Partisan Gerrymandering
      and the Efficiency Gap," *University of Chicago Law Review* 82 (March 2015): 831–900,
      https://uchicagolawjournalsmshaytiubv.devcloud.acquia-sites.com/sites/lawreview
      .uchicago.edu/files/04%20; Eric McGhee, "Measuring Partisan Bias in Single-Member
      District Electoral Systems," *Legislative Studies Quarterly* 39, no. 1 (2014): 55–85.

242n **There are also mathematical:** Daryl DeFord, et. al, "Implementing Partisan Symmetry:
      Problems and Paradoxes," *Political Analysis* 31, no. 3 (2021): 1–20, doi:10.1017/pan.2021.49.

244   **to foul play.** Benjamin Cover, "Quantifying Partisan Gerrymandering: An Evaluation
      of the Efficiency Gap Proposal," *Stanford Law Review* 70 (2018), https://digitalcommons
      .law.uidaho.edu/cgi/viewcontent.cgi?article=1126&context=faculty_scholarship.

244   **them, of course).** Nathaniel Rakich, "Did Redistricting Commissions Live up to Their
      Promise?" FiveThirtyEight, January 24, 2022, https://fivethirtyeight.com/features/did
      -redistricting-commissions-live-up-to-their-promise.

247n **This and the previous issue can be recast:** Benjamin Cover, "Quantifying Partisan
      Gerrymandering: An Evaluation of the Efficiency Gap Proposal," *Stanford Law Review*
      70 (2018), https://digitalcommons.law.uidaho.edu/cgi/viewcontent.cgi?article
      =1126&context=faculty_scholarship; Mira Bernstein and Moon Duchin, "A Formula
      Goes to Court: Partisan Gerrymandering and the Efficiency Gap," *Notices of the Ameri-
      can Mathematical Society* 64, no. 9 (2017): 1020–1024, https://doi.org/10.1090/noti1573.

248   **the efficiency gap.** Nicholas Stephanopoulos and Eric McGhee, "The Measure of a
      Metric: The Debate over Quantifying Partisan Gerrymandering," *Stanford Law Review*
      70 (2018): 1503–1568.

248   **this practical experience.** Cover, "Quantifying Partisan Gerrymandering."

248   **conditions for voting.** Michael Waldman, Patrick Berry, Robyn Sanders, and Sara Loving, "Voting Laws Roundup: December 2021," Brennan Center for Justice, March 1, 2023, https://www.brennancenter.org/our-work/research-reports/voting-laws-roundup-december-2021.

248   **ID were mandatory.** Christina A. Cassidy and Ivan Moreno, "Wisconsin Voter ID Law Proved Insurmountable for Many," *Milwaukee Journal Sentinel*, May 14, 2017, https://www.jsonline.com/story/news/politics/2017/05/14/wisconsin-voter-id-law-proved-insurmountable-many/321680001/.

249   ***Gill v. Whitford* case.** For discussion of improvements to the efficiency gap, see Kristopher Tapp, "Measuring Political Gerrymandering," *American Mathematical Monthly* 126, no. 7 (2019): 593–609, https://doi.org/10.1080/00029890.2019.1609324.

254   **paper on compactness.** Daniel D. Polsby and Robert D. Popper, "The Third Criterion: Compactness as a Procedural Safeguard against Partisan Gerrymandering," *Yale Law & Policy Review* 9, no. 2 (1991): 301–353.

254   **you guys have."** Christopher Ingraham, "Trudeau Says Canada Does Redistricting Better than We Do. Is He Right?" *Washington Post*, September 25, 2021, https://www.washingtonpost.com/business/2018/09/25/trudeau-says-canada-does-redistricting-better-we-do-is-he-right/.

259   **of the scores.** This and related issues are nicely elaborated in Moon Duchin and Bridget Eileen Tenner, "Discrete Geometry for Electoral Geography," *ArXiv:1808.05860 [Physics]*, August 15, 2018, https://arxiv.org/pdf/1808.05860.pdf.

259   **Minimum Convex Polygon."** Brief of Amicus Curiae Fair Democracy, League of Women Voters of Pennsylvania v. The Commonwealth of Pennsylvania, Supreme Court of Pennsylvania, 159 MM 2017, https://www.pacourts.us/Storage/media/pdfs/20211215/165722-feb.19,2018-briefofamicuscuriae(fairdemocracy).pdf.

260   **could have been.** Jowei Chen and Jonathan Rodden, "Unintentional Gerrymandering: Political Geography and Electoral Bias in Legislatures," *Quarterly Journal of Political Science* 8, no. 3 (2013): 239–69.

263   **roughly equal populations.** Benjamin Fifield, Michael Higgins, Kosuke Imai, and Alexander Tarr, "Automated Redistricting Simulation Using Markov Chain Monte Carlo," *Journal of Computational and Graphical Statistics* 29, no. 4 (2020): 715–28.

264   **and two coauthors.** Gregory Herschlag, Robert Ravier, and Jonathan C. Mattingly, "Evaluating Partisan Gerrymandering in Wisconsin," *ArXiv:1709.01596 [Physics, Stat]*, September 5, 2017, https://arxiv.org/pdf/1709.01596v1.pdf.

265   **coauthored by Mattingly,** Gregory Herschlag, Han Sung Kang, Justin Luo, Christy Vaughn Graves, Sachet Bangia, Robert Ravier, and Jonathan C. Mattingly, "Quantifying Gerrymandering in North Carolina," *Statistics and Public Policy* 7, no. 1 (2020): 30–38, https://doi.org/10.1080/2330443x.2020.1796400.

265n   **Another recent and promising:** Marion Campisi Thomas Ratliff, Stephanie Somersille, and Ellen Veomett, "Geography and Election Outcome Metric: An Introduction," *Election Law Journal: Rules, Politics, and Policy* 21, no. 3 (2022): 200–219, https://doi.org/10.1089/elj.2021.0054.

267   **in the future."** Vieth v. Jubelirer, 541 U.S. 267 (2004) (Kennedy, J., concurring opinion), https://www.law.cornell.edu/supct/html/02-1580.ZC.html.

266n   **Incorporating the Voting Rights Act:** Amariah Becker, Moon Duchin, Dara Gold, and Sam Hirsch, 2021. "Computational Redistricting and the Voting Rights Act." *Election Law Journal: Rules, Politics, and Policy*, 20 (4). https://doi.org/10.1089/elj.2020.0704.

267   **of unconstitutional partisanship."** Bernard Grofman and Gary King, "The Future of Partisan Symmetry as a Judicial Test for Partisan Gerrymandering after LULAC v. Perry," *Election Law Journal* 6, no. 1 (2007): 2–35, https://gking.harvard.edu/files/jp.pdf.

267   **state of affairs."** A compelling argument rebuking this concern can be found in Grofman and King, "The Future of Partisan Symmetry as a Judicial Test for Partisan Gerrymandering after LULAC v. Perry." Grofman and King are two of the authors of an amicus brief proposing partisan symmetry as a tool to detect gerrymandering.

267   **gap of 11.46%).** Cover, "Quantifying Partisan Gerrymandering."

268   **worth your time.** Video can be found at *"Gill v. Whitford Oral Argument,"* C-SPAN, March 28, 2018, https://www.c-span.org/video/?432595-1%2Fsupreme-court-hears -oral-argument-wisconsin-gerrymandering-case. Transcript is available at https://www .wsj.com/public/resources/documents/Scotus20171003GillvWhitford.pdf.

268   **him publicly afterward.)** Colleen Flaherty, "Chief Justice John Roberts Calls Data on Gerrymandering 'Sociological Gobbledygook.' Sociology Fires Back," Inside Higher Ed, October 11, 2017, https://www.insidehighered.com/news/2017/10/12/chief-justice -john-roberts-calls-data-gerrymandering-sociological-gobbledygook.

268   **were exceptionally gerrymandered.** Herschlag et al., "Quantifying Gerrymandering in North Carolina."

268   **of its kind.** Amicus Brief of Mathematicians, Law Professors, and Student in Support of Appellees and Affirmance, Robert A. Rucho v. Common Cause, Linda H. Lamone v. O. John Benisek, nos. 18-422, 18-726 (March 8, 2019), https://mggg.org/SCOTUS -MathBrief.pdf.

269n   **One of the signatories:** Jordan Ellenberg, "The Supreme Court's Math Problem," *Slate*, March 29, 2019, https://slate.com/news-and-politics/2019/03/scotus-gerryman dering-case-mathematicians-brief-elena-kagan.html.

269   **mathematics is troublesome.** For more on the Supreme Court's innumeracy, see Oliver Roeder, "The Supreme Court Is Allergic to Math," FiveThirtyEight, October 17, 2017, https://fivethirtyeight.com/features/the-supreme-court-is-allergic-to-math/.

270   **O'Connor in 1993,** Sandra Day O'Connor, Ruth O. Shaw, et al. v. Janet Reno, Attorney General, et al., 509 US 630 (1993) (No. 92-357) (Justice O'Connor for the Court), Sandra Day O'Connor Institute Digital Library, accessed March 31, 2023, https://oconnorlibrary .org/supreme-court/shaw-v-reno-1992.

270   **Reock score it.** Aaron R. Kaufman, Gary King, and Mayya Komisarchik, "How to Measure Legislative District Compactness If You Only Know It When You See It," *American Journal of Political Science* 65, no. 3 (2021): 533–50, https://doi.org/10.1111/ajps .12603.

270   ***Rucho v. Common Cause.*** Nice summaries of how mathematicians are aiding the judicial system can be found in Scott Hershberger, "Courts, Commissions, and Consultations:

How Mathematicians Are Working to End Gerrymandering," *Notices of the American Mathematical Society* 69, no. 4 (2022): 1, https://doi.org/10.1090/noti2461; Heidi Opdyke, "Mathematicians' Work Helps Change How People Vote," Phys.org, November 8, 2019, https://phys.org/news/2019-11-mathematicians-people-vote.html.

270   **as an example.** Maria Chikina, Alan Frieze, and Wesley Pegden, "Assessing Significance in a Markov Chain without Mixing," *Proceedings of the National Academy of Sciences* 114, no. 11 (2017): 2860–2864, https://doi.org/10.1073/pnas.1617540114.

271   *Gill v. Whitford.* Brief of Political Geography Scholars as *Amici Curiae* in Support of Appellees Beverly R. Gill v. William Whitford (September 5, 2017) (No. 16-1161), https://www.scotusblog.com/wp-content/uploads/2017/09/16-1161-bsac-political-geography.pdf.

271   *Rucho v. Common Cause.* Brief of *Amici Curiae* Professors Wesley Pegden, Jonathan Rodden, and Samuel S.-H. Wang in Support of Appellees Robert A. Rucho v. Common Cause (March 8, 2019) (No. 18-422), https://www.supremecourt.gov/DocketPDF/18/18-422/91394/20190308165646674_2019.03.08%20Amicus%20in%20Rucho%20v.%20Common%20Cause.pdf.

271   **just as gerrymandered.** Mark Joseph Stern, "Pennsylvania Governor Rejects Latest GOP Gerrymander, Ensuring Court-Drawn Map for Midterms," *Slate*, February 13, 2018, https://slate.com/news-and-politics/2018/02/pennsylvania-governor-rejects-new-gop-gerrymander-ensuring-court-drawn-map-for-2018.html. For the full account of the story, see David Daley, *Unrigged: How Americans Are Battling Back to Save Democracy* (New York: Liveright, 2020).

271   **2021 congressional map.** Ian Millhiser, "A New Supreme Court Case Could Make It Nearly Impossible to Stop Racial Gerrymanders," *Vox*, February 1, 2022, https://www.vox.com/2022/2/1/22910909/supreme-court-racial-gerrymander-alabama-merrill-singleton-milligan.

272   **for that case.** Brief of Computational Redistricting Experts as *Amici Curiae* in Support of Appellees and Respondents, John H. Merrill v. Evan Milligan, John H. Merrill v. Marcus Caster (Nos. 21-1086, 21-1087), https://www.supremecourt.gov/DocketPDF/21/21-1086/230272/20220718153650363_21-1086%2021-1087%20bsac%20Computational%20Redistricting%20Experts.pdf.

Chapter Thirteen

Page

276   **bucked that trend).** Aaron Blake, "Why the GOP's Popular-Vote Edge Hasn't Translated to More House Seats," *Washington Post*, November 15, 2022, https://www.washingtonpost.com/politics/2022/11/14/republican-popular-vote-seats/.

276   **32%, and 3.5%.** Roderick McInnes, "General Election 2019: Turning Votes into Seats," House of Commons Library, January 10, 2020, https://commonslibrary.parliament.uk/general-election-2019-turning-votes-into-seats/.

276   **us, the voters."** "Cleese on PR (Full Length)," 1987, YouTube video, https://www.youtube.com/watch?v=NSUKMa1cYHk.

277   **the world's democracies.** Thea Ridley-Castle, "How Many Countries around the World Use Proportional Representation?" Electoral Reform Society, March 20, 2023, https://www.electoral-reform.org.uk/how-many-countries-around-the-world-use-proportional-representation/.

277   **Parliament before Brexit.** Neil Johnston. 2019. "How Do European Parliamentary Elections Work?" Commonslibrary.parliament.uk, May, https://commonslibrary.parliament.uk/how-do-european-parliamentary-elections-work/.

278   **electorate more faithfully.** J. Vowles, "Introducing Proportional Representation: The New Zealand Experience," *Parliamentary Affairs* 53, no. 4 (2000): 680–96.

280   **representation and plurality.** Matthew Soberg Shugart and Martin P. Wattenberg, "Mixed-Member Electoral Systems: A Definition and Typology," in *Mixed-Member Electoral Systems*, ed. Matthew Soberg Shugart and Martin P. Wattenberg (Oxford: Oxford University Press, 2003), 9–24.

280   **a compensatory seat.** For more on the compensatory seats in Bosnia (in Bosnian), see F. H., "Zašto Na Izborima Nije Bitno Samo Koliko, Nego I Gdje Osvojite Glasove," KLIX, September 12, 2022, https://www.klix.ba/vijesti/bih/zasto-na-izborima-nije-bitno-samo-koliko-nego-i-gdje-osvojite-glasove/220905073.

283   **and the same.** Michel L. Balinski and H. Peyton Young, *Fair Representation* (Washington, DC: Brookings Institution Press, 2010).

285   **Denmark, for example).** Nicholas R. Miller, "Election Inversions under Proportional Representation," *Scandinavian Political Studies* 38, no. 1 (2014): 4–25.

288   **along with gerrymandering.** Shelby County v. Holder, 570 U.S. 529 (2013) (Justice Ginsburg, with whom Justice Breyer, Justice Sotomayor, and Justice Kagan join, dissenting), https://supreme.justia.com/cases/federal/us/570/529/#tab-opinion-1970751.

288   **the school committee.** Shannon Dooling, "A Lawsuit Challenged Lowell's Voting System. 2 Years Later, the City Agreed to Change It," *WBUR*, May 29, 2019, https://www.wbur.org/news/2019/05/29/lowell-voting-rights-lawsuit-at-large-system.

288   **in Lowell's history.** "Lowell City Council Election Results 2021," KhmerPost USA, November 8, 2021, https://khmerpostusa.com/lowell-city-council-election-results-2021. A detailed account of the Lowell case can be found in Chapter 21 of Moon Duchin and Olivia Walch, eds., *Political Geometry* (Cham: Birkhäuser, 2022).

288   **their state legislature.** "State Legislative Chambers that Use Multi-Member Districts," n.d., Ballotpedia, https://ballotpedia.org/State_legislative_chambers_that_use_multi-member_districts.

293n   **Here's a strange thing:** David McCune and Adam Graham-Squire, "Monotonicity Anomalies in Scottish Local Government Elections," ArXiv.org. May 28, 2023. https://doi.org/10.48550/arXiv.2305.17741.

296   **on Representative Government.** John Stuart Mill, *Considerations on Representative Government* (1861; repr., George Routledge & Sons, 1991).

296   **therefore of civilization."** John S. Mill, *The Later Letters of John Stuart Mill, 1849–1873*, vol. 2, ed. Francis E. Mineka and Dwight N. Lindley (London: Routledge, 1972).

296   **dabble in it.** For a good historical account, see Douglas J. Amy, "The Forgotten History of the Single Transferable Vote in the United States," *Representation* 34, no. 1 (1996): 13–20.

297   **a Kremlin import.** Robert J. Kolesar, "Communism, Race, and the Defeat of Proportional Representation in Cold War America," paper presented at New England Historical
Association Conference, Amherst College, Amherst, April 20, 1996, https://fairvote.org
/archives/communism-race-and-the-defeat-of-proportional-representation-in-cold-war
-america/.

297n   **The cumulative voting story is even shorter:** "Effectiveness of Fair Representation
Voting Systems for Racial Minority Voters," FairVote, January 2015, https://fairvote.app
.box.com/v/fair-rep-voting-rights.

297   **three to five.** Gerdus Benade, Ruth Buck, Moon Duchin, Dara Gold, and Thomas
Weighill, "Ranked Choice Voting and Minority Representation," unpublished paper,
February 18, 2021, https://doi.org/10.2139/ssrn.3778021.

298   **counts is lacking.** Douglas Amy, "PR Library: Common Criticisms of PR and Re-
sponses to Them," FairVote, accessed April 7, 2023, https://fairvote.org/archives
/common-criticisms-of-pr-and-responses-to-them.

299   **for fewer seats.** Nikhil Garg, Wes Gurnee, David Rothschild, and David Shmoys, "Com-
batting Gerrymandering with Social Choice," in *EC '22: Proceedings of the 23rd ACM
Conference on Economics and Computation* (New York: Association for Computing Machinery, 2022), 560–561, https://doi.org/10.1145/3490486.3538254; Douglas Amy, "PR
Library: How Proportional Representation Would Finally Solve Our Redistricting and
Gerrymandering Problems," FairVote, accessed April 7, 2023, https://fairvote.org/archives
/how-proportional-representation-would-finally-solve-our-redistricting-and-gerry
mandering-problems.

300   **are treated equally.** Emily Badger, "How the Rural-Urban Divide Became America's
Political Fault Line," *New York Times,* May 21, 2019, https://www.nytimes.com/2019/05
/21/upshot/america-political-divide-urban-rural.html.

300   **close to call.** Mark Bauer, "How Multi-Member Districts Can Make a More Balanced
Electorate (and Put an End to Gerrymandering)," Rank The Vote, February 3, 2022,
https://rankthevote.us/how-multi-member-districts-can-make-a-more-balanced
-electorate-and-put-an-end-to-gerrymandering/.

300   **better for women.** Richard E. Matland and Donley T. Studlar, "The Contagion of
Women Candidates in Single-Member District and Proportional Representation Electoral Systems: Canada and Norway," *Journal of Politics* 58, no. 3 (1996): 707–733.

301   **in plurality systems.** Anna Fahey, "This Voting Reform May Get More Women into
Elected Office," Sightline Institute, October 30, 2018, https://www.sightline.org/2018
/10/25/getting-more-women-elected-us-canada-proportional-representation/; Haly
Jungwirth, "International Women's Day and Proportional Representation," FairVote,
March 8, 2022, https://www.fairvote.org/international_women_s_day_and_proportional
_representation.

301   **their own party.** Steven Hill and Robert Richie, "The Case for Proportional Representation," *Boston Review,* March 1, 1998, https://bostonreview.net/articles/robert-richie
-steven-hill-case-proportional-representation.

301   **to do so).** "Women Winning: Electoral Reforms Drive Change," RepresentWomen,
accessed April 7, 2023, https://www.representwomen.org/women_winning.

301   **of minority voters.** Benade et al., "Ranked Choice Voting and Minority Representation."

301   **increase by 40%.** Benjamin Oestericher, "To Reflect All of Its People, America Must Become a Proportional Democracy," FairVote, December 4, 2020, https://fairvote.org /to_reflect_all_of_its_people_america_must_become_a_proportional_democracy.

301   **for underrepresented groups.** MGGG Redistricting Lab, "Modeling the Fair Representation Act," Metric Geometry and Gerrymandering Group, July 7, 2022, https://mggg .org/fra-report.

302   **would achieve this.** HR 3863—Fair Representation Act, 117th Congress (2021–2022), Congress.gov, https://www.congress.gov/bill/117th-congress/house-bill/3863 ?s+1&r+1.

302   **But things change."** Patrick Gavin, "Novoselic Plays for Proportional Vote," *POLITICO*, March 18, 2014, https://www.politico.com/story/2014/03/krist-novoselic-fairvote -proportional-representation-104781; "FairVote Chair Krist Novoselic on Democracy and Proportional Representation," FairVote, December 12, 2012, https://fairvote.org /fairvote-chair-krist-novoselic-on-democracy-and-proportional-representation/.

Chapter Fourteen

Page

308   **came to be."** Jordan Ellenberg, *Shape: The Hidden Geometry of Information, Biology, Strategy, Democracy, and Everything Else* (New York: Penguin Press, 2021), 354.

312   **of state sovereignty.** Jesse Wegman, *Let the People Pick the President: The Case for Abolishing the Electoral College* (New York: St. Martin's Griffin, 2021).

313   **the national legislature.** Drew DeSilver, "Among Democracies, U.S. Stands out in How It Chooses Its Head of State," Pew Research Center, November 22, 2016, https://www .pewresearch.org/fact-tank/2016/11/22/among-democracies-u-s-stands-out-in-how-it -chooses-its-head-of-state/.

314   **no enforcement mechanism.** Congressional Research Service, "Supreme Court Clarifies Rules for Electoral College: States May Restrict Faithless Electors," CRS Legal Sidebar, July 10, 2020, https://crsreports.congress.gov/product/pdf/LSB/LSB10515.

314   **affected the outcome.** "Presidential Elections," FairVote, accessed April 7, 2023, https:// fairvote.org/resources/presidential-elections/#faithless-electors.

314   **party over another.** Robert S. Erikson, Karl Sigman, and Linan Yao, "Electoral College Bias and the 2020 Presidential Election," *Proceedings of the National Academy of Sciences* 117, no. 45 (2020): 27940–27944.

317   **to national outcomes.** Edward B. Foley, Jeremy B. White, Sam Sutton and Carly Sitrin, and Bill Mahoney and Josh Gerstein, "An Idea for Electoral College Reform that Both Parties Might Actually Like," *POLITICO*, January 12, 2019, https://www.politico.com /magazine/story/2019/01/12/electoral-college-reform-conservatives-223965/.

318   **votes at all.** Christopher Klein, "Here's How Third-Party Candidates Have Changed Elections," History, May 31, 2018, https://www.history.com/news/third-party-candidates -election-influence-facts.

319*n*   **An argument could be made:** Theo Lippman Jr., "Electors Complicated 1960 Vote," *Baltimore Sun*, December 19, 2000, https://www.baltimoresun.com/news/bs-xpm-2000 -12-19-0012190382-story.html.

320   **of each other.** Michael Geruso, Dean Spears, and Ishaana Talesara, "Inversions in US Presidential Elections: 1836–2016," working paper 26247, National Bureau of Economic Research, September 9, 2019, https://www.nber.org/papers/w26247.

322   **total votes cast.** Danielle Kurtzleben, "How to Win the Presidency with 23 Percent of the Popular Vote," NPR, November 2, 2016, https://www.npr.org/2016/11/02/500112248 /how-to-win-the-presidency-with-27-percent-of-the-popular-vote.

326   **the popular vote.** Rebecca Salzer and Jocelyn Kiley, "Majority of Americans Continue to Favor Moving Away from Electoral College," Pew Research Center, August 5, 2022, https://www.pewresearch.org/fact-tank/2022/08/05/majority-of-americans-continue -to-favor-moving-away-from-electoral-college.

326   **with abolishing it.** Robert Alexander, "Republicans Were against the Electoral College before They Were for It," *CNN*, October 26, 2020, https://www.cnn.com/2020/10/26 /opinions/gop-electoral-college-abolish-opinion-alexander/index.html.

326   **the Electoral College.** "Past Attempts at Reform," FairVote, n.d., https://fairvote.org /archives/the_electoral_college-past_attempts_at_reform/.

326   **Cohen of Tennessee).** Mr. Cohen et al., Joint Resolution Proposing an Amendment to Abolish the Electoral College, H. J. Res.14, 117th Congress, 1st Session, January 11, 2021, https://www.congress.gov/117/bills/hjres14/BILLS-117hjres14ih.pdf.

326   **less than 1%.** "A Constitutional Amendment to Abolish the Electoral College," History, Art & Archives, September 18, 1969, https://history.house.gov/HistoricalHighlight /Detail/25769816548.

326   **supermajorities and filibusters).** Jesse Wegman, "The Filibuster That Saved the Electoral College," *New York Times*, February 8, 2021, https://www.nytimes.com/2021/02 /08/opinion/filibuster-electoral-college.html; Dave Roos, "How the Electoral College Was Nearly Abolished in 1970," History, August 25, 2020, https://www.history.com/news /electoral-college-nearly-abolished-thurmond.

328   **way they vote.** Wegman, *Let the People Pick the President*.

328   **up small states.** Wegman, *Let the People Pick the President*.

329   **matter more politically.** Meilan Solly, "Why Do Maine and Nebraska Split Their Electoral Votes?" *Smithsonian Magazine*, November 5, 2020, https://www.smithsonianmag .com/smart-news/why-do-maine-and-nebraska-split-their-electoral-votes-180976219.

329   **to Clinton's 248.** 270toWin Staff, "The 2016 Election under Alternate Electoral Allocation Methods," 270toWin, February 7, 2017, https://www.270towin.com/news/2017/02 /07/the-2016-election-under-alternate-electoral-allocation-methods_446.html.

330   **over Al Gore.** Claire Daviss and Rob Richie, "Fuzzy Math: Wrong Way Reforms for Allocating Electoral Votes," FairVote Policy Perspective, January 2015, https://fairvote .app.box.com/v/fuzzy-math-wrong-way-reforms.

330   **traditionally swing states.** "Analysis of the Congressional-District Method of Awarding Electoral Votes," National Popular Vote, https://www.nationalpopularvote.com/analysis -congressional-district-method-awarding-electoral-votes.

330   **in twelve states.**) "Analysis of the Congressional-District Method of Awarding Electoral Votes."

332   **say is necessary.** Jeffrey M. Jones, "Support for Third U.S. Political Party at High Point," Gallup, February 15, 2021, https://news.gallup.com/poll/329639/support-third-political-party-high-point.aspx.

332   **and Nader 13.** Daviss and Richie, "Fuzzy Math."

334   **can be changed.** "Summary: State Laws Regarding Presidential Electors," National Association of Secretaries of State, 2020, https://www.nass.org/sites/default/files/surveys/2020-10/summary-electoral-college-laws-Oct20.pdf.

334   **of court challenges.** Alexander Keyssar, *Why Do We Still Have the Electoral College?* (Cambridge, MA: Harvard University Press, 2020).

334   **"legal train wreck."** Keyssar, *Why Do We Still Have the Electoral College?*

INDEX

Page numbers in *italics* refer to figures.

O'Connor, Sandra Day, 270

Ohio, 219, 220, 222, 224, 231, 241, 300, 320, 334

Oklahoma, 182

Oliver, John, 210

Olympic Games: medal counts at, 79; scoring at, 85; site selection for, 62, 68n

One Nation Party, 42

One Person, One Vote (Seabrook), 221

one person, one vote doctrine, 4, 168, 215, 328; Electoral College incompatible with, 323, 327, 335; interstate malapportionment incompatible with, 162; ranked choice voting compatible with, 67

open list proportional representation, 279, 349

open primaries, 132

optimization, 5, 14, 97–98, 105

Oregon, 116, 123, 221, 222

packing, in gerrymandering, 209–10, 215, 224, 242–49, 349

pairwise (sequential) voting, 93–95, 136

Palin, Sarah, 64–65, 73, 110

papal succession, 24, 128

paradox of positive association, 71

Paraguay, 313

Pareto, Vilfredo, 106n

Pareto criterion, 106n

parity, as voting method, 13, 19, 349

partisan bias, 225, 238, 239–40, 266, 349

partisan symmetry, 235–37, 240, 340

party list proportional representation, 278–81, 285–86, 349;

Paul, Ron, 314

Pegden, Wesley, 270–71

Pelosi, Nancy, 225n

Peltola, Mary, 64–65, 73

Pennsylvania, 116, 309, 334; gerrymandering in, 219, 222, 231, 250, 251, 259, 263, 267, 271; in 2016 election, 37, 318, 320n

Perdue, David, 45

Perot, Ross, 36–37, 117, 127, 318

Perron-Frøbenius theorem, 16, 260

Perry, Rick, 129

plurality voting (first-past the post; relative majority; winner take all), 28, 96, 276, 277; defined, 349; manipulability of, 129; shortcomings of, 29–31, 34–38, 42, 44, 143, 204

plurality bloc voting, 286–89, 349

+2 effect, 324, 330, 349

Political Parties (Duverger), 39

Polk, James, K., 36, 187

poll taxes, 214

Polsby-Popper compactness score, 253–54, 258–60, 270, 349

popular vote, 349

population monotonicity, 201, 202, 349

population paradox, 181, 186, 190, 201, 202, 349

population polygon score, 257, 259, 350

Populist Party, 41n

Portugal, 46, 279

positional voting, 78–79

positive political theory, 107

Poundstone, William, 120, 129

Powell, Colin, 314

precinct, 350

preference order (preference ballot), 53

preferential method (ranked choice method), 53, 132–33, 350; growing use of, 4, 58. *See also* instant runoff method

primaries: in one-party districts, 2n; open, 132; presidential, 276; spoiler effect in, 37, 132; top-two, 29, 46, 351

Princeton Gerrymandering Project, 258

prison gerrymandering, 226, 350

probability distribution function, 264, 350

profile, in ranked choice voting, 54–56, 350

Progressive Party, 41n

Prohibition Party, 41n

proportional ranked choice voting (choice voting; generalized Hare method; single transferable vote), 289–96, 300–2, 340, 351

proportional representation, 208, 234–35; benefits of, 41–42, 278, 298–303; bloc

A NOTE ON THE TYPE

This book has been composed in Arno, an Old-style serif typeface in the classic Venetian tradition, designed by Robert Slimbach at Adobe.

GPSR Authorized Representative: Easy Access System Europe - Mustamäe tee
50, 10621 Tallinn, Estonia, gpsr.requests@easproject.com